Plant Galls

Proceedings of a Joint International Symposium of the British Plant Gall Society and the
Systematics Association held in London, July 1992.

The Systematics Association
Special Volume No. 49

Plant Galls

Organisms, Interactions, Populations

Edited by

MICHÈLE A. J. WILLIAMS

International Mycological Institute

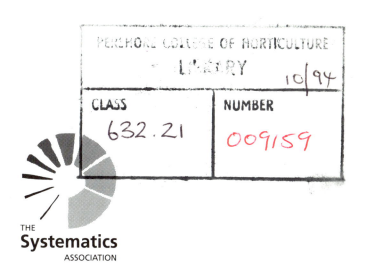
THE
Systematics
ASSOCIATION

Published for the SYSTEMATICS ASSOCIATION by
CLARENDON PRESS · OXFORD
1994

Oxford University Press, Walton Street, Oxford OX2 6DP
Oxford   New York   Toronto
Delhi   Bombay   Calcutta   Madras   Karachi
Kuala Lumpur   Singapore   Hong Kong   Tokyo
Nairobi   Dar es Salaam   Cape Town
Melbourne   Auckland   Madrid
and associated companies in
Berlin   Ibadan

Oxford is a trade mark of Oxford University Press

Published in the United States
by Oxford University Press Inc., New York

A catalogue record for this book is available from the British Library

Library of Congress Cataloging in Publication Data
Plant galls: organisms, interactions, populations/edited by Michèle A. Williams.
(The Systematics Association special volume; no. 49)
"Proceedings of a joint international symposium of the British
Plant Gall Society and the Systematics Association held in London,
July 1992"—P.
Includes bibliographical references (p. 000) and index.
1. Galls (Botany)–Congresses.   2. Gall insects–Congresses.
3. Insect-plant relationships–Congresses.   I. Williams, Michèle A. J.
II. British Plant Gall Society.   III. Systematics Association.   IV. Series.
SB767.P58 1994   632'.2–dc20   93–38148

ISBN 0 19 8577699

Typeset by Selwood Systems, Midsomer Norton
Printed in Great Britain by
Biddles Ltd
Guildford & King's Lynn

Preface

This volume has as its basis papers presented at an international symposium held 15–17 July 1992, at the Royal Entomological Society of London. The meeting was convened jointly by the Systematics Association and the British Plant Gall Society. The original concept of the symposium was to involve workers from as many disciplines as possible related by their common interest in the study of plant galls, and to reflect the current huge range of subject approaches and wide variety of organisms studied by cecidologists.

The main themes of the symposium were 'Structure, Systematics and Evolution', 'Ecological Aspects', and 'Physiological Aspects'. Many of the papers presented impinged on more than one of these subject areas and this volume is ordered according to a loose taxonomic framework for the gall-causing organisms, progressing from bacteria and fungi through to the rotifers, mites, and insects.

Thanks are due to many people who helped to make the meeting a success. The staff of the Royal Entomological Society were most helpful, as were Dr Keith Harris and his staff at the International Institute of Entomology who provided a magnificent venue for the reception. I am grateful to all those who contributed papers and posters, and to the chairmen of the five paper-reading sessions: Professors Joe Shorthouse, Michael Claridge, Laurence Mound, and Drs Francis Gilbert and Chris Leach. Dr Chris Leach also kindly provided valuable concluding remarks.

The background administration and day to day organization of the conference was largely in the hands of three of my colleagues from the International Mycological Institute: Miss Caroline Lowe, Miss Janet Pryse, and Mr John David, all three of whom also worked tirelessly on the preparations for the reception. I owe them an enormous debt of gratitude.

Contributions to this volume were reviewed by Dr Roger Blackman, Mr John David, Dr Keith Harris, Dr Mark Holderness, Dr Eric Hollowday, Dr Chris Leach, Dr Margaret Redfern, and Dr Brian Spooner. I am grateful to all the referees for their careful work.

I should also like to express my gratitude to fellow members of the Council of the Systematics Association and to various members of the British Plant Gall Society, notably Drs Chris Leach and Brian Spooner, for their support and advice during the organization of the meeting. The financial contributions of the Systematics Association and the Royal Society are gratefully acknowledged.

Finally, thanks are due to my husband, Robert, for his support and encouragement whilst I was working on this volume.

Lisbon   M.A.J.W.

January 1993

Contents

Contributors

J. ARNOLD-RINEHART
Department of Animal Ecology, University of Bayreuth, D-95440 Bayreuth, Germany.

CHARLOTTE ASTBURY
Department of Life Science, University of Nottingham, Nottingham NG7 2RD, UK.

JENNY BEDINGFIELD
Department of Life Science, University of Nottingham, Nottingham NG7 2RD, UK.
GIUSEPPE BAGATTO
Department of Biology, Laurentian University, Sudbury, Ontario, P3E 2C6, Canada.

MARGRET H. BAYER
Fox Chase Cancer Center, Institute for Cancer Research, Philadelphia, PA 19111, USA.

JAN BOCZEK
Agricultural University of Warsaw, 02-766 Warsaw, Poland.

ROBERT CAMERON
School of Continuing Studies, University of Birmingham, Birmingham B15 2TT, UK.

MICHAEL F. CLARIDGE
School of Pure and Applied Biology, University of Wales, Cardiff, CF1 3TL, UK.

B. COCKRELL
c/o S. Malcolm, Department of Biological Sciences, Western Michigan University, Kalamazoo, M 49008, USA.

MARGARET E. COLLINSON
Department of Geology, Royal Holloway University of London, Egham, Surrey TW20 0EX, UK.

G. CSÓKA
Department of Forest Protection, Forest Research Institute, P.O. Box 49, 2100 Gödöllő, Hungary.

M. J. CRAWLEY
Department of Biology and NERC Centre for Population Biology, Imperial College at Silwood Park, Ascot, Berkshire SL5 7PY, UK.

I.S. CURTIS
Plant Genetic Manipulation Group, Department of Life Science, University of Nottingham, Nottingham NG7 2RD, UK.

M.R. DAVEY
Plant Genetic Manipulation Group, Department of Life Science, University of Nottingham, Nottingham NG7 2RD, UK.

HASSAN ALI DAWAH
School of Pure and Applied Biology, University of Wales, Cardiff CF1 3TL, UK.

LUC DE BRUYN
Department of Biology, Evolutionary Biology Group, University of Antwerp (RUCA), Groenenborgerlaan 171, 2020 Antwerpen, Belgium.

BRUCE ENNIS
Department of Life Science, University of Nottingham, Nottingham NG7 2RD, UK.

W.A. FOSTER
Department of Zoology, University of Cambridge, Downing Street, Cambridge CB2 3EJ, UK.

K.M.A. GARTLAND
Department of Molecular and Life Sciences, Dundee Institute of Technology, Dundee DD1 1HG, UK.

FRANCIS GILBERT
Department of Life Science, University of Nottingham, Nottingham NG7 2RD, UK.

D.A. GRIFFITHS
Bunting Biological Control Ltd, Great Horkesley, Colchester, Essex CO6 4AJ, UK.

ROSEMARY S. HAILS
Imperial College at Silwood Park, Ascot, Berkshire SL5 7PY, UK.

KEITH M. HARRIS
International Institute of Entomology, 56 Queen's Gate, London SW7 5JR, UK.

DAVID L. HAWKSWORTH
International Mycological Institute, Bakeham Lane, Egham, Surrey TW20 9TY, UK.

A.J. HICK
Department of Pure and Applied Biology, University of Leeds LS2 9JT, UK.

R.N. HIGTON
Lord Williams's School, Thame, Oxon OX9 2AQ, UK.

ROSMARIE HONEGGER
Institut für Pflanzenbiologie, Universität Zürich, Zollikerstrasse 107, CH-8008 Zürich, Switzerland.

BRUCE ING
Department of Biology, Chester College of Higher Education, Cheyney Road, Chester CH1 4BJ, UK.

R.M. JENKINS
Department of Plant Sciences, University of Oxford, South Parks Road, Oxford OX1 3RB, UK.

SAMANTHA LAWSON
Department of Life Science, University of Nottingham, Nottingham NG7 2RD, UK.

I.J. LEITCH
Jodrell Laboratory, Royal Botanic Gardens, Kew, Richmond, Surrey TW9 3AB, UK.

D.J. MABBERLEY
Department of Plant Sciences, University of Oxford, South Parks Road, Oxford OX1 3RB, UK.

I.F.G. McLEAN
109 Miller Way, Brampton, Huntingdon, Cambridgeshire PE18 8TZ, UK.

LAURENCE A. MOUND
The Natural History Museum, London SW7 5BD, UK.

P.A. NORTHCOTT
Department of Zoology, University of Cambridge, Downing Street, Cambridge CB2 3EJ, UK.

I. DAVID R. PERIES
Department of Biological Sciences, University of Durham, South Road, Durham DH1 3LE, UK.

J.B. POWER
Plant Genetic Manipulation Group, Department of Life Science, University of Nottingham, Nottingham NG7 2RD, UK.

T.F. PREECE
Kinton, Turners Lane, Lynclys Hill, Near Oswestry, Shropshire SY10 8LL, UK.

MARGARET REDFERN
School of Continuing Studies, University of Birmingham, Birmingham B15 2TT, UK.

K. SCHÖNROGGE
Department of Biology and NERC Centre for Population Biology, Imperial College at Silwood Park, Ascot, Berkshire SL5 7PY, UK.

ANDREW C. SCOTT
Department of Geology, Royal Holloway University of London, Egham, Surrey TW20 0EX, UK.

JOSEPH D. SHORTHOUSE
Department of Biology, Laurentian University, Sudbury, Ontario, P3E 2C6, Canada.

TRACEY SITCH
Department of Life Science, University of Nottingham, Nottingham NG7 2RD, UK.

B.M. SPOONER
31 Balmoral Crescent, West Molesey, Surrey KT8 1QA, UK.

JONATHAN STEPHENSON
Department of Geology, Royal Holloway University of London, Egham, Surrey TW20 0EX, UK.

G.N. STONE
Department of Biology and NERC Centre for Population Biology, Imperial College at Silwood Park, Ascot, Berkshire SL5 7PY, UK.

PAUL J. SUNNUCKS
Institute of Zoology, Regent's Park, London NW1 4RY, UK.

MICHÈLE A.J. WILLIAMS
International Mycological Institute, Bakeham Lane, Egham, Surrey TW20 9TY, UK.

H. ZWÖLFER
Department of Animal Ecology, University of Bayreuth, D-95440 Bayreuth, Germany.

1. Plant galls: a perspective

MICHÈLE A.J. WILLIAMS

International Mycological Institute, Bakeham Lane, Egham, Surrey, UK

Abstract

Galls have been of interest to man from historical times to the present day. Despite or because of this fact, the term 'plant gall' has been applied to many different systems. Definitions of the term are mostly inadequate and its employment as a general expression rather than a term suggesting scientific precision is preferred. Manifestations which have been called plant galls are caused by a wide variety of organisms interacting with a broad range of hosts. These interactions in turn provide opportunities for complex assemblages of organisms. Cecidology, the study of plant galls, uses a diverse range of biological techniques and involves many disciplines, a representative selection of which are drawn upon to provide the material for this volume.

Introduction

In seventeenth-century England on 29 May each year people wore sprigs of oak with gilded oak-apples [the gall caused by the cynipid *Biorhiza pallida* (Olivier)]. Commanded by Act of Parliament, 'Oak Apple Day' was observed as a day of thanksgiving for the restoration of the monarchy (Evans 1981; Robbins 1992 and references therein). Nowadays plant galls have lost this significance but continue to hold a fascination for many professional scientists and laymen alike.

Certain common plant galls are highly visible, brightly coloured structures and as such have caught the eye of commentators on natural subjects through the ages. Swanton (1912) noted that galls were mentioned in the writings of Theophrastus (in the third century BC) and, subsequently, by such famous authors as Dioscorides and Pliny (writing in the first century AD) and, more recently, John Evelyn (in the seventeenth century). Throughout history, galls have played a part in the affairs of man. In medieval times, plant galls were used as medicines and for

Plant Galls (ed. Michèle A. J. Williams), Systematics Association Special Volume No. 49, pp. 1–7. Clarendon Press, Oxford, 1994. © The Systematics Association, 1994.

dyestuffs and were the subject of much superstition and folklore. In the recent past they have been important commercial commodities used industrially as a source of tannins and in ink production (Connold 1908; Swanton 1912; Briggs 1986). Malpighi, in the seventeenth century, is generally credited with having made the first scientific investigations into galls. Cecidology, the science of the study of galls, made great progress in the late nineteenth and early twentieth centuries. Mani (1992) has listed the major early contributors to the field. Today, scientific interest is broadly based. Researchers value galls for their intrinsic biological interest, their use as model systems for physiological and other studies, and their potential application in disciplines such as biological control and plant breeding. Many galls are not, however, viewed favourably. Certain galls, particularly some caused by nematodes and fungi, are manifestations of crop plant pests and diseases which continue to cause severe economic losses.

Definition of the term 'plant gall'

The term 'plant gall' is widely used by scientists and laymen. All workers in the field of cecidology must have a concept of what they understand by the expression plant gall and most works dealing with the subject make some attempt to define the term. There is, however, no current widely accepted definition. Recently, several articles have discussed the definition of the term 'gall' (Stubbs 1987; Spooner 1990; Redfern 1992) and it is not appropriate to rehearse here all the arguments for and against different definitions. The major areas of dissent are outlined below.

Definitions attempt to categorize the interaction between a gall causer, which may be any one of a variety of types of organisms and its host, which historically was a plant, but now that fungi are recognized as a distinct Kingdom, the host definition must be modified to include members of both Kingdoms. Definitions may include benefit(s) the causer (or the host) may sustain, the nature of the morphological manifestation of the interaction, and/or may attempt to confine a broad concept in a few words, without listing specific attributes.

Fundamental difficulties arise when too many specific attributes are placed in the definition. Perceived benefits are often confused by anthropocentric ideas, such as a 'home' or 'shelter'. Definitions which stress nutritional relationships are inappropriate when the physiological processes of many galls have been so little investigated. Likewise, changes in cell numbers and size (used, for example, in the definition provided by Darlington (1975)) have been accurately researched in very few gall interactions. Photomicroscopic studies and electron microscopy may reveal alterations which are not visible to the naked eye, but few gall systems have been investigated using even routine techniques. Emphasis on the abnormality of the gall interaction constricts the definition, as in some instances, for example, root nodules, the gall is a manifestation of the 'normal', that is, usual, development. Thus, the more rigorous definitions tend to exclude examples of interactions which are classically considered to be galls.

However, broad definitions such as that advanced by Meyer (1987 and see discussion in Spooner (1990)) and simple definitions are generally regarded as imprecise ('non-scientific') and tend to allow the inclusion of systems which are not conventionally regarded as galls.

Terminology often has to be modified as knowledge increases. Cecidiologists should accept that the conceptual interaction which is a 'plant gall' has had a broad interpretation for many years. The term is best loosely applied and treating it as rigorous is erroneous. A universally acceptable and accepted definition may well be an impossibility and cecidology might profit more if the scientists concentrated on studying the galls rather than playing with words.

The symposium from which this volume is derived covered a selection of subjects relating to systems which have been traditionally regarded as plant galls and, perhaps more importantly, which the scientists themselves consider to be galls.

Gall-causing organisms

Manifestations which have been termed plant galls originate from the specific interaction of the gall causer with the host organism. Causer groups are not restricted to insects. Table 1.1 shows principal causer groups which have been adopted by the British Plant Gall Society in compiling a database for the preparation of a *Checklist of British gall-causing organisms* (see Spooner and Bowdrey 1993).

On a world-wide scale several more insect groups rank as major gall causers (see, for example, the list provided by Dreger-Jauffret and Shorthouse (1992)).

Hosts for galls

Although perhaps the greatest diversity of galls is seen within particular flowering plant families, excrescences or malformations which have been considered to be galls have been recorded from angiosperms, gymnosperms, pteridophytes, bryophytes, algae, and fungi. However, different host groups, genera, and even species show very different

Table 1.1.   Principal groups of gall-causing organisms in the UK

Viruses		
Bacteria		
Mycoplasmas		
Actinomycetes		
Fungi: (16 orders)		
Angiosperms		
Protozoa		
Rotifera		
Nematoda		
Arachnida:	Acarina:	Eriophyidae
		Tarsonemidae
		Tetranychidae
Insecta:	Thysanoptera:	Thripidae
	Homoptera:	Adelgidae
		Aphalaridae
		Aphididae
		Cercopidae
		Coccidae
		Pemphigidae
		Phylloxeridae
		Psyllidae
		Triozidae
	Lepidoptera:	(17 families)
	Coleoptera:	Apionidae
		Cerambycidae
		Chrysomelidae
		Curculionidae
		Scolytidae
	Diptera:	Agromyziae
		Anthomyidae
		Cecidomyiidae
		Chloropidae
		Tephritidae
	Hymenoptera:	Blasticotomidae
		Cynipidae
		Tenthredinidae

propensities for being galled. Algae and bryophytes have few galls recorded from them. Gymnosperms have markedly fewer than angiosperms. Insect-induced galls occur much more commonly on dicotyledonous than monocotyledonous angiosperms (Mani 1964). Certain angiosperm families, for example, Compositae, Rosaceae, and Fagaceae are particularly prone to galling by cynipids (Mani 1964). Furthermore, host–causer patterns vary geographically.

Usually the interaction of one causer with one host is distinct and characteristic. In some cases the distinct nature of the galls may be used to assist the rapid identification of very similar causer organisms. Complexity in the host–galler relationship may include either the production of different galls on the same host by different generations of the causer (for example, some cynipids) or the galling of two alternate hosts at different stages in the causer's life cycle, as exemplified by certain cynipid gall wasps and rust fungi.

Different parts of the host may be galled to lesser or greater degrees. Galls have been recorded from organs including leaf and flower buds, leaves, flowers, fruits, stems, petioles, and roots, the fronds of ferns, and the thalli of bryophytes, fungi, and algae. Certain causer groups may predominantly gall one particular organ. Often the niche is very specific, for example, the pinnule of a fern frond, the vein of a leaf, or the anthers of a flower. Many galls are only associated with one part of the host, but others can occur at different sites on the same host. Some galls may be localized, affecting only a few cells whilst others distort the whole of an organ and in some cases may distort the whole growth form of the host organism, such as the witches' brooms of woody plants caused by mites, viruses, and fungi.

Such complex interactions are undoubtedly the result of intricate evolutionary processes. However, the coevolutionary and biogeographical relationships between gall causers and their hosts is, in most cases, little studied and very poorly understood.

Types of gall

The variety of interactions considered to be galls has resulted in several sets of terminology for types of gall. The classifications adopted by Küster (1911) recognized galls as being organoid, where the plant organ is modified or being histoid, where tissue modification produces a novel organ. Prosoplasmatic galls are complex, with finite growth and may be highly differentiated, whilst kataplasmatic galls show indeterminate growth and tissues little differentiated from the host. Other descriptive names have been given to the morphology of the galls: 'pouch galls', 'scroll galls', 'blister galls', etc. Swanton (1912) used a set of terms based on the complexity, position, and appearance of galls. A more recently adapted set of these categories has been provided by Dreger-Jauffret and Shorthouse (1992). The majority of these names for galls are often employed imprecisely and if not qualified by detail should be used with caution.

Galls as communities

Very few galls are restricted solely to the interaction of a causer with a host. Most galls at some stage in their existence provide opportunities for a community of associated organisms. Inquilines exist inside the gall alongside the causer and feed on the gall tissue. They may cause apparently little harm but in some cases may starve the causer. Parasitoids prey upon the larvae in the gall, either those of the causer or of other inhabitants of the gall, including inquilines and other parasitoids. These organisms often cause death of the causer larvae, so cannot be considered true parasites. Other gall-related organisms include predators which seek out, kill, and eat the larva of insect galls, and vagrants, which are organisms that may be found passing through the gall, perhaps using it for shelter. The complexity of certain interactions between insects inhabiting cynipid galls has been extensively reviewed by Wiebes-Rijks and Short-house (1992). Larger animals, such as rodents and birds, may use galls or their inhabitants for food material. Many saprobic fungi take advantage of insect galls in the litter layer or indeed on trees and sometimes specialized communities grow on galls (see, for example, Palmer 1990). Galls may also provide opportunities for biotrophic parasitic fungi and even symbiotic fungi (Bissett and Borkent 1988). Other microorganisms such as algae and bacteria may be found on and within plant galls.

Disciplines of study

Plant galls provide opportunities for studies of taxonomy, evolution, biogeography, all branches of ecology, developmental and nutritional physiology, morphology, developmental anatomy, and other disciplines. Galls are also studied in more applied fields such as plant pathology, plant breeding and genetics, and biological control. The interactive nature of the systems and the fact that the galls themselves are static (unlike many of the causers) and many are readily visible (and therefore easily collected) further enhance their practicality as models for study. In all these areas there is still a vast amount of work to be done. Our knowledge of galls is so poor that, in addition to the studies of a vast body of professional workers, the opportunity for amateur naturalists to make valuable scientific contributions remains.

This volume reports and reviews work from only a small proportion of the fields where galls are studied. The authors are unanimous in their view that gall biology is a profitable field of endeavour and that cecidology should attract wider participation.

References

Bissett, J. and Borkent, A. (1988). Ambrosia galls: the significance of fungal nutrition in the evolution of the Cecidomyiidae (Diptera). In *Coevolution of fungi with plants and animals* (ed. K.A. Pirozynski and D.L. Hawksworth), pp. 204–25. Academic Press, London.

Briggs, J. (1986). Historical uses of plant galls. *Cecidology*, **1,** 6–7.

Connold, E.T. (1908). *British oak galls*. Adlard and Son, London.

Darlington, A. (1975). *The pocket encyclopaedia of plant galls in colour*, (revised edn). Blandford Press, Poole.

Dreger-Jauffret, F. and Shorthouse, J.D. (1992). Diversity of gall-inducing insects and their galls. In *Biology of insect-induced galls* (ed. J.D. Shorthouse and O. Rohfritsch), pp. 8–33, Oxford University Press, New York.

Evans, I.H. (1981). *Brewer's dictionary of phrase and fable*, (2nd edn). Cassell, London.

Küster, E. (1911). *Die Gallen der Pflanzen; ein Lehrbuch der Botaniker und Entomologen*. S. Hirzel, Leipzig.

Mani, M.S. (1964). *Ecology of plant galls*. Dr W. Junk, The Hague.

Mani, M.S. (1992). Introduction to cecidology. In *Biology of insect-induced galls* (ed. J.D. Shorthouse and O. Rohfritsch), pp. 3–7. Oxford University Press, New York.

Meyer, J. (1987). *Plant galls and gall inducers*. Borntraeger, Berlin.

Palmer, J.T. (1990). Sclerotiniaceous cup fungi on oak galls. *Cecidology*, **5,** 31–4.

Redfern, M. (1992). What are galls? *Cecidology*, **7,** 81–3.

Robbins, S. (1992). Oak-apples in history and folklore. *Cecidology*, **7,** 5–9.

Spooner, B.M. (1990). Some problems in defining the word 'gall'. *Cecidology*, **5,** 51–2.

Spooner, B.M. and Bowdrey, J.P. (1993). Checklist of British gall causing organisms: progress report 1992. *Cecidology*, **8,** 31–5.

Stubbs, F.B. (1987). What is a gall? *Cecidology*, **2,** 49–50.

Swanton, E.W. (1912). *British plant-galls. A classified textbook of cecidology*. Methuen, London.

Wiebes-Rijks, A.A. and Shorthouse, J.D. (1992). Ecological relationships of insects inhabiting cynipid galls. In *Biology of insect-induced galls*, (ed. J.D. Shorthouse and O. Rohfritsch), pp. 238–57. Oxford University Press, New York.

2. *Agrobacterium*-induced crown gall and hairy root diseases: their biology and application to plant genetic engineering

M.R. DAVEY*, I.S. CURTIS*, K.M.A. GARTLAND†, and J.B. POWER*

*Plant Genetic Manipulation Group, Department of Life Science, University of Nottingham, Nottingham, UK † Department of Molecular and Life Sciences, Dundee Institute of Technology, Dundee, UK

Abstract

The Gram-negative soil bacteria *Agrobacterium tumefaciens* and *A. rhizogenes* infect a wide range of dicotyledonous plants at wound sites, where they incite the development of tumours or roots characteristic of crown gall and hairy root diseases. The molecular basis of these diseases involves the transfer and integration of the T-DNA region of the bacterial Ti (tumour-inducing) and Ri (root-inducing) plasmids into the genome of recipient plant cells, followed by expression of bacterial genes. The precise mechanism of T-DNA transfer is not clear, although it involves interaction between the T-DNA border sequences and plasmid virulence genes. Tumours induced by agrobacteria carrying wild-type Ti plasmids generally remain undifferentiated following transfer to culture. In contrast, cultured transformed roots of several species produce transgenic shoots either spontaneously or following stimulation with the appropriate growth regulators. Ri-transformed plants have characteristic phenotypes, including wrinkled leaves, dwarfism, and reduced fertility. Such effects result from an alteration in cell physiology.

The natural ability of *Agrobacterium* to genetically engineer plant cells has been exploited through the development of cointegrate and binary vectors. Chimaeric genes, of varying complexity, have been constructed and inserted between the T-DNA borders. Removal of the bacterial oncogenicity genes from the T-DNA has enabled phenotypically normal, fertile transgenic plants to be produced. This technology has permitted studies of tissue- and organ-specific gene expression. In addition, agronomically useful characteristics have been inserted into dicotyledonous crop plants, including resistance to viruses, insects, and herbicides.

Plant Galls (ed. Michèle A. J. Williams), Systematics Association Special Volume No. 49, pp. 9–56. Clarendon Press, Oxford, 1994. © The Systematics Association, 1994.

Introduction

Events which modify plant growth and development have always intrigued botanists. The formation of galls on higher plants is an excellent example of a phenomenon on which considerable research has been focused. Crown gall disease, incited by *Agrobacterium tumefaciens*, has been studied for over a century, with major advances being made in understanding the molecular interactions between bacterial and plant cells. The hairy root syndrome, induced by *A. rhizogenes*, has also received attention, particularly during the last two decades, but, to some extent, studies of this disease have been overshadowed by the momentum generated by the crown gall system. Nevertheless, the ability to clone bacterial genes involved in hairy root disease and to introduce them, either separately or in combination, into target plants, is also allowing this disease to be characterized at the molecular level. Although several of the events which occur during the interaction of *Agrobacterium* with plant cells are still not fully elucidated, the knowledge which is currently available has been utilized to develop methods for inserting a variety of foreign genes into plants.

Characteristics of crown gall and hairy root diseases

1. Infection of plants under field conditions

Crown gall and hairy root diseases are characterized by unlimited and rapid host cell proliferation following invasion of fresh wound sites by *Agrobacterium*. Many dicotyledons are affected, although some plants are more susceptible than others. A recent survey showed that in angiosperms, 641 plants belonging to 331 genera and 93 families are infected by *A. tumefaciens* (Erwin and Stuteville 1990). Tumours have also been reported in lower plants (De Cleene and De Ley 1976). *Agrobacterium rhizogenes* has a more restricted host range than *A. tumefaciens*. In monocotyledons, only a few plants, mainly in the Liliaceae (Bytebier *et al.* 1987) and Amaryllidaceae, are weakly susceptible to the bacterium.

In nature, infection by *Agrobacterium* usually occurs at ground level near the junction of the stem and roots, which, in the case of invasion by *A. tumefaciens*, may result in tumours in the crown of the plant. Corms, stems, and leaves of woody and herbaceous plants are also vulnerable if damaged. Subterranean chewing insects may aid *Agrobacterium* infection. Galls vary in diameter from a few millimetres to 5–15 cm in older trees and can be up to 22.7 kg in weight (Walker 1969). Deeper wounds facilitate infection and usually result in larger galls. Anatomically, a gall is composed of disorganized vascular and parenchymatous tissues, although some tumours, such as those induced by the nopaline strain T37 of *A. tumefaciens*, produce abnormal, stunted shoots (teratomata), with deformed leaves and fasciated stems. Galls often reduce plant growth and, in some cases, cause host death. Crown gall disease occurs world-wide in nurseries and orchards, especially where apples and pears are grown, when 80 per cent of more of the stock may have to be removed following an epidemic (Moore *et al.* 1990).

Hairy root or woolly knot disease (Riker 1930) is considered to be special type of tumorous response (White and Nester 1980). It is common on grafted apple trees and is characterized by numerous small, cone-like roots, which usually emerge from hard swellings at the graft union. Where plants are infected at ground level, proliferation of adventitious roots may produce a secondary rhizosphere near the soil surface. In contrast to crown galls, the additional root system may be beneficial (Tepfer 1984) and enhance the drought tolerance of apple seedlings (Moore *et al.* 1979).

Cells of crown galls and hairy roots synthesize opines which are absent from non-transformed cells. Opines support the growth of the *Agrobacterium* strain or a related strain, that inflicts the disease. This confers a competitive advantage on the pathogen over other unrelated strains, as only the inciting bacteria can utilize the opines induced as carbon and nitrogen sources (Moore *et al.* 1990). In addition to serving as an energy source, opines can stimulate conjugative transfer of bacterial plasmids. Because plasmid-encoded genes are involved in opine catabolism, this ensures a population of bacteria with the metabolic potential to utilize the opines. Agrobacteria can be cultured on media with opines as the sole carbon and nitrogen sources.

2. Induction of crown galls and hairy roots in the laboratory

Axenic stems and leaves provide suitable explants for inoculation by *Agrobacterium* using a sharp instrument to induce wound sites. Tumours and hairy roots proliferate rapidly under the high humidity conditions of culture. In the case of *A. tumefaciens*, secondary tumours frequently develop at sites remote from the point of inoculation. The hairy root phenotype is clearly apparent following infection of plants such as tobacco, a mass of roots emerging from the point of inoculation to give a typical 'hairy' appearance. Such roots often exhibit prolific root hair development. In other cases, as in legumes such as *Glycine canescens* (Rech *et al.* 1988), relatively few roots are produced at the inoculation site. Consequently, several authors prefer to use the term 'transformed' rather than 'hairy' to describe these roots.

Both crown galls and transformed roots can be excised from the host plant, freed of the inciting bacteria by antibiotic treatment, and maintained indefinitely in culture with regular transfer on nutrient media lacking auxins and cytokinins. Transformed roots are plagiotropic and negatively geotropic with extensive branching. They grow more rapidly *in vitro* than non-transformed roots (Tepfer 1983). In contrast to crown galls, transformed roots are capable of producing whole plants, the regeneration of shoots occurring spontaneously in some species. Roots of other plants require induction by auxins and cytokinins (Tepfer 1984). Tempé and Casse-Delbart (1989) and Tepfer (1990) listed the production of transformed roots of 116 species from 26 plant families, with plant regeneration from 37 species. Plants regenerated from *A. rhizogenes*-transformed roots exhibit aberrant phenotypes, including wrinkled leaves, reduced apical dominance of stems and roots, shorter internodes, a change from biennial to annual flowering and change in flower morphology, reduced fertility, and seed set (Tepfer 1984). In Belgium endive (*Cichorium intybus*), transformation circumvented the need for vernalization to induce flowering (Sun *et al.* 1991).

Historical background and classification of agrobacteria

1. Early experiments on crown gall disease

Several investigators studied crown gall disease between 1892 and 1904 and demonstrated that healthy trees developed galls when macerated tumours were placed near their roots. This suggested that the disease was communicable, although the mechanism was not understood. The Italian worker Cavora isolated the bacterium in 1897 from stem tumours of grape and succeeded in reproducing disease symptoms by inoculating healthy plants. Subsequently, Smith and Townsend isolated a bacterium from Marguerite which they transferred to agar plates. These workers were able to induce galls on healthy plants by pricking the latter with needles dipped in the bacterial culture. In 1907 they reported *A. tumefaciens* as being the causative agent of crown gall disease (Smith and Townsend 1907). Young plant tissues showed symptoms of gall formation within 4–5 days of inoculation.

Jensen (1910) isolated galls from field grown plants of sugar beet and grafted them onto red beet. He demonstrated that the galls exhibited autonomous growth, since some were free of inciting agrobacteria after serial grafting, but continued to grow in the absence of the bacteria. This important observation provided evidence that plant cells had undergone a permanent change or transformation. Later, Smith (1917) detected differences in the host range of different *Agrobacterium* strains and in the growth and morphology of the tumours induced by these strains.

2. Characteristics of agrobacteria and their classification

Agrobacteria are normally rod shaped (0.6–1.0 μm by 1.5–3.0 μm), Gram-negative, aerobic, soil bacteria which occur singly or in pairs. They are motile by one to six peritrichous flagella, are non-sporing, and produce copious extracellular polysaccharide when grown on carbohydrate-containing medium. Frequently, they bear cellulose fibrils which are thinner and longer than their flagella. In culture, the bacterial colonies are non-pigmented and smooth, becoming striated with age. Classification of agrobacteria is based on several characters, such as biochemical behaviour *in vitro* and in habitat. They belong, together with *Rhizobium* and *Phillobacterium* to the Rhizobiaceae and carry plasmids larger than 100 kb in size.

When numerical taxonomy was applied to the genus, it became clear that the early classification system was unreliable (Kersters *et al.* 1973) and that inclusion of certain bacteria in the genus *Agrobacterium* was unjustified. The classification of *Agrobacterium* species is based on their phytopathogenic behaviour (Kersters and De Ley 1984). Thus, species which incite crown galls are *A. tumefaciens*, those which cause hairy root are *A. rhizogenes*, the agent inducing cane galls on *Rubus* species is *A. rubi*, while *A. radiobacter* is non-pathogenic (Moore *et al.* 1990). Species-specific strains are further divided according to the opines synthesized by the galls or roots which the bacteria incite (Table 2.1). The difficulty with this nomenclature is that pathogenicity is based on the presence of a plasmid which can be lost or transferred to a non-pathogenic strain which subsequently becomes pathogenic. Since plasmids move freely between bacteria, the tumour- and root-inducing properties move with them. Problems of this nature have resulted in current nomenclatures and taxonomic structures based on morphology, physiology, and genotypic traits to have no correlation (Kersters and De Ley 1984). Studies of phenotypic characters, chromosomal DNA, and comparisons of electrophoretic protein patterns, together with tests such as the production of ketolactose from lactose, growth at 29, 35, or 37°C, the sensitivity to sodium chloride, growth on erythritol as carbon source, and serological analyses (Alarcon *et al.* 1987), have enabled agrobacteria to be classified into the three biotypes I, II, or III according to their chromosomal characteristics (Kerr and Brisbane 1983). It is thought such a nomenclature objectively differentiates the genus, although it has not yet been officially accepted for distinguishing strains at the species level (Moore *et al.* 1990).

Table 2.1. Classification of the Ti and Ri plasmids according to the opines synthesized by tumours and transformed roots

Ti plasmid type	Representative strains	Opines synthesized by tumours	Ri plasmid type	Representative strains	Opines synthesized by roots
Octopine-type	A6, ACH5, B6, R10, 15955	Octopine, octopinic acid (ornopine), agropine, agropinic acid, mannopine, mannopinic acid, lysopine, histopine	Agropine-type	A4, 1855, 15834, TR105, HR1	Agropine, mannopine, mannopinic acid, agropinic acid, agrocinopine A
Nopaline-type	C58, T37, H100	Nopaline, nopalinic acid (ornaline), agrocinopine A	Mannopine-type	8196, TR107	Mannopine, mannopinic acid, agropinic acid, agrocinopine C
Agropine-type	A281, AT1, AT4, Bo542	Agropine, agropinic acid, mannopinic acid, agrocinopine C	Cucumopine-type	NCPPB2657, 2659	Cucumopine
Succinamopine-type	Eu6, AT181	Succinamopine			
Grapevine-type	K305, K308	Octopine, new unknown opines			

Physiology of *Agrobacterium* transformed cells and plants

Studies of the physiology of transformed cells are few compared to those directed at molecular aspects of transformation and the application of this knowledge to plant genetic engineering.

1. Opine synthesis

One of the most significant biochemical differences between *Agrobacterium*-transformed cells and their non-transformed counterparts is that transformed cells synthesize opines. Opines are low molecular weight nitrogenous- or phosphorus-containing compounds, which are derivatives of amino acids and sugars (Tempé and Goldmann 1982).

Three main classes of opines have been detected in crown galls. Those of the octopine class are *N*-a-(D-1-carboxyethyl) derivatives of L-arginine (octopine), L-ornithine (octopinic acid), L-lysine (lysopine), and L-histidine (histopine), induced by strains of *A. tumefaciens* such as ACH5 and B6. The nopaline group are *N*-a-(1,3-dicarboxypropyl) derivatives of L-arginine (nopaline) and L-ornithine (nopalinic acid), induced by strains such as T37 and C58. The agropine family includes derivatives of L-glutamine such as agropine, mannopine, mannopinic acid, and agropinic acid, which are present in Bo 542- and A281-induced tumours. Crown galls incited by strains AT181, Eu6, and T10/73, synthesize succinamopine. The latter has a structure analogous to that of nopaline, but with asparagine replacing arginine. Some octopine tumours also synthesize arginine and related compounds.

The agrocinopines are phosphorylated sugars whose structure has not been fully determined. Agrocinopines A and B are present in some nopaline tumours, while agrocinopines C and D have been detected in agropine tumours. Succinamopine has been found in tumours induced by bacterial strains Eu6 and AT181, while galls of the grape vine type also synthesize a number of unidentified opines, in addition to octopine.

Three classes of opines have been detected in roots transformed by *A. rhizogenes* (Table 2.1). In general, the concentration of opines present in transformed roots permits them to be visualized on paper electropherograms following staining with silver nitrate. In contrast, opines are present in tumours in much lower concentrations, necessitating the use of the more sensitive reagent phenanthrenequinone for detection.

2. Nutrient requirements, growth regulators, and proteins of transformed cells

A characteristic which has been known for many years is that transformed tissues have less exacting nutrient requirements than non-transformed tissues. Likewise, transformed roots can be grown on simple nutrient media. Analysis has shown that crown galls contain auxins and cytokinins at concentrations sufficient to support continuous growth and cell division, with the level of indole acetic acid being 2–500-fold higher than in non-transformed cells. Cytokinins, such as transzeatin riboside, have also been detected at levels 1620-fold higher than in non-transformed tissues. Other studies of growth regulators in relation to gene function are discussed later under the section on plasmid organization. There is evidence that transformation can affect the permeability and uptake capacity of cells, enabling octopine-type tumours to utilize lactose as a carbon source. Radiolabelled potassium and orthophosphate are also accumulated more rapidly than by non-transformed cells.

Comparisons have been made of the proteins of tumours and non-transformed cells. Increases in RNA synthesis during tumour induction in potato tuber discs were paralleled by an increase in chromatin-bound DNA-dependent polymerases I and II. The phosphorylation of proteins also rose during tumour development, with phosphorylation of low molecular weight proteins occurring mainly following cessation of mitotic activity.

Studies with transformed roots have shown that transformation modifies the metabolic equilibrium of plant cells, with changes in the major groups of peroxidases. The peroxidases present in the roots of Ri-derived plants were the same as those in roots of non-transformed tobacco plants (Gᴵᴵ, Gᴵᴵᴵ, and Gᴵᵛ). In contrast, leaves of transformed plants exhibited these groups, together with Gᴵ peroxidases, whereas leaves of non-transformed plants contained only peroxidases of groups Gᴵ and Gᴵᴵᴵ. The levels of pathogenesis-related proteins were similar in transformed and non-transformed tobacco plants. There is evidence that Ri-transformed plants of tobacco produce more ethylene, which is related to their more vigorous growth in sealed containers compared with non-transformed plants (Tepfer 1984). Collectively, these biochemical differences contribute to the enhanced growth of transformed cells and roots in culture and are probably reflected in phenotypic aberrations, such as wrinkled leaves, in plants regenerated from transformed roots.

Evidence for the involvement of bacterial plasmids in crown gall and hairy root diseases

The fact that crown gall tumours synthesized opines suggested that a 'tumour-inducing principle' might move from the bacterium to control the synthesis of these compounds. For many years, the nature of this principle was investigated using different approaches. Sterile fractions of the bacteria were tested for pathogenicity, but without positive, reproducible results (Gribnau and Veldstra 1969). The role of growth substances was examined, as were DNA preparations from pathogenic *Agrobacterium* strains, but the preparations failed to induce crown galls (Kado *et al.* 1972). Interestingly, virulence could be transferred from *A. tumefaciens* into the non-pathogenic *A. radiobacter* when water-splashed soil containing *A. radiobacter* made contact with stems of tomato plants inoculated with *A. tumefaciens* (Kerr 1969).

Concurrent with some of these studies was the recognition that bacterial cells carry extrachromosomal plasmids, the latter being large, covalently closed, circular DNA molecules which replicate autonomously and which encode specific bacterial functions. The use of buoyant density gradient centrifugation permitted plasmids to be isolated, while the use of restriction endonucleases to cleave DNA at specific base sequences, combined with techniques such as agarose gel electrophoresis, Southern blotting, and autoradiography, enabled restriction maps of isolated plasmids to be constructed. Whilst emphasis was on plasmids of *Escherichia coli*, DNA hybridization studies indicated that *A. tumefaciens* DNA sequences were present in genomic DNA from crown galls, suggesting that bacterial DNA was transferred into the host plant during transformation. These bacterial genes were expressed, since bacterial proteins could be detected immunologically in tumours (Schilperoort *et al.* 1969). Such investigations also suggested that bacterial DNA was involved in pathogenicity. Interestingly, the ability of the highly virulent nopaline strain C58 of *A. tumefaciens* to induce tumours was lost when the bacterium was grown for 4 weeks at 36°C. In other strains, the loss of pathogenicity was more rapid and began to occur within 48 h.

The application of alkaline and neutral sucrose gradient and dye-buoyant density centrifugation to DNA preparations from *A. tumefaciens* enabled plasmids to be isolated from this bacterium (Zaenen *et al.* 1974). In spite of the fact that only microgram quantities could be prepared, even with the introduction of improved ultracentrifugation technology employing vertical rotors (Draper *et al.* 1982), a single copy, non-amplifiable mega plasmid was identified in *A. tumefaciens*. The plasmid was visualized by electron microscopy and characterized by restriction enzyme analysis. Studies also correlated the loss of virulence of *A. tumefaciens* to loss of this large plasmid of approximately 200 kb in size. Furthermore, transfer of plasmid from tumourigenic strains of *Agrobacterium* to plasmid-free strains enabled the latter to induce tumours (Van Larabeke *et al.* 1975), confirming the involvement of the plasmid in tumour induction. Thus, the term tumour-inducing (Ti) plasmid was coined. Subsequent studies, involving DNA–DNA hybridization, showed that a specific segment of the Ti plasmid, the transferred or T-DNA carrying the genes for tumour induction, became integrated into the higher plant genome, where it was expressed (Chilton *et al.* 1977).

The type of Ti plasmid carried by the bacterial strains was defined by the opines synthesized by transformed plant cells. Likewise, virulence in *A. rhizogenes* was also shown to be encoded by large root-inducing (Ri) plasmids (White and Nester 1980).

Organization of Ti and Ri plasmids

1. Size of Ti and Ri plasmids

The Ti plasmid of oncogenic strains of *A. tumefaciens* is a single copy molecule, approximately 90–115×10^6 daltons (180–220 kb) in size. All strains of *A. rhizogenes* have at least one plasmid of a size comparable to that of the Ti plasmid, with some strains harbouring three plasmids. For example, strain 15843 carries plasmids a (107×10^6 daltons), b (154×10^6 daltons) and c (258×10^6 daltons). Virulence is carried on plasmid b, while plasmid c is a cointegrate of a and b.

The octopine-type Ti plasmids, such as pTi ACH5, have been well characterized at the molecular level. Approximately 180 genes reside on the 200 kb plasmid, with loci for plasmid replication (*ori*), plasmid stability, virulence (*vir*), incompatibility (*inc*), and transfer (*tra*) between bacteria by conjugation. The *ape* gene confers on bacterial cells resistance to or exclusion of the bacteriophage AP1. Genes are also present which code for a permease and a membrane-bound oxidase to enable the bacterial cells to take up and to catabolize opines. A third gene is responsible for the degradation of the resulting arginine, the activity of such genes being induced by opines. The transfer of Ti plasmids between bacteria is also influenced by opines.

2. T-DNA regions of Ti and Ri plasmids

The T-DNA regions of the Ti and Ri plasmids carry the genes which are transferred to, integrated, and expressed in recipient plant cells. Several authors (for example, Ooms 1992) have published T-DNA restriction maps. The T-DNA in a nopaline-type Ti plasmid is present as a contiguous segment of approximately 23 kb, while in octopine-type plasmids it consists of two regions, the left (TL)-DNA and right (TR)-DNA of 13.1 and 7.9 kb, respectively. Transfer of the TL- and TR-DNAs occurs independently. Although the TL-DNA is present in all transformed plant cells and is essential for transformation, the TR-DNA is frequently absent. However, when present, the TR-DNA may be linked to the TL-DNA or integrated independently into the plant genome with a similar or different copy number. The nopaline T-DNA and the octopine TL-DNA carry the genes for synthesis of phytohormones, together with the genes for synthesis and secretion of opines. These genes are located between the 24 bp (stated as 25 bp by some authors) direct imperfect repeat left and right borders of the T-DNA. Both TL- and TR-DNAs have these borders. The border sequences flanking the T-DNA are essential for transformation. These sequences contain the incorporation sites for cleavage and excision of the T-DNA during transfer to plant cells (Yanofsky *et al.* 1986). On the TL-DNA, nine open reading frames (ORFs) larger than 380 bp have been identified, eight of which correspond to RNA transcripts are low in abundance, although some of their functions are known. Six genes have been identified, of which three are involved in tumour growth. Two of these genes, *iaaM (tms1, aux-1,* transcript 1) and *iaaH (tms2, aux-2,* transcript 2) encode an oxygenase and a hydrolyase, respectively that convert tryptophan to the auxin indoleacetic acid through the inactive intermediate indole 3-acetamide (Thomashow *et al.* 1987). Indole 3-acetamide hydrolase is also able to hydrolyse indole 3-acetonitrile and esters of indole acetic acid to active auxins. The way in which different dicotyledons respond to the T-DNA auxin genes is remarkably variable. For example, the same auxin biosynthetic gene can induce auxin independence in *Nicotiana glutinosa,* but fail to do so in *N. tabacum.* Under some circumstances, it is even possible for the transfer of a gene involved in cytokinin biosynthesis to be involved in auxin autonomy (Binns *et al.* 1987). The ratio of auxin:cytokinin in tumour tissue is, however, likely to be a major determinant of the gall's phenotype. Differences in the way auxin is sensed and auxin biosynthesis regulated may also be important in determining the final phenotype exhibited. At the present time, our understanding of these factors and their interaction in plants is far from complete.

The third gene of major importance on the TL-DNA, *ipt (tmr, cyt,* transcript 4), encodes dimethylallyl pyrophosphate 5'-adenosine monophosphate transferase or isopentenyltransferase. This is involved in the condensation of 5'-adenosine monophosphate with 2-isopentenyl pyrophosphate to form *n*6-(2-isopentenyl)-5'-adenosine monophosphate, which may thereafter be hydroxylated into adenine- or zeatin-type cytokinins (Barry *et al.* 1984). Ti plasmids deleted for the auxin biosynthetic genes, but with a functional isopentenyl transferase, tend to incite shooty teratomas. The isopentenyl transferase gene may also play a role in extending the host range of some *Agrobacterium* strains. Expression of T-DNA genes 1,2, and 4 tends to lead to increased concentrations of auxins and cytokinins. The extent of these elevations vary tremendously and may be influenced by the copy number of T-DNA insertions, insertion position with the plant genome, methylation patterns, and the growth receptor status of the transformed tissue. These factors may all have an effect on the physiology and final phenotype of transformed cells. Results using transformed plants provide an indication of the plant's metabolic state, but it must be remembered that it is only a snapshot of an exceedingly complex system, in one cell of which, for example, there may be 10 mRNA transcripts with an abundance of 4500 molecules per cell and as many as 11 300 mRNAs with an abundance of 17 copies per cell (Goldberg *et al.* 1978). The gene for octopine or nopaline synthesis (*nos* or *ocs*, transcript 3) is located near the right border sequence of the T-DNA. Two genes, for transcripts 6a and 6b, determine octopine and nopaline secretion (*ons*) by crown galls (Messens *et al.* 1985) and a protein which may modulate cytokinin and auxin activities in transformed cells (Hooykaas *et al.* 1988).

The Ti TR-DNA encodes five transcripts of which 1' and 2' are involved in mannopine biosynthesis and transcript 0' in the subsequent biosynthesis of agropine (Komro *et al.* 1985). The functions of transcripts 3' and 4' are not known. The TR-DNA genes are not essential for tumourigenesis.

Left (TL)- and right (TR)-DNA regions have been identified in agropine-type Ri plasmids. The two DNAs are each approximately 20 kb in length and are separated by approximately 15 kb of DNA which is not transferred to the plant genome. The TR-DNA contains genes involved in auxin synthesis, which are homologous to *aux-1* and *aux-2* from the T-DNA of Ti plasmids, as well as genes involved in agropine synthesis (*ags*) (Huffman *et al.* 1984). Transposing insertions into these genes of the Ri TR-DNA make the *A. rhizogenes* strains avirulent (Offringa *et al.* 1986). This avirulence can be negated by exogenous application of auxin during transformation, indicating that auxins produced by TR-DNA expression provide the initial stimulus for the growth of Ri TL-DNA-transformed cells (Cardarelli *et al.* 1987). The Ri TL-DNA does not appear to be closely related to any other characterized Ti plasmid, although limited homology has been reported to the T-DNA of the nopaline-type plasmids, presumably to the region involved in agrocinopine synthesis (Huffman *et al.* 1984).

In the Ri TL-DNA, 11 ORFs and at least 10 transcripts have been identified, with four loci designated *rol* A,B,C, and D being shown to affect the transformation response in some plants (White *et al.* 1985). In transformed roots, the TL-DNA encoded RNAs vary in abundance, with levels up to 0.01 per cent of the poly A-RNA in roots from the most abundant transcript 8 (Durand-Tardif *et al.* 1985; Ooms *et al.* 1986). Most of the transcripts are differentially expressed in different organs of transgenic plants regenerated from Ri-transformed roots. Ri TL-DNA gene products may modify plant cell membranes, possibly through modulation of auxin receptor activity (Shen *et al.* 1988). Mutations of the *rol* loci have demonstrated that the TL-DNA is responsible for the aberrant phenotype of transgenic plants regenerated from Ri transformed roots, which suggests that this region is either involved in cytokinin synthesis or in altering cytokinin metabolism in such plants (Sinkar *et al.* 1988). Cardarelli *et al.* (1987) have shown that *rol A, B,* and *C* act synergistically both in rhizogenesis and in generating plant growth abnormalities.

The importance of the *rol B* locus in hairy root formation has been borne out by the work of Shen *et al.* (1988) and Maurel *et al.* (1990), which demonstrated that *rol B* expression increases auxin sensitivity by at least 100-fold in tobacco protoplasts. The *rol B* encoded protein hydrolyses indole 3-glucosides, releasing active auxin from the inactive glucoside conjugates (Estruch *et al.* 1991*a*). The *rol C* encoded protein acts similarly to release free cytokinins from their glucoside conjugates and is localized in the cytosol (Estruch *et al.* 1991*b,c*). At the phenotypic level, *rol C* expression leads to the regeneration of dwarf plants, with increased shoot formation. As yet, little is known regarding the functions of the two remaining *rol A* and *rol D* loci. The Ri plasmid *rol* loci appear to be able to interact with the TR-DNA auxin biosynthetic loci to influence free IAA levels in *Solanum dulcamara* (Gartland *et al.* 1991). Free IAA levels in cells expressing the TR-DNA auxin genes only were significantly greater than the concentrations found in cells expressing the *rol A, B,* and *C* genes as well as the TR-DNA auxin genes. This may suggest some form of feedback interaction between the *rol* genes and the TR-DNA encoded auxin biosynthetic machinery. Support for this may come from Dominov *et al.* (1992), who have shown that cytokinins and auxins can influence gene expression in *N. plumbaginifolia* by feedback regulation of one another. This may involve activation by phosphorylation of critical proteins in the growth regulator signalling pathway and dephosphorylation during feedback inhibition. It is clear that the *rol* loci can exert a profound influence on plant growth and development.

Transformed roots have been obtained by infection with agrobacteria-carrying Ri plasmids that carry only the TR-DNA or only the TL-DNA, suggesting that agropine-type Ri plasmids can induce root proliferation by two independent mechanisms. However, neither the TR-DNA nor the TL-DNA alone induce a response comparable to that of wild-type strains, implying that the two regions cooperate in the wild-type response. Auxin production may be responsible for root induction by the TR-DNA alone (Vilaine and Casse-Delbart 1987*b*). Bouchez and Camilleri (1990) identified a gene homologous to the TL-DNA *rolB* gene in the TR-DNA of the agropine Ri plasmid A4 (designated *rolB^{TR}*) which may contribute to the root-inducing activity of the Ri TR-DNA.

Less information is available for the mannopine-type and cucumopine-type Ri plasmids and only one T-DNA region has been identified (Combart *et al.* 1987). Investigations of the T-DNA of the mannopine-type plasmid 8196 showed the absence of the *aux* genes, but homologies with the *rol* loci of the TL-DNA. Vilaine and Casse-Delbart (1987*b*) proposed a similarity between the molecular mechanisms of root induction by mannopine-type Ri T-DNA and agropine Ri TL-DNA.

Genetic interaction between *Agrobacterium* and plants

1. Attraction and attachment of Agrobacterium to plant cells

Agrobacteria infect their hosts at wound sites. Wounded plant tissues exude a conglomerate of chemicals, including biosynthetic precursors or breakdown products of lignin (aromatics) and wound hormones (aliphatics) (Shaw *et al.* 1988). The chemotactic response between the bacteria and the plant attractant can be achieved at concentrations as low as 10^{-7} molar.

Attachment of bacteria to the plant surface appears to be necessary for tumour induction, since non-attaching mutants are avirulent (Douglas *et al.* 1982). Matthysee (1984) proposed that the host plant cell has a receptor site of protein and pectin, which is exposed on its surface and a second binding site, of unknown composition, on the plasma membrane. The bacteria have a binding site composed of protein and lipopolysaccharide exposed on their outer membranes. Both plant receptor and bacterial binding sites appear to exist prior to any contact between the bacteria and the plant. After binding, the bacteria synthesize cellulose fibrils, some of which anchor the micro-organisms to the surface of plant cells. Others entrap unattached bacteria, creating a large aggregate of bacteria and fibrils on plant cell surfaces.

2. Involvement of bacterial chromosomal genes in bacterial attachment to plant cells

Loci on the bacterial chromosome are important in tumour formation. Thus, the neighbouring, constitutively expressed, chromosomal virulence genes *chvA* and *chvB* are required for attachment of bacteria to plant cells (Douglas *et al.* 1985). Zorreguieta *et al.* (1988) found that *chvB* codes for a 235 kDa protein involved in the formation of a cyclic beta-1,2 glucan, while Cangelosi *et al.* (1989) reported that *chvB* determines a protein located in the bacterial inner membrane. This protein is necessary for transport of beta-1,2 glucan into the periplasm, indicating a possible role of this compound in the attachment of agrobacteria to plant cell walls. Another chromosomal locus, *psca*, is involved in the synthesis of major neutral and acidic extracellular polysaccharides (Thomashow *et al.* 1987). Other genes, *attC43* and *attC69*, are thought to be involved in bacterial attachment to plant cells.

3. Plant phenolic compounds and the virulence region of Ti and Ri plasmids

Two plant phenolic compounds, acetosyringone (4-acetyl-2,6-dimethoxyphenol) and alpha-hydroxy-acetosyringone (4-[2-hydroxyacetyl]-2,6-demethoxyphenol) activate the Ti and Ri plasmid virulence (*vir*) genes (Stachel *et al.* 1985) at concentrations two orders of magnitude higher than required for chemotaxis (Shaw *et al.* 1988). Interestingly, these plant phenolic compounds are not produced by some of the monocotyledons which fail to respond to *Agrobacterium* infection (Usami *et al.* 1987). Nucleotide sequence and DNA hybridization have suggested a functional similarity between the *vir* regions of Ti and Ri plasmids (Sinkar *et al.* 1987). Indeed, all the known Ti plasmid *vir* loci are present on agropine type Ri plasmids (Huffman *et al.* 1984).

The *vir* region of Ti plasmids, sited outside the T-DNA, is a segment of approximately 40 kb consisting of at least 22 genes in seven operons (*virA* to *virG*), which are involved in the excision and transfer of the T-DNA to plant cells (Stachel and Nester 1986). Detailed reviews have been published of the *vir* region, including restriction maps of this part of the Ti plasmids (Melchers and Hooykaas 1987).

VirA, B, D, and *G* are essential for tumour formation in all plants assessed; *virC, E,* and *F* are important in some cases and, consequently, determine host specificity. Transcription of the *vir* genes is regulated by the constitutively expressed genes *virA* and *virG* (Winans *et al.* 1988). These regulatory genes are homologous to other prokaryotic regulatory proteins (Ronson *et al.* 1987). *VirA* acts as a histidine kinase (Huang *et al.* 1990) and its protein, which is present on the bacterial inner membrane, recognizes the presence of plant cells by interacting with plant phenolic compounds in the periplasmic space. The phenolic compounds induce autophosphorylation of *virA* which, in turn, phosphorylates *virG*, the response regulator (Jin *et al.* 1990). Both *virA* and *virG* are regulators, whose action is required to induce the other *vir* genes. Hence, mutants in *virA* and *virG* are avirulent (Lewin 1990). Phosphorylation of the *virG* protein induces *virG* expression to a higher level, through transcription from a different location, compared to the site used for constitutive expression. *VirG* is a sequence-specific DNA-binding protein which attaches to DNA sequences, called *vir* boxes, located upstream of the promoters of the virulence genes. The binding protein may function by activating transcription of the *vir* regions (Pazour and Das 1990). Recent studies have shown that one or more *vir* box sequences are required for expression of *virB, virE,* and *virG*, although the number and location of *vir* box sequences vary with different Ti plasmids. Closely related Ri plasmids have a *vir* box sequence equivalent to the 3' half of the Ti plasmid *vir* box (Aoyama and Oka 1990).

The *vir* genes B, C, D, and E are important in excision of the T-DNA and its transfer. Any sequence between the T-DNA repeated flanking sequence of 24 nucleotides is transferred to plant cells. The *virD* locus has four ORFs, of which two have been characterized. *VirD1* encodes topisomerase and *virD2* an endonuclease, which process the repeats by a covalent complex to produce single- and double-stranded breaks at the right border (Yanofsky *et al.* 1986). This initiates the production of free, single-stranded T-DNA molecules or T-strands (Stachel *et al.* 1987), corresponding to the 'bottom strand' of the T-DNA. The start or 5' end of a T-strand maps to the T-DNA right border, with the transfer being from the right to the left border. Deletion of the right border prevents T-DNA transfer; deletion of the left border results in transfer past the T-DNA to a pseudoborder or, in some cases, all the way round to the right border (Rubin 1986). One of the *virC* proteins binds to a sequence next to the right border repeat, called 'overdrive'. This sequence is a natural enhancer, which ensues high efficiency T-strand synthesis and which can function up to 6714 bp away from the 24 bp repeat (Van Haaren *et al.* 1987).

The *virE* locus encodes 7.0 and 60.5 kDa polypeptides in octopine Ti plasmids (Winans *et al.* 1987) and a 64 kDa polypeptide in nopaline Ti plasmids (Hirooka *et al.* 1987), which are single-stranded DNA-binding proteins. It is thought that these proteins form a complex with T-strand DNA, protecting the latter against bacterial and plant nucleases during T-DNA transfer to plant cells. The protein(s) may also be involved in T-DNA integration into recipient plant cells (Citovsky *et al.* 1989). Other *virE* and *virF* genes are probably host range determinants, since their functions are necessary for tumour induction on certain plants (Melchers *et al.* 1990). Some of the 11 proteins encoded by the *virB* locus are involved in T-strand transfer to plant cells. Indeed, they may determine a pilus or pore for T-DNA mobilization via a conjugation-like mechanism (Christie *et al.* 1988; Thompson *et al.* 1988), although it is still not clear how T-DNA passes across the bacterial and plant membranes. In nopaline-type Ti plasmids, an accessory *vir* gene (*tzs*) is responsible for the synthesis and secretion of transzeatin by the bacteria, which may condition plant cells for transformation, possibly by inducing cell division (Powell *et al.* 1988).

Integration of T-DNA into plant genomes

The mechanism of T-DNA integration into the plant genome is unclear. Molecular analysis of the genome of transformed plants and their progeny shows that foreign genes are often truncated, rearranged, or present in tandem. Frequently, single or on average three tandem copies of the T-DNA integrate into the genomes of dicotyledons. Integration is random at one or more genomic locations (Wallroth *et al.* 1986) and is predominately in the form of inverted repeats, although direct repeats are also found (Jorgensen *et al.* 1987). Insertion of the T-DNA is more precise on the right side than the left, suggesting that its integration, like the generation of the transferable T-strand copy, is directed by the right T-DNA border. T-DNA transfer is thought to be stimulated by the 'overdrive' region of the Ti plasmid (Lewin 1990).

A variety of T-DNA as well as target DNA rearrangements, including deletions, inversions, and duplications have been found (Mayerhofer *et al.* 1991), indicating that T-DNA integration may be a multistep process involving several different types of recombination, replication, and repair activities, most probably mediated by enzymes encoded by the host plant. Gheysen *et al.* (1987) have proposed that T-DNA integration is a four-stage process following transfer of the T-strand into the plant nucleus, although it is not clear how the T-strands traverse the cytoplasm and enter the nucleus. They proposed that

(1) a protein at the 5' end of the T-strand interacts with a nicked sequence in the plant DNA;

(2) attachment of the T-strand to one strand of the plant DNA results in a torsional strain which is relieved by producing a second nick on the opposite strand of the target site, the position of this nick varying with the structure of the plant chromatin at the insertion site;

(3) each end of the T-strand is ligated to the plant DNA and its homologous strand is copied by cellular enzymes;

(4) repair and replication of the staggered nicks in the plant DNA result in a repeated sequence of variable length and additional rearrangements at the ends of the inserted T-DNA.

Host functions and the *virD2* protein mediate the formation of the initial synapsis, partial homologous pairing, and DNA repair of the junctions between T-DNA inserts and target plant DNA sequences (Mayerhofer *et al.* 1991). Foreign DNA integration into the plant genome probably occurs during the first two division cycles, usually preceding endomitotic chromosome doubling, preferentially during the DNA replication phase. Once integrated, the T-DNA genes are expressed, resulting in the typical crown gall or transformed root phenotypes.

Agrobacterium for genetic engineering of plants

The most successful way of genetically engineering dicotyledonous plants is by employing the Ti and Ri plasmids of *Agrobacterium* as gene vectors.

Other bacteria such as *Rhizobium* and *Rhodococcus* may transform some plants in nature, but have not been exploited as vector systems (Crespi *et al.* 1992). Vectors based on viral genomes have been constructed, but these have limitations due to their pathogenic nature, restrictions in the size of DNA that can be packaged, and host range. In addition, the transgene is not stably transmitted through seed progeny. Transformation by isolated DNA generally requires a protoplast-to-plant system, which still does not exist for many species.

1. Early experiments using Ti plasmids as gene vectors

The first experiments in transferring foreign DNA into plants using the Ti plasmid as a vector involved insertion of the bacterial transposon Tn7 into the nopaline synthase gene of pTiT37 (Hernalsteens *et al.* 1980). Crown galls initiated on tobacco carried Tn7 as demonstrated by DNA hybridization. Tn7 encodes resistance to streptomycin, spectinomycin, and trimethoprim in *Escherichia coli*, but crown galls failed to survive in culture on medium containing these antibiotics. This indicated that bacterial genes failed to express in plant cells, because they lacked the correct regulatory sequences. Shortly after this demonstration of gene transfer, improvements in recombinant DNA technology enabled chimaeric genes, consisting of a suitable eukaryotic promoter, the gene sequence of interest, and a termination sequence, to be constructed. Such chimaeric genes expressed in transformed plant cells. For example, the gene for neomycin phosphotransferase II (*npt*II), which confers resistance to aminoglycoside antibiotics such as kanamycin sulphate, was cloned in place of the nopaline synthase (*nos*) gene within the T-DNA of a Ti plasmid. Galls initiated by agrobacteria carrying this engineered plasmid grew *in vitro* on medium containing kanamycin sulphate, demonstrating that the *nos* promoter and terminator sequences were suitable for efficient transgene expression (Herrera-Estrella *et al.* 1983*a*). Other genes, such as the octopine synthase (*ocs*) and the chloramphenicol acetyltransferase (*cat*) genes, were also expressed in transformed plant cells using the *nos* regulatory system. At least 50 kb of DNA could be introduced into plant cells using the Ti plasmid as a vector. These experiments formed a basis for the transfer of other, more important genes into plants.

2. Disarmed Ti plasmids for plant transformation

A major limitation of using wild-type Ti plasmids for gene transfer was that transformed cells either failed to produce shoots or regenerated shoots were phenotypically abnormal and did not develop roots. Indeed, such shoots had to be grafted onto non-transformed stock plants in order to maintain them *ex vitro* to flowering. When it was discovered that the oncogenic (*onc*) genes encoded by the Ti plasmid were neither required for the transfer of T-DNA to plant cells, nor its integration into the nuclear DNA, the assembly of gene transfer agents was directed to the construction of disarmed vectors lacking the *onc* genes. Although removal of the *onc* genes resulted in a failure of neoplastic outgrowths to develop on infected explants, transformed cells could be identified by the expression of chimaeric bacterial antibiotic resistance genes (for example, the *npt*II gene) under selection conditions (Draper *et al.* 1988). An advantage of disarmed vectors was that regenerated plants were phenotypically normal and non-oncogenic T-DNAs present in regenerated plants were transmitted to seed progeny in a Mendelian fashion.

Irrespective of the nature of the DNA to be transferred into plants, the sequence must be flanked by T-DNA borders and be stably maintained in the *Agrobacterium* strain harbouring a complete *vir* region, either *in cis* or *in trans*. Additionally, chromosomal genes associated with virulence must be present for transformation. The large size of Ti plasmids causes difficulty for gene manipulation, especially in locating single restriction enzyme sites into which foreign DNA can be cloned. When it was found that the *vir* region will act *in trans* on T-DNA carried on another plasmid, it offered the opportunity for small plasmids, more amenable to manipulation, to be incorporated into the transformation system. Overall, vectors require a selectable marker which functions in *Agrobacterium*, a selectable marker for expression in plants, and the border sequences for T-DNA transfer to plants. The recombinant DNA needs to be linked to the correct plant regulatory signals, which, in early experiments, were T-DNA-derived, such as *nos* and *ocs*. For constitutive gene expression studies, such sequences have been superseded by more powerful promoters, such as the 35S promoter from cauliflower mosaic virus (CaMV). Where agronomically useful genes are being introduced into plant cells, it is essential to ensure that the foreign DNA of interest and the marker DNA sequence are transferred into the same cell. Thus, it is preferable if both sequences are tightly linked. Currently, there are two types of disarmed Ti plasmid vectors in use. Specifically, the *cis* or cointegrate vectors and the *trans* or binary vectors.

3. Cointegrate Ti and Ri vectors

The construction of disarmed, cointegrate vectors involves deletion of the *onc* genes and their replacement with a DNA sequence having homology to a small cloning vector, such as pBR322 from *E. coli*. An example is pGV3850, which was one of the first disarmed Ti vectors to be produced (Zambryski *et al.* 1983). When the disarmed Ti plasmid, with its intact *vir* region and a small *E. coli* cloning (intermediate) vector, carrying a plant selectable marker gene together with sites for cloning of foreign DNA, are introduced into the same bacterial cell, they undergo homologous recombination. This results in the foreign gene(s) of interest being inserted between the T-DNA borders. The intermediate vector is introduced into *Agrobacterium* by conjugation and transconjugants selected by appropriate antibiotic treatment. In other constructs, the T-DNA borders are present on the intermediate vector, as in pGV2260 (Deblaere *et al.* 1985), while in split-end vectors (Fraley *et al.* 1985) the right border is on the intermediate vector and the left border is on the disarmed Ti plasmid. However, after cointegration, the foreign DNA is again inserted between the T-DNA border sequences on the modified Ti plasmid.

In contrast to wild-type Ti plasmids, which must be disarmed by removal of the *onc* genes to allow regeneration of plants, wild-type, non-disarmed Ri plasmids can be used as recipients for the intermediate vector. The latter possesses a homologous region for recombination within the T-DNA, which, in agropine-type strains, is usually within the TL-DNA (Morgan *et al.* 1987). Disarmed Ri plasmids have also been constructed in which the TL-DNA *rol* genes have been placed by pBR322 sequences (McInnes *et al.* 1989). These vectors have been used to insert chimaeric antibiotic resistance genes into plants and also to study the expression of *rol* genes, either alone or in combination, in plants (McInnes *et al.* 1991). In general, strains of *A. rhizogenes* carrying disarmed Ri plasmids are not as virulent as strains carrying wild-type Ri plasmids.

4. Binary vectors

Binary vectors can replicate in both *E. coli* and *Agrobacterium* and possess the left and right T-DNA borders, which flank a polylinker sequence into which foreign genes can be cloned. The vectors are usually engineered in *E. coli* and transferred by conjugation into *Agrobacterium* strains which contain a disarmed Ti plasmid lacking border sequences, but an intact *vir* region. The foreign DNA inserted into the cloning vector is transferred to plant cells by the activity of the *vir* region acting *in trans*. Many cloning vectors, such as pBIN19, cannot be directly conjugated into *Agrobacterium* as they lack the *tra* (transfer) and *mob* (mobility) functions, but have a *bom* site where conjugative transfer can be initiated. However, conjugation can occur by triparental mating, provided these missing functions are provided *in trans* by helper plasmids such as pRK2013.

The knowledge that Ti and Ri *vir* functions can complement each other has permitted the use of Ti- and Ri-based vectors in *A. rhizogenes* strains carrying wild-type Ri plasmids, as well as in disarmed *A. tumefaciens* strains (Hooykaas *et al.* 1984). Thus, disarmed Ti-derived binary vectors can be introduced into *A. rhizogenes* and will replicate, provided that the strain is grown under antibiotic selection. Infection of explants results in transformed roots and regenerated shoots, which acquire both the disarmed T-DNA from pTi, as well as the T-DNA from pRi (Simpson *et al.* 1986). This indicates that the *vir* region in *A. rhizogenes* is able to act upon the *A. tumefaciens*-derived border sequences located on the binary vector to effect T-DNA transfer. Some experiments have demonstrated that only the T-DNA from the binary vector is transferred (Shahin *et al.* 1986), suggesting an independent transfer of the T-DNAs. The latter is desirable, especially in crop plants, since the absence of Ri plasmid sequences permits the regeneration of phenotypically normal plants.

Another binary strategy involves the construction of vectors containing either the border sequences of a pRi T-DNA (disarmed micro-Ri plasmids, for example, pMRK62) or the border sequences, together with portions of the pRi T-DNA (root-inducing mini-Ri plasmids, such as pMRKE15) (Vilaine and Casse-Delbart 1987*a*). These binary vectors possess the origins of replication of ColE1 and of agropine-type Ri plasmids, bacterial selection markers, and a plant selectable marker between the T-DNA borders. They can be introduced into an *Agribacterium* strain harbouring either a Ti or a Ri plasmid lacking its entire T-DNA region. An additional binary strategy involves vectors carrying only the replication origin of agropine plasmids, allowing their replication in *A. tumefaciens* (Jouanin *et al.* 1985). These origins of replication have been used to construct vectors which contain the T-DNA borders of Ti plasmids. Such vectors have been used to introduce genomic libraries into plant cells (Simoens *et al.* 1986).

In general, vectors are designed for specific purposes and the system employed depends on the aim of the experiment and the target plant. Thus, expression vectors, cosmid vectors, plasmid–cosmid rescue vectors, and gene-tagging vectors have been constructed (Draper *et al.* 1988). Mobilizing of vectors into *Agrobacterium* can be variable, but not difficult. The frequency of transconjugant selection in *Agrobacterium* is higher for *trans* (10^0–10^{-1}) than for *cis* vectors (10^{-5}–10^{-7} (Fraley *et al.* 1985). Binary vectors can be stabilized in any chromosomal background in *Agrobacterium*, although antibiotic selection is required to maintain the vector in the host strain. In contrast, cointegrate vectors are generally stable in *Agrobacterium* in the absence of antibiotics and they have been reported to transform plant cells at higher frequencies than binary vectors (McCormick *et al.* 1986).

Chimaeric marker genes for plant transformation

A number of chimaeric genes have been constructed to monitor gene expression and to facilitate the selection of transformed tissues and plants.

In general, stably transformed plant cells have been selected by their resistance to antibiotics and herbicides.

1. Genes conferring antibiotic resistance

The gene for neomycin phosphotransferase type II (NPTII; synonym aminoglycoside phosphotransferase (3') II [APH(3')II]), originally isolated from the bacterial transposon Tn5, confers resistance to aminoglycoside antibiotics such as kanamycin, neomycin, Geneticin 418 (G418), and paramomycin. The NPTII enzyme catalyses the orthophosphorylation of these antibiotics and, by inhibiting the antibiotics from interacting with their ribosomal targets, overcomes the inhibition of protein synthesis. Since the first reports of the *npt*II gene being employed for tobacco and petunia (Bevan *et al.* 1983), this gene has been the marker used most extensively for selecting transformed tissues and plants (Klee and Rogers 1989). The coding sequence of the gene has been fused to strong constitutively expressed promoters, such as the *nos* and CaMV promoters, to maximize expression in plant cells. Enzymatic assays have been developed for NPTII expression, based on the ability of the enzyme to catalyse the phosphorylation of kanamycin sulphate *in vitro*, using ATP labelled at the gamma phosphate (McDonnell *et al.* 1987). A non-radioactive ELISA method has also been developed to detect and to quantify NPTII (5 Prime → 3 Prime, Inc. 1990). The 'tightness' of selection, by the resistance of cells to kanamycin sulphate or G418 may vary considerably from one plant to another.

Whilst cells of *Nicotiana* and *Petunia* are easily selected by their kanamycin resistance, the same does not hold for *Brassica*. Additionally, the selection can vary in the same plant according to the source material. Exposure of cells to kanamycin sulphate for extended periods may also inhibit shoot regeneration from transformed tissues.

Methotrexate is an inhibitor of the enzyme dihydrofolate reductase. Plant cells are very sensitive to low concentrations of this compound, and a chimaeric bacterial methotrexate-insensitive dihydrofolate reductase gene, under the *nos* or the CaMV 35S promoters, has been used to select transformed plant cells. In general, transformation frequencies for some species are not as high as with the *npt*II gene, which may be due to the greater toxicity of methotrexate relative to kanamycin (Klee and Rogers 1989).

Hygromycin B is an aminoglycoside antibiotic which inhibits protein synthesis by disrupting ribosomal function. Resistance in *E. coli* is encoded by a plasmid-borne hygromycin phosphotransferase gene, which also expresses in plant cells under a suitable promoter. Although hygromycin gives 'tight' selection, it is highly toxic to mammalian cells. Another antibiotic which has been used for selection is the polypeptide bleomycin, which interacts with DNA to produce single- and double-stranded breaks. A gene from Tn5 codes for a polypeptide which confers resistance to bleomycin. Although this gene has been transferred into tobacco (Hille *et al.* 1986), it has not been used extensively in plant transformation.

A gene from Tn9, coding for chloramphenicol acetyltransferase (CAT), has also been employed as a selectable marker (Herrera-Estrella *et al.* 1983*b*). However, CAT expression offers poor selection, since non-transformed plant tissues often exhibit endogenous CAT activity, as in *Brassica* (Balazs and Bonneville 1987).

2. Genes conferring resistance to herbicides

Knowledge of the mechanism of action of many herbicides, together with current interest to generate herbicide-resistant crop plants, has stimulated the use of such compounds as selectable markers for plant transformation. Transgenic tobacco and tomato plants tolerant to glyphosate, an inhibitor of the enzyme enolpyruvylshikimate-3-phosphate (EPSP) synthase, have been obtained following transformation by either a bacterial mutant gene (Fillatti *et al.* 1987) or a gene which overexpresses the plant wild-type enzyme (Shah *et al.* 1986). Resistance to chlorsulphuron, a sulphonyl urea herbicide which blocks the synthesis of branched-chain amino acids by inhibiting the enzyme acetolactate synthase (ALS), has been reported in transformed tobacco and pea tissues (Mullineaux *et al.* 1987), following introduction of a mutant gene coding for a resistant enzyme. Tomato, tobacco, and potato plants have also been obtained which are resistant to the non-selective herbicide phosphinotricin, a potent inhibitor of glutamine synthase in plant tissues. In these studies, a chimaeric *bar* gene encoding phosphinotricin acetyltransferase, which confers resistance to these herbicides in bacteria, was introduced into plant cells.

3. Marker genes for rapid assays of transformation

Scorable markers permit verification of the transformed nature of cells and tissues, with opines being used initially to screen for transformation. Assays for the presence of opines in transformed tissues involves electrophoretic separation, followed by detection with reagents such as phenanthrenequinone (octopine and nopaline) or silver nitrate (agropine and mannopine). However, opines are difficult to quantify, which has limitations for studying levels of gene expression. In addition, the assay must include adequate controls, since damaged non-transformed plant tissues can produce compounds which interfere with the assay (Klee and Rogers 1989).

Chloramphenicol acetyltransferase (CAT) activity has been used by many workers to monitor gene expression. The assay involves the acetylation of chloramphenicol with ^{14}C-labelled acetyl co-enzyme A. Reaction products are separated by thin layer chromatography, auto-radiographed, and individual acetylation products measured by scintillation counting.

Beta-glucuronidase (GUS), with a monomer molecular weight of 68 200 daltons, is a hydrolase that cleaves a wide range of glucuronides. The gene encoding this enzyme, isolated from *E. coli* (*uid A* locus), has been engineered to express in a variety of organisms, including plants. It has been employed to investigate tissue-specific gene expression (Jefferson *et al.* 1987), often within 36 hours of transformation. The enzyme can cleave commercially available substrates, such as X-Gluc (X-glucuronide) for histochemical analysis to indicate the sites of GUS activity and 4-MUG (4-methylumbelliferyl glucuronide) for fluorometric analysis to study levels of *gus* gene expression. The enzyme is very stable and can be assayed over a wide pH range (optimum 5.2–8.0), with little or no background activity in several biological systems. The sensitivity of the assays and the ability to localize the enzyme in transformed tissues, has led to the GUS system being applied extensively to plants.

Other scorable markers for plant transformation, although not used extensively as CAT and GUS, are the luciferase genes isolated from fireflies and bioluminescent bacteria (*Vibrio harveyii* and *V. fischeri*). Insect luciferase is encoded by a single gene and requires luciferin, ATP, and oxygen as substrates, while the bacterial *lux* gene encodes two polypeptides which require oxygen, reduced flavin mononucleotide, and the aldehyde decanal as substrates. Gene expression can be monitored by rapid, sensitive, and non-destructive *in vivo* methods using a luminometer. The fusion of the *luxA* and *luxB* genes into a single functional unit, allowing the production of light mediated by a fused protein in tobacco cells (Kirchner *et al.* 1989), has simplified the luciferase system for plants.

Experimental systems for gene transfer to plants

The introduction of genes into plants depends on several factors, including the species and variety, the anatomy and physiology of target cells, the competence of cells for transformation, and the vector system. Several experimental approaches have been developed, most of which utilize axenic seedlings and cultured shoots as a source of explants. Glasshouse-grown plants have also been used as a source of explants. Inoculation of explants and their incubation in closed plastic or glass containers in designated growth rooms, provide the containment conditions required when handling agrobacteria, especially those carrying recombinant DNA.

1. Inoculation of stems, leaves, and roots with Agrobacterium

The simplest approach involves inoculation of stems and leaves with a sharp instrument, such as a scalpel blade or toothpick, previously immersed in bacteria from a recently subcultured agar plate or an actively growing, overnight bacterial suspension. Alternatively, stems can be decapitated and the cut surface inoculated with a loopful of bacterial suspension. Oncogenic strains of *Agrobacterium* induce crown galls and transformed roots which appear at the inoculation sites within 7–21 days, depending on the virulence of the strains.

The leaf-disc transformation system described by Horsch *et al.* (1985) has been used extensively, particularly for members of the Solanaceae. Leaf explants are removed using a suitable instrument (for example, a cork borer) and immersed for a few seconds or minutes in an overnight culture of *Agrobacterium* harbouring the vector carrying the gene(s) of interest. After blotting on sterile filter paper to remove excess bacteria, the explants are transferred to the surface of an appropriate agar culture medium for approximately 48 h. Subsequently, they are cultured on a medium containing an antibiotic, such as cefotaxime, to kill the bacteria and the appropriate combination of growth regulators to induce shoot production from dividing cells at the cut surface of the explants. This method is successful with disarmed vectors carrying a selectable marker, such as the *npt*II gene. Shoots which continue to grow in the presence of antibiotic (for example, kanamycin sulphate), included in the regeneration medium at concentrations which normally inhibit the growth of non-transformed shoots, are putative transformants. The latter are excised from the explant, characterized at the molecular level, and grown to maturity.

Similar approaches have been developed for transgenic plant production from root explants of *Arabidopsis thaliana* (Valvekens *et al.* 1988) and from epidermal segments from flowering stems of tobacco (Trinh *et al.* 1987). Incubating the explants over a layer of actively growing nurse cells, such as those from a suspension of the same or a different plant, during infection with the bacteria, may improve the transformation frequency. The nurse cells probably provide amino acids and other compounds which stimulate growth of target cells and which increase the competence of the latter to receive foreign DNA. Likewise, treatment of *Agrobacterium* with acetosyringone may promote transgenic shoot production.

2. Co-cultivation of protoplast-derived cells and suspension-cultured cells with Agrobacterium

The co-cultivation of protoplast-derived cells with *Agrobacterium* works well, but has seen limited application. Protoplasts from leaves or cell suspensions are cultured until they are actively dividing, at which time the cultures are infected with *Agrobacterium* (400–1000 bacteria per plant cell) for 36–48 h. During co-cultivation (infection), the plant cells and bacteria aggregate. Subsequently, the bacteria are removed by washing with culture medium containing antibiotics such as carbenicillin, cefotaxime, vancomycin, or augmentin. Protoplast-derived cells are cultured in the presence of antibiotics to remove any remaining bacteria and transformed tissues recovered using the appropriate selection encoded by the chimaeric marker gene. An additional advantage of the use of isolated protoplasts and protoplast-derived cells as recipients for foreign DNA is that there is a greater potential for exposure of somaclonal variation. Some variant traits may be beneficial and may be superimposed on transformation.

Dividing suspension-cultured cells can also be transformed by an approach similar to that employed for protoplast-derived cells (An 1985). Layering of recipient cells onto filter paper over a nurse culture of cells plated in semi-solid medium has been used to stimulate gene transfer into proembryogenic cell suspensions (Scott and Draper 1987). An advantage of this method is that the cells undergoing transformation are easily transferred to the appropriate media for bacterial elimination, callus growth, and shoot regeneration.

3. Transformation of seeds by Agrobacterium

Most of the approaches for gene insertion rely upon reproducible methods to regenerate plants from cultured explants, callus, and protoplast-derived cells. In order to minimize culture procedures, Feldmann and Marks (1987) incubated germinating seeds of *Arabidopsis thaliana* for 24 h with agrobacteria harbouring a disarmed Ti plasmid carrying the *npt*II gene. Transformed seedlings were selected by their resistance to kanamycin sulphate and resistant plants carrying T-DNA were recovered. The *npt*II gene was transmitted to seed progeny in a Mendelian manner. Such an approach is attractive because of its simplicity. However, the mechanism by which transformation occurs has not been elucidated and the general applicability of this approach to other plants has not been confirmed.

Stability and expression of foreign genes in plant tissues

There are numerous examples of the expression of both wild-type T-DNA genes and chimaeric genes and their inheritance in plants transformed by *Agrobacterium*. Stable and transient expression of bacterial, plant, and animal genes, under the control of *nos*, CaMV 35S, and other promoters carried on chimaeric non-oncogenic vectors, have been assayed by examination of phenotypic features, the use of enzymatic assays, and by Southern, Northern, and Western blotting (Weising *et al.* 1988). The level of gene expression varies widely between transformed plants and may be associated with copy number of the transferred gene (Morgan *et al.* 1987). The variability of expression may also reflect differences between insertion sites, since different regions of the plant genome are expressed at different levels and at different stages of development. Additionally, the effects may be associated with differential methylation (Peerbolte *et al.* 1986*a*,*b*). Integrated foreign genes can be transmitted to seed progeny in a Mendelian manner, although non-Mendelian inheritance may be associated with T-DNA integration at high copy number (Deroles and Gardner 1988). Foreign genes have been shown to have a high degree of meiotic stability, although there are examples of loss of the transformed phenotype during meiosis, because of deletion of the gene or its inactivation by methylation. Reversible methylation and inactivation of genes in one T-DNA has been reported to result from the presence of a second, unlinked T-DNA in the genome of the progeny of tobacco plants which have undergone a second transformation (Matzke *et al.* 1989).

The precise length of T-DNA inserted into plant cells may vary, with the left end of the TL-DNA often being truncated. Abnormally short T-DNAs may be responsible for phenotypic diversity in cell clones resulting from the co-cultivation of protoplast-derived cells with *Agrobacterium*. Inactivation or loss of T-DNA has also been reported during culture of crown galls and transformed roots, which may be related, after long periods of stability, to rearrangements of DNA sequences. Spontaneous deletion of pRi TL and TR-DNAs has been observed in potato root clones which had been expressing the foreign trait during prolonged culture and in a number of their regenerated plants (Hanisch Ten Cate *et al.* 1990). Although the causes of these T-DNA deletions have not been elucidated, it is unlikely that they are due to variation in chromosome number, as plants and roots transformed by *A. rhizogenes* usually retain their diploid chromosome complement during sexual propagation, as well as during long periods in culture.

Application of the natural gene transfer system of *Agrobacterium* to the genetic engineering of crop plants

The last 5 years have witnessed a number of agronomically important genes being inserted into crop plants using *Agrobacterium*-mediated gene delivery. Such genes include those for herbicide, insect, and virus resistances, with many biotechnology companies focusing their attention in these areas.

1. Induction of herbicide resistance in plants

Herbicides provide more efficient weed control compared to manual or mechanical methods. The requirement to apply such compounds in limited amounts to reduce cost and environmental pollution has stimulated efforts to generate herbicide-resistant plants. Improved understanding of the mode of action of herbicides has also contributed to success in this area (Stalker 1991). The fact that many of the herbicide resistance genes are single dominant traits makes them amenable to current gene transfer technologies. To date, two strategies have been developed for engineering herbicide resistance in plants.

a. Modification of the target of herbicide action Glyphosate (*N*-phosphonomethylglycine) is an analogue of glycine and is the active ingredient of Roundup, a wide-spectrum herbicide marketed by the Monsanto Company. The compound inhibits the enzyme 5-enol-pyruvylshikimate-3-phosphate (EPSP) synthase. The latter catalyses the condensation of phosphoenolpyruvate and shikimate-3-phosphate to 5-enolpyruvylshikimate-3-phosphate, a precursor of a range of aromatic secondary metabolites and the amino acids tryptophan, phenylalanine, and tyrosine. Ethylmethanesulphate mutagenization of the bacterium *Salmonella typhimurium* produced two-types of glyphosate-resistant mutants at the *aroA* gene, specifically, those which overproduced EPSP synthase by an up-promoter mutation and others which showed a structurally altered EPSP synthase, making the bacterial cells insensitive to the herbicide. Both *aroA* gene mutants have been used to produce herbicide-tolerant plants. Thus, an *aroA* gene, with a single proline to serine substitution resulting in a decreased affinity for glyphosate in the bacterium, was transferred into tobacco (Comai *et al.* 1985) and tomato (Fillatti *et al.* 1987) under the control of the *ocs* and *mas* promoters. Transgenic plants were three times more tolerant to glyphosate compared to non-transformed plants, herbicide tolerance being correlated with the level of expression of the gene. However, agronomically, the level of tolerance was still inadequate. In these experiments, EPSP synthase from the introduced gene was expressed in the cytoplasm, whereas the enzyme is normally produced in chloroplasts. Consequently, a transit peptide-encoding sequence of the Rubisco small subunit gene was fused upstream of the promoter of the *aroA* gene. This resulted in a 1000-fold increase in resistance (Della Cioppa *et al.* 1986). Encouraging results have been obtained with field-grown transgenic tomato plants with the chloroplast-directed glyphosate-resistant bacterial EPSP synthase. Such transgenic plants gave a normal yield of fruit, but were stunted. As stunting is the result of the herbicide being concentrated in meristematic regions, an increased and localized expression of EPSP synthase and/or a kinetically efficient glyphosate-resistant enzyme could prevent this effect (Stalker 1991). Other workers have attempted to hyperexpress the wild-type EPSP synthase by fusing the strong, constitutive CaMV 35S promoter upstream of the gene (Shah *et al.* 1986). However, when the construct was tested in *Petunia*, only a 20–40-fold increase in activity of the enzyme resulted, corresponding to a 10-fold increase in herbicide resistance.

The herbicides Glean and Oust have the sulphonylurea-type compounds chlorsulphuron and sulphometuron methyl as their active ingredients. These herbicides inhibit acetolactate synthase (ALS). Chlorsulphuron- and sulphometuron-resistant plants have been obtained from haploid tobacco protoplasts (Chaleff and Ray 1984) by selection during culture, while resistant seedlings of *A. thaliana* were obtained following seed mutagenization (Haughn and Somerville 1986). Resistance in these cases was not due to overproduction of ALS, but was the result of a single base substitution in the gene. Such genes have been isolated and introduced into other plants as chimaeric constructs. For example, tobacco plants transformed with the cloned *Arabidopsis crs1* gene (Haughn *et al.* 1988) and tobacco protoplasts transformed with the *surB-Hra* gene (Falco *et al.* 1987), have shown resistance to field application rates of sulphonylurea herbicides. In the case of linseed, sulphonylurea resistant plants have been grown in field trials and have expressed a variable degree of resistance, although most were tolerant of field dose applications and gave normal seed yields (McHughen *et al.* 1990). ALS is also the target enzyme for imidazolanone and triazopyrimidine herbicides, but the precise mode of action of sulphonylurea, imidazolanone, and triazopyrimidine herbicides is still unresolved.

b. Detoxification or degradation of herbicides   Herbicide-degrading bacteria are attractive as a source of desirable genes, because their metabolic pathways are easier to elucidate and the identification and purification of degradative/detoxification enzymes are simpler, when compared to plants (Stalker 1991). *Klebsiella ozaenae* has been isolated from soil treated with bromoxynil (3,5-dibromo-4-hydroxybenzonitrile), a weed killer used for broad leaved plants. The bacterium used the chemical as its nitrogen source, the herbicide being degraded by the enzyme nitrilase encoded by the *bxn* gene. Both tobacco and tomato have been transformed with the *bxn* gene, using the tobacco Rubisco small subunit and the CaMV 35S promoters in *Agrobacterium* binary vectors. Levels of resistance up to 10-fold the field application rate of bromoxynil were observed in both plants (Stalker *et al.* 1988). In cotton, the nitrilase gene had no detrimental effect on the growth and yield of transgenic plants.

The gene *tfdA*, which encodes for 2,4-dichlorophenoxyacetate mono-oxygenase, has been isolated from *Alcaligenes eutrophus* strain JMP134 (Streber *et al.* 1987) and has been shown to degrade the herbicide 2,4-dichlorophenoxyacetic acid (2,4-D). Glasshouse trials have demonstrated a three- to five-fold resistance to 2,4-D by tobacco plants transformed by the *tfdkA* gene (Streber and Willmitzer 1989).

L-Phosphinothricin (PPT) was discovered as an antibiotic produced by *Streptomyces viridochromogenes* and is an analogue of glutamate. PPT is a competitive inhibitor of glutamine synthase, which converts glutamate to glutamine. Glutamine synthase is the only enzyme able to detoxify ammonia. The *bar* gene from *S. hygroscopicous* encodes the enzyme phosphinothricin acetyltransferase (PAT), which acetylates the free amino acid groups of PPT. This gene, has been transferred into tobacco, tomato, and potato under the CaMV 35S promoter to give high levels of resistance to the herbicide (De Block *et al.* 1987). Under field conditions, De Greef *et al.* (1989) observed resistance to the herbicide Basta in transgenic tobacco and potato plants and the latter had the same agronomic characters as unsprayed plants. For example, a glutamate synthase cDNA clone was obtained from a PPT-tolerant suspension culture of *Medicago sativa* and linked to the CaMV 35S promoter. However, this gene, when introduced into tobacco, gave a lower level of tolerance to PPT than the *bar* gene.

Atrazine and the chemically unrelated ureas bind to chloroplast thylakoid membranes at the *psbA* gene, which encodes for the Q_B protein, an important component in electron transport of photosystem II. Genes conferring resistance to atrazine have been isolated, but their introduction into chloroplasts has not been successful, because of the lack of a reproducible system for transforming these organelles (Cornelissen *et al.* 1987). However, limited atrazine tolerance was observed in tobacco, when the plant-derived *psbA* gene from *Abutilon hybridus* was converted into a nuclear gene by fusing to the Rubisco small subunit promoter and small subunit transit peptide sequence (Cheung *et al.* 1988).

2. Insect resistance in transgenic plants

It is not yet feasible to produce transgenic plants capable of synthesizing insecticides, because the enzymes for the reactions are unknown. Additionally, the transfer of polygenes is difficult and the regulation of such genes to enable pathways to function is beyond current technology (Gatehouse *et al.* 1991). If a plant lacked one enzyme in a pathway to make an effective insecticide, as is the case for *Trifolium repens* in relation to phaseollin, it might be possible to genetically engineer this metabolic pathway. However, there are few examples where this might be achieved.

Insecticidal proteins, being products of single genes, are highly desirable for gene transfer, although ribosome-inactivating proteins must be avoided because of their toxicity to both plants and animals. To date, two insecticidal proteins have been transferred to and evaluated in transgenic plants, namely the Bt toxin from *Bacillus thuringiensis* and protease inhibitors.

a. Expression of Bt toxins in transgenic plants For more than 20 years, the crystal protein produced by the soil bacterium, *B. thuringiensis*, has been used in insecticidal sprays, but high production costs and instability of the protein under field conditions, have focused attention on transfer of the *Bt* gene to plants. The intact protein is non-toxic to insects, but alkaline conditions of the midgut of some insects permit hydrolysis, while further proteolytic processing releases toxic fragments which disrupt insect gut membranes. The protein is non-toxic to animals, since it cannot be processed in the mammalian gut. Different strains of *B. thuringiensis* produce different toxins; the latter are each effective against a limited range of insects. In total, over 50 lepidopteran and a few coleopteran species are affected.

Since only the N-terminal end of the *Bt* gene is necessary for toxicity, studies have been made of the effects of different truncated Bt toxins as a means of controlling insects (Fischhoff *et al.* 1987). The Bt toxin-encoding sequences have been cloned into plant expression vectors (Velten *et al.* 1984) under the *mas* or the CaMV 35S promoters, along with the *nptII* gene. When transferred into tomato and tobacco using *Agrobacterium*, the level of Bt toxin was highest when the coding sequence had been truncated, although the toxin amounted to only 0.02 per cent of total leaf soluble protein. The entire Bt sequence had adverse effects on transgenic plants. Plants transformed with the truncated *Bt* gene were insecticidal towards tobacco hornworm (*Manduca sexta*), with up to 100 per cent insect mortality within a few days. These results emphasized the potency of Bt toxin when expressed at low levels. More serious pests belonging to the Noctuid Lepidopterans, such as tobacco budworm (*Heliothis virescens*) and corn earworm (*Heliothis zea*), were controlled, although they were less sensitive to the toxin. Experiments have been performed to elevate the expression of Bt toxin in transgenic plants, to increase the range of insects controlled, by linking the minimum *Bt* sequence necessary for toxicity to enhanced promoters (Fuchs *et al.* 1989).

Field trials of Bt-transformed plants have been of limited nature, but have given encouraging results (Gatehouse *et al.* 1991). A heavy infestation of tomato pinworm (*Keiferia lycopersicella*) in Florida caused severe damage to control plants. In contrast, plants transformed with the *Bt* gene, which was expressed in all tissues, were substantially protected (Fischoff, 1989).

b. Transfer of protease inhibitors into plants   Plants have evolved physical and chemical defence barriers against insects. Such barriers result from the expression of many genes and these traits are impossible to transfer into plants (Dawson *et al.* 1989). However, the cowpea trypsin inhibitor (CpTI), a single-gene product, is ideal for genetic engineering. Accession TVu2027 of cowpea (*Vigna unguiculata*), showed resistance to larvae of the bruchid beetle (*Callosobruchus maculatus* F.), due to elevated levels of trypsin inhibitor. The inhibitor affected serine, trypsin, and chymotrypsin proteases of insects, resulting in abnormal development and death due to deficiency of essential amino acids (Hilder *et al.* 1991). The inhibitor is also active as an antimetabolite in a wide range of insects, including members of the orders Lepidoptera and Coleoptera. Although CpTI is also inhibitory to mammalian trypsin, it is degraded by pepsin in the gut of animals and so is non-toxic.

A cDNA clone encoding a trypsin inhibitor was inserted into a *Sma1* site in the binary vector pRoK2, under the control of the CaMV 35S promoter and the *nos* terminator, using a *nos-neo* gene, conferring kanamycin resistance, as a selectable marker (Hilder *et al.* 1987). The construct was transferred into tobacco by leaf disc transformation using *A. tumefaciens* strain LBA4404. Transgene expression in leaves of transformed plants was variable from below the limit of detection to approximately 0.9 per cent of the total soluble protein using dot immunobinding assays of rabbit anti-CpTi antiserum. There appeared to be no correlation between T-DNA copy number and expression of CpTI, which was considered to be due to a 'position effect'. Transformed tobacco plants were tested for insect resistance using first instar larvae of the tobacco budworm (*Heliothis virescens*) in sealed plantaria. Plants which showed the greatest resistance to the pest expressed CpTI at the highest levels, with a transgene expression threshold of 0.5 per cent soluble protein as CpTI.

3. Virus resistance in transgenic plants

Viruses cause major economic losses through reduction in yield (Zaitlin and Hull, 1987). The nucleic acids of viruses can be either DNA or RNA and may be single- or double-stranded. Viruses enter plant tissue through damaged cell walls and plasma membranes. Infection by insect-borne viruses occurs through the insect stylet during feeding. Once inside plant cells, viral particles are uncoated, releasing their genetic material which replicates without integrating into the host genome.

The manipulation of virus resistance by conventional breeding relies on resistant cultivars, but this approach is limited by a small gene pool. 'Cross-protection' has been used to confer virus resistance on plants, in which susceptible crops are infected with a mild strain of a virus which does not induce severe symptoms. This confers resistance against virulent strains of the same or closely related viruses. Limitations of this approach are that mild strains may mutate to become more virulent, they may act synergistically with other viruses to affect plant growth and they may reduce crop yield. Increased knowledge of the molecular biology of viral function has resulted in four strategies being proposed to control viruses using gene transfer technology.

a. Modified cross-protection   This is based on identifying and separating viral genes and gene products, which confer protection, from genes encoding proteins responsible for symptom formation. When the tobacco mosaic virus (TMV) coat protein was transferred into tobacco under the control of the CaMV 35S promoter using *A. tumefaciens* as a vector (Abel *et al.* 1986), transgenic plants developed symptoms more slowly than controls and 10–60 per cent of the transgenics failed to show symptoms. Generally, protection is specific against the virus whose coat protein has been introduced. For example, when the coat protein gene of alfalfa mosaic virus (AMV) was transferred into tobacco, protection was specific against AMV and the gene offered no protection against TMV (Loesch-Fries *et al.* 1988). The mechanism of cross-protection is unclear, but expression of a viral coat protein may prevent the uncoating of other invading viruses, inhibiting genome replication and protein synthesis. Alternatively, the coat protein may compete for factors required by invading viruses for the uncoating process.

b. Use of satellite nucleic acids   Satellite RNAs are found in some plant RNA viruses and are replicated and packaged normally, but they are not required for viral replication and spread. If DNA copies of satellite RNA are transferred into a susceptible plant, the transcribed satellite RNA is able to inhibit symptom formation. Thus, when a DNA copy of a satellite RNA of cucumber mosaic virus (CMV) under the CaMV 35S RNA promoter was introduced into tobacco using *Agrobacterium*, the level of satellite RNA was low in uninfected transgenic plants, but the RNA level was greatly increased upon viral infection (Harrison *et al.* 1987). Accumulation of satellite RNA reduced symptom formation compared with the symptoms seen in non-transformed, viral infected plants. A closely related virus, tomato aspermy (TAV), also induced the synthesis of CMV satellite RNA, with a reduction in disease symptoms.

Investigations have been made on the effects of cDNA copy number of satellite RNA on plant protection (Gerlach *et al.* 1987). When tobacco ringspot virus (TobRV) satellite cDNA was made into a trimer (three copies in tandem) and transferred into tobacco, the level of transcription of the transgene increased considerably compared to when a monomer was employed. A reduction in ringspot symptoms and a decrease in the replication of infectious virus was observed for the trimer. However, there was no reduction in symptoms with the monomeric form. Interestingly, the CaMV 35S promoter used in these experiments was induced only by the presence of an infective virus. It is not known how satellite RNA protects transgenic plants. It may be that it interacts directly or indirectly with genomic RNA or the symptom-producing process. The satellite RNA approach does present problems. For example, the sequence can be protective in one plant and be virulent in another host, while the sequence may mutate to a virulent form.

c. Use of antisense RNA   This strategy is based on transferring an antisense RNA (minus-strand RNA) which, through base pairing, binds to the sense strand (plus-strand or messenger) RNA, preventing translation of a specific viral sequence and conferring viral resistance. Transgenic plants expressing antisense RNA to the 3' region, including the coat protein gene of TMV (Powell *et al.* 1989), potato virus X (PVX) (Hemenway *et al.* 1988), or CMV (Cuozzo *et al.* 1988) RNAs, showed some protection against the respective viruses, although it was lower compared to plants expressing viral coat protein genes. In general, protection using antisense RNA constructs has been observed only when transgenic plants were infected with low inoculum levels (Buck 1991). Experiments have been performed to study the regions of the antisense RNA which caused protection. Constructs were designed which had antisense RNA to either the TMV coat protein gene or the 3' untranslated region of TMV RNA. Protection was seen only with the antisense 3' untranslated region, suggesting that the transgene interacted with the replicase binding site and so reduced the synthesis of the negative RNA strand of the virus. The efficiency of gene suppression by this method could be improved using additional enhancer elements together with stronger promoters. This kind of protection has been limited to RNA viruses, but it is possible that the technique may be more effective against DNA viruses, such as the gemini viruses.

d. Use of ribozymes as virus resistance genes   Ribozymes are RNA molecules that cleave RNA in a sequence-specific manner and these enzymes could be targeted against many different regions of a viral genome. Although the full potential of ribozymes in transgenic plants remains to be evaluated, it is another approach for producing virus resistant plants (Buck 1991).

4. Transfer of other important genes into plants

Several genes of agronomic and economic value have been transferred into plants using *Agrobacterium* vectors. The isolation of specific promoters has permitted transgene expression to be targeted to tissues and organs.

a. Flower specific genes   The TA 29 promoter, isolated during studies of anther-specific genes in tobacco, has been linked to a synthetic *Aspergillus oryzae* T1 RNase gene or to an RNase gene from *Bacillus amyloliquefaciens* and transferred into tobacco (Mariani *et al.* 1990). The promoter retained its specificity, since the coding sequence was only transcribed in the tapetum at a particular stage of floral development. Transgenic plants were phenotypically normal, but failed to produce functional pollen, resulting in male sterility. The same construct has been transferred into oilseed rape with similar results (Peacock 1990). These observations are of relevance to plant breeding as this technology permits the production of hybrid seed without the need for costly hand emasculation.

b. Antisense genes   The use of *Agrobacterium* to introduce into plants chimaeric genes, which inhibit the expression of other specific genes, has been possible through an antisense approach. Recombinant DNA technology has made it feasible to invert DNA sequences and to link them to suitable gene promoters, generating chimaeric antisense genes which are transcribed to generate the 'wrong' (antisense) strand of mRMA. The antisense RNA inhibits the accumulation of normal mRNA by forming an unstable RNA–RNA sense–antisense duplex, preventing translation of the sense strand. This approach has been successful for delaying fruit ripening by inhibiting expression of the polygalacturonase gene. An antisense polygalacturonase gene, based on the first half of the polygalacturonase cDNA, was transferred into tomato using a Ti plasmid vector (Smith *et al.* 1988). Consequently, the fruit was more resistant to squashing and cracking due to cells being more firmly bonded together (Bird *et al.* 1991). The antisense gene was stably inherited and the level of polygalacturonase mRNA was significantly lower for those plants possessing two copies of the construct compared to those plants having one or less copies. Antisense RNA technology has also been used to down-regulate flavonoid-specific genes, such as the chalcone synthase gene, to produce novel colours and patterns in flowers and to generate male sterile plants (Mol 1991).

c. Modification of biochemical pathways; antibody and secondary product synthesis   *Agrobacterium*-mediated transformation has enabled new biochemical pathways to be developed. For example, the maize gene for dihydroflavonol-4-reductase (DFR) has been transferred into *Petunia*.

Normally, the *Petunia* DFR gene product is unable to use pelargonidin as a substrate. However, the transgene-encoded enzyme could utilize this compound, resulting in novel petunias with brick-red coloured flowers (Meyer *et al.* 1987).

Complementary DNA from mouse hybridoma mRNA has been successfully expressed in transgenic tobacco (Hiatt *et al.* 1989). Plants expressing single heavy or light antibody chains were crossed to yield progeny which expressed both chains simultaneously. Functional antibody amounted to 1.3 per cent of total leaf protein. It has been postulated that the yield of antibody could be increased to over 10 per cent of total leaf protein using stronger promoter and enhancer sequences. Economically, this technology could be competitive compared to producing antibodies in hybridoma cells. Of note is the potential of Ti and Ri plasmids in secondary metabolism, where transformed roots and shooty teratomas have been used for the overproduction and biotransformation of metabolites in medicinal plants (Saito *et al.* 1992).

Conclusion

Whilst considerable effort has been directed to studies of the phenotypic, molecular, and biochemical changes associated with the interaction of *Agrobacterium* with plants, several of the events which accompany transformation, particularly the mechanisms of T-DNA transfer and integration into recipient genomes, are still not understood. Many of the genes on the Ti and Ri plasmids remain to be defined. Indeed, it is probable that additional genes on the T-DNA and *vir* regions will be characterized in due course. Future work seems likely to be aimed primarily at the *vir* region, since its activity determines, to a large extent, the success of gene transfer to plants.

Several agronomically important genes have already been introduced into plants and some of the latter field trialled. In many cases, gene expression still needs to be maximized. Constitutive expression of these genes may be detrimental to the host plant, with the need to use promoters which function at defined stages of plant development and which respond to specific stimuli, such as virus infection. In the case of insect and virus resistance, a combination of different genetic engineering approaches may be required to protect plants adequately against these agents. Many questions remain to be answered concerning the environmental safety and consumer acceptance of genetically engineered crops.

Currently, *Agrobacterium*-mediated gene delivery is the method of choice for introducing foreign DNA into plants. Extension of this technology to a wider range of species will depend upon reproducible shoot regeneration from cultured cells and explants. However, in the immediate future, most monocotyledons will remain recalcitrant to this approach until the reasons have been elucidated for their inability to respond to *Agrobacterium*.

References

Abel, P.P., Nelson, R.S., De, B., Hoffmann, N., Rogers, S.G., Fraley, R.T. *et al.* (1986). Delay of disease resistance in transgenic plants that express the tobacco mosaic virus coat protein gene. *Science*, **232,** 738–43.

Alarcon, B., Lopez, M.M., Cambra, M., and Ortiz, J. (1987). Comparative study of *Agrobacterium* biotypes 1, 2 and 3 by electrophoresis and serological methods. *Journal of Applied Bacteriology*, **62,** 295–308.

An, G. (1985). High efficiency transformation of cultured tobacco cells. *Plant Physiology*, **79,** 568–70.

Aoyama, T. and Oka, A. (1990). A common mechanism of transcriptional activation by the three positive regulators, VirG, PhoB and OmpR. *FEBS Letters*, **363,** 1–4.

Balazs, E. and Bonneville, J.M. (1987). Chloramphenicol acetyl transferase activity in *Brassica* spp. *Plant Science*, **50,** 65–8.

Barry, G., Rogers, S., Fraley, R., and Brand, D. (1984). Identification of a cloned cytokinin biosynthetic enzyme. *Proceedings of the National Academy of Sciences USA*, **81,** 4776–80.

Bevan, M.W., Flavell, R.B., and Chilton, M.D. (1983). A chimaeric antibiotic resistance gene as a selectable marker for plant cell transformation. *Nature*, **304,** 184–7.

Binns, A.N., Labriola, J., and Black, R.C. (1987). Initiation of auxin autonomy in *Nicotiana glutinosa* cells by the cytokinin biosynthesis gene from *Agrobacterium tumefaciens*. *Planta*, **171,** 539–48.

Bird, C.R., Ray, J.A., Fletcher, J.D., Boniwell, J.M., Bird, A.S., Teulieres, C., *et al.* (1991). Using anti-sense RNA to study gene function: inhibition of carotenoid biosynthesis in transgenic tomatoes. *Bio/Technology*, **9,** 635–9.

Bouchez, D. and Camilleri, C. (1990). Identification of a putative *rolB* gene on the T$_R$-DNA of *Agrobacterium rhizogenes* A4 Ri plasmid. *Plant Molecular Biology*, **14,** 617–19.

Buck, K.W. (1991). Virus-resistant plants. In *Plant genetic engineering*. (ed. D. Grierson), pp. 136–78. Blackie, Glasgow, London.

Bytebier, B., De Boeck, F., De Greve, H., Van Montagu, M., and Hernalsteens, J.P. (1987). T-DNA organisation in tumour cultures and transgenic plants of the monocotyledon *Asparagus officinalis*. *Proceedings of the Academy of Sciences USA*, **84,** 5345–9.

Cangelosi, G.A., Martinetti, G., Leigh, J.A., Lee, C.C., Theines, C., and Nester, E.W. (1989). Role of *Agrobacterium tumefaciens* ChvA protein in export of β-1, 2 glucan. *Journal of Bacteriology*, **171,** 1609–15.

Cardarelli, M., Mariotti, D., Pomponi, M., Spano, L., Capone, I., and Constantino, P. (1987). *Agrobacterium rhizogenes* T-DNA genes capable of inducing hairy root phenotype. *Molecular and General Genetics*, **209,** 475–80.

Chaleff, R.S. and Ray, T.B. (1984). Herbicide-resistant mutants from tobacco cell cultures. *Science*, **223**, 1148–51.

Cheung, A.Y., Bogorad, L., Van Montagu, M., and Schell, J. (1988). Relocating a gene for herbicide tolerance: a chloroplast gene is converted into a nuclear gene. *Proceedings of the National Academy of Sciences USA*, **85**, 391–5.

Chilton, M.D., Drummond, M.H., Merlo, D.J., Saiky, D., Montoya, A.L., Nester, E.W. *et al.* (1977). Stable incorporation of plasmid DNA into higher plant cells; the molecular basis of crown gall tumorigenesis. *Cell*, **11**, 263–71.

Christie, P.J., Ward, J.E., Winans, S.C., and Nester, E.W. (1988). The *Agrobacterium tumefaciens VirE2* gene product is a single-stranded-DNA-binding protein that associates with T-DNA. *Journal of Bacteriology*, **170**, 2659–67.

Citovsky, V., Wong, M.L., and Zambryski, P. (1989). Cooperative interaction of *Agrobacterium VirE2* protein with single-stranded DNA: Implications for the T-DNA transfer process. *Proceedings of the National Academy of Sciences USA*, **86**, 1193–7.

Comai, L., Facciotti, D., Hiatt, W.R., Thompson, G., Rose, R.E., and Stalker, D.M. (1985). Expression in plants of a mutant *aroA* gene from *Salmonella typhimurium* confers tolerance to glyphosate. *Nature*, **317**, 741–4.

Combart, A., Brevet, J., Borowski, D., Cam, K., and Tempé, J. (1987). Physical map of the T-DNA region of *Agrobacterium rhizogenes* NCPPB 2659. *Plasmid*, **18**, 70–5.

Cornelissen, M.J., De Block, M., Van Montagu, M., Leemans, J., Schrier, P.H., and Schell, J. (1987). Plasmid transformation: a progress report. In *Plant gene research: plant DNA infectious agents*, (ed. T.H. John and J. Schell), pp. 311–20. Springer-Verlag, Vienna.

Crespi, M., Messens, E., Caplan, A.B., Van Montagu, M., and Desomer, J. (1992). Fasciation induced by the phytopathogen *Rhodococcus fascians* depends upon a linear plasmid encoding a cytokinin synthase gene. *EMBO Journal*, **11**, 795–804.

Cuozzo, M., O'Connell, K.M., Kaniewski, W., Farg, R.X., Chua, N.H., and Tumer, N.E. (1988). Viral protection in transgenic tobacco plants expressing the cucumber mosaic virus coat protein or its antisense RNA. *Bio/Technology*, **6**, 549–57.

Dawson, G.W., Hallahan, D.L., Mudd, A., Patel, M.M., Pickett, J.A., Wadhans, L.A. *et al.* (1989). Secondary plant metabolites as targets for genetically modifying crops for pest resistance. *Pesticide Science*, **27**, 191–201.

Deblaere, R., Bytebier, B., DeGreve, H., Schell, J., Van Montagu, M., and Leemans, J. (1985). Efficient octopine Ti plasmid-derived vectors for *Agrobacterium*-mediated gene transfer to plants. *Nucleic Acids Research*, **13**, 4777–88.

De Block, M., Botterman, J., Vandewiele, M., Dockx, J., Thoen, C., Gosselé, V. *et al.* (1987). Engineering herbicide resistance in plants by expression of a detoxifying enzyme. *EMBO Journal*, **6**, 2513–18.

De Cleene, M. and De Ley, J. (1976). The host range of crown gall. *Botanical Review*, **42**, 389–466.

De Greef, W., Delon, R., De Block, M., Leemans, J., and Bolterman, J. (1989).

Evaluation of herbicide resistance in transgenic crops under field conditions. *Bio/Technology*, **7**, 61–4.

Della-Cioppa, G., Bauer, S.C., Klein, B.K., Shah, D.M., Fraley, R.T. and Kishore, G.M. (1986). Translocation of the precursor of 5-enolpyruryl-shikimate-3-phosphate synthase into chloroplasts of higher plants *in vitro*. *Proceedings of the National Academy of Sciences USA*, **83**, 6873–7.

Deroles, S.C. and Gardner, R.C. (1988). Analysis of the T-DNA structure in a large number of transgenic petunias generated by *Agrobacterium*-mediated transformation. *Plant Molecular Biology*, **11**, 365–77.

Dominov, J.A., Stenzler, L., Lee, S., Schwarz, J.J., Leisner, S., and Howell, S.H. (1992). Cytokinins and auxins control the expression of a gene in *Nicotiana plumbaginifolia* cells by feedback regulation. *The Plant Cell*, **4**, 451–61.

Douglas, C.J., Halperin, W., and Nester, E.W. (1982). *Agrobacterium tumefaciens* mutant affected in attachment to plant cells. *Journal of Bacteriology*, **152**, 1265–75.

Douglas, C.J., Staneloni, R.J., Rubin, R.A., and Nester, E.W. (1985). Identi-fication and genetic analysis of an *Agrobacterium tumefaciens* chromosomal virulence region. *Journal of Bacteriology*, **161**, 850–60.

Draper, J., Davey, M.R., Freeman, J.P., and Cocking, E.C. (1982). Isolation of plasmid DNA from *Agrobacterium* by isopycnic density gradient centrifugation in vertical rotors. *Experimentia*, **38**, 101–2.

Draper, J., Scott, R., Armitage, P., and Walden, R. (1988). *Plant genetic trans-formation and gene expression. A laboratory manual*. Blackwell Scientific Publications, Oxford.

Durand-Tardiff, M., Broglie, R., Slightom, J., and Tepfer, D. (1985). Structure and expression of Ri T-DNA from *Agrobacterium rhizogenes* in *Nicotiana tabacum*. *Journal of Molecular Biology*, **186**, 557–64.

Erwin, D.C. and Stuteville, D.L. (1990). *Compendium of alfalfa diseases*, (2nd edn). APS Press, Minnesota.

Estruch, J.J., Schell, J., and Spena, A (1991*a*). The protein encoded by the *rol* B plant oncogene hydrolyses indole glucosides. *EMBO Journal*, **10**, 3125–8.

Estruch, J.J., Parets-Soler, T., Schmulling, A., and Spena, A. (1991*b*). Cytosolic localisation in transgenic plants of the *rol* C peptide from *Agrobacterium rhizogenes*. *Plant Molecular Biology*, **17**, 547–50.

Estruch, J.J., Chriqui, D., Grossmann, K., Schell, J., and Spena, A. (1991*c*). The plant oncogene *rol* C is responsible for the release of cytokinins from glucoside conjugates. *EMBO Journal*, **10**, 2889–95.

Falco, S.C., Knowlton, S., Larossa, R.A., Smith, J.K., and Mazur, B.J. (1987). Herbicides that inhibit amino acid biosynthesis in the sulphonylureas—a case study. In *1987 British crop protection conference—weeds*, p. 149–58. BCPC Publications, Farnham, Surrey.

Feldmann, K.A. and Marks, M.D. (1987). *Agrobacterium*-mediated transformation of germinating seeds of *Arabidopsis thaliana*. A non-tissue culture approach. *Molecular and General Genetics*, **208**, 1–9.

Fillatti, J. J., Kiser, J., Rose, B., and Comai, L. (1987). Efficient transformation of tomato and introduction and expression of a gene for herbicide tolerance.

48   M.R. Davey, I.S. Curtis, K.M.A. Gartland, and J.B. Power

In UCLA symposium in plant biology, Vol. 4—tomato biotechnology, ed. D. J.
Nevins and R.A. Jones), pp. 199–210. Allan Liss, New York.
Fischhoff, D.A. (1989). Plants as delivery systems for biopesticides. Agbiotech,
1989, p. 373. Conference Proceedings, Arlington.
Fischhoff, D.A., Bowdish, K.S., Perlak, F.J., Marrone, P.G., McCormick, S.M.,
Niedermeyer, J.G. et al. (1987). Insect tolerant transgenic tomato plants.
Bio/Technology, 5, 807–13.
5 Prime → 3 Prime, Inc. (1990). Non-radioactive neomycin phosphotransferase
II (NPT-II) ELISA. Prime Report, 2,(1), 4–5.
Fraley, R.T., Rogers, S.G., Horsch, R.B., Eichholtz, D.A., and Flick, J.S. (1985).
The SEV system: a new disarmed Ti plasmid. Bio/Technology, 3, 629–35.
Fuchs, R., MacIntosh, S., Kishore, G., Perlak, F., Dean, D., Stone, T. et al.
(1989). Enhanced expression/efficiency of transgenic plants which express the
Bacillus thuringiensis insect control protein. Agbiotech, 1989, p. 210. Conference
Proceedings, Arlington.
Gartland, K.M.A., McInnes, E., Hall, J.F., Mulligan, A.J., Elliott, M.C., and
Davey, M.R. (1991). Effects of Ri plasmid rol gene expression on the IAA
content of transformed roots of Solanum dulcamara L. Plant Growth Regulation,
10, 235–41.
Gatehouse, J.A., Hilder, V.A., and Gatehouse, A.M.R. (1991). Genetic engin-
eering of plants for insect resistance. In Plant genetic engineering, (ed. D.
Grierson, pp. 105–35. Blackie, Glasgow, London.
Gerlach, W.L., Llewellyn, D., and Haseloff, J. (1987). Construction of a plant
disease resistance gene from the satellite RNA of tobacco ringspot virus.
Nature, 328, 802–5.
Gheysen, G., Van Montagu, M., and Zambryski, P. (1987). Integration of
Agrobacterium tumefaciens T-DNA involves rearrangements of target plant
sequence. Proceedings of the National Academy of Sciences USA, 84, 6169–73.
Goldberg, R.B., Hoschek, G., Kamalay, T.C., and Timberlake, W.E. (1978).
Sequence complexity of nuclear and polysomal RNA in leaves of the
tobacco plant. Cell, 14, 123–31.
Gribnau, A.G.M. and Veldstra, H. (1969). The influence of mitomycin C on
the induction of crown gall tumors. FEBS Letters, 3, 115–17.
Hanish Ten Cate, C.H., Loonen, A.E.H.N., Ottaviani, M.P., Ennik, L., Van
Eldick, J., and Stiekema, W.J. (1990). Frequent spontaneous deletions of Ri
T-DNA in Agrobacterium rhizogenes transformed potato roots and regenerated
plants. Plant Moleculer Biology, 14, 735–41.
Harrison, B.D., Mayo, M.A., and Baulcombe, D.C. (1987). Virus resistance in
transgenic plants that express cucumber mosaic virus satellite RNA. Nature,
328, 799–802.
Haughn, G.W., Smith, J., Mazur, B., and Somerville, C. (1988). Transformation
with a mutant Arabidopsis acetolactate synthase gene renders tobacco resistant
to sulfonylurea herbicides. Molecular and General Genetics, 211, 266–71.
Haughn, G.W. and Somerville, C. (1986). Sulfonylurea-resistant mutants of
Arabidopsis thaliana. Molecular and General Genetics, 204, 430–4.
Hemenway, C., Fang, R-X., Kaniewska, W.K., Chua, N.H., and Tumer,
N.E. (1988). Analysis of the mechanisms of protection in transgenic plants

expressing the potato virus X coat protein or its antisense RNA. *EMBO Journal*, **7,** 1273–80.

Hernalsteens, J.P., Van Vliet, F., De Beuckeleer, M., Depicker, A., Engler, G., Lemmers, M. *et al*. (1980). The *Agrobacterium tumefaciens* Ti plasmid as a host vector for introducing DNA in plant cells. *Nature*, **287,** 654–6.

Herrera-Estrella, L., Depicker, A., Van Montagu, M., and Schell, J. (1983*a*). Expression of chimaeric genes transferred to plants using a Ti plasmid-derived vector. *Nature*, **303,** 209–13.

Herrera-Estrella, L., De Block, M., Messens, E., Hernalsteens, J.P., Van Montagu, M., and Schell, J. (1983*b*). Chimaeric genes as dominant selectable markers in plant cells. *EMBO Journal*, **4,** 2987–95.

Hiatt, A., Cafferkay, R., and Bowdish, K. (1989). Production of antibodies in transgenic plants. *Nature*, **342,** 76–8.

Hilder, V.A., Gatehouse, A.M.R., and Boulter, D. (1991). Genetic engineering of crops for insect resistance using genes of plant origin. In *Genetic engineering of crop plants*, (ed. G.W. Lycett and D. Grierson). Butterworths, London.

Hilder, V.A., Gatehouse, A.M.R., Sheerman, S.E., Barker, R.F., and Boulter, D. (1987). A novel mechanism of insect resistance engineered into tobacco. *Nature*, **330,** 160–3.

Hille, J., Verheggen, F., Roevink, P., Franssen, H., Van Kammen, A., and Zabel, P. (1986). Bleomycin resistance: a new dominant selectable marker for plant cell transformation. *Plant Molecular Biology*, **7,** 171–6.

Hirooka, T., Rogowski, P.M., and Kado, C.I. (1987). Characterization of the *vir*E locus of *Agrobacterium tumefaciens* plamid pTiC58. *Journal of Bacteriology* **169,** 1529–36.

Hooykaas, P.J.J., den Dulk-Ras, M., and Schilperoort, R.A. (1988). The *Agrobacterium tumefaciens* T-DNA gene 6^b is an *onc* gene. *Plant Molecular Biology*, **11,** 791–4.

Hooykaas, P.J.J., Hofker, M., den Dulk-Ras, M., and Schilperoort, R.A. (1984). A comparison of virulence determinants in an octopine Ti plasmid, a nopaline Ti plasmid and a Ri plasmid by complementation analysis of *Agrobacterium tumefaciens* mutants. *Plasmid*, **11,** 195–205.

Horsch, R.B., Fry, J.E., Hoffman, N.L., Eichholtz, D., Rogers, S.G., and Fraley, R.T. (1985). A simple and general method for transferring genes into plants. *Science*, **227,** 1121–31.

Huang, Y., Mord, P., Powell, B., and Kado, C.I. (1990). *VirA*, a coregulator of Ti-specified virulence genes, is phospohorylated *in vitro*. *Journal of Bacteriology*, **172,** 1142–4.

Huffman, G.A., White, F.F., Gordon, M.P., and Nester, E.W. (1984). Hairy-root inducing plasmid: physical map and homology to tumor-inducing plasmids. *Journal of Bacteriology*, **157,** 269–76.

Jefferson, R.A., Kavanagh, T.A., and Bevan, M.W. (1987). GUS fusions: beta-glucuronidase as a sensitive and versatile gene fusion marker in higher plants. *EMBO Journal*, **6,** 3901–7.

Jensen, C.O. (1910). Von echten Geschwulsten bei Pflanzen. *Rapp Conference International Etude Cancer*, **2,** 214–54.

Jin, S., Roitsch, T., Christie, P.J., and Nester, E.W. (1990). The regulatory VirG protein specifically binds to a *cis*-acting regulatory sequence involved in transcriptional activation of *Agrobacterium tumefaciens* virulence genes. *Journal of Bacteriology*, **172**, 531–7.

Jorgensen, R., Snyder, C., and Jones, J.D.G. (1987). T-DNA is organized predominantly in inverted repeat structures in plants transformed with *Agrobacterium tumefaciens* C58 derivatives. *Molecular and General Genetics*, **207**, 471–7.

Jouanin, L., Vilaine, F., D'Enfert, C., and Casse-Delbart, F. (1985). Localization and restriction maps of replication origin regions of the plasmids of *Agrobacterium rhizogenes* strain A4. *Molecular and General Genetics*, **210**, 370–4.

Kado, C.I., Heskett, M.G., and Langley, R.A. (1972). Studies on *Agrobacterium tumefaciens*: characterisation of strains 1D135 and B6, and analysis of the bacterial chromosome, transfer RNA and ribosomes for tumour-inducing ability. *Physological Plant Pathology*, **2**, 47–57.

Kerr, A. (1969). Transfer of virulence between strains of *Agrobacterium*. *Nature*, **223**, 1175–6.

Kerr, A. and Brisbane, P.G. (1983). *Agrobacterium*. In: *Plant bacterial diseases: a diagnostic guide*, (ed. P.C. Fahy and G.J. Persley, pp. 27–43. Academic Press, Australia.

Kersters, K. and De Ley, J. (1984). *Agrobacterium* Conn 1942. In: *Bergey's manual of systematic bacteriology* 1, ed. N.R. Krieg pp. 244–54. Williams and Wilkins Co., Baltimore.

Kersters, K., De Ley, J., Sneath, P.H.A., and Sackin, M. (1973). Numerical taxonomic analysis of *Agrobacterium*. *Journal of General Microbiology*, **78**, 227–39.

Kirchner, G., Roberts, J.L., Gustafson, G.D., and Ingoloa, T.D. (1989). Active bacterial luciferase from a fused gene: expression of a *Vibrio harveyi lux*AB translational fusion in bacteria, yeast and plant cells. *Gene*, **81**, 349–54.

Klee, H.J. and Rogers, S.G. (1989). Plant gene vectors and genetic transformation systems based on the use of *Agrobacterium tumefaciens*. In *Cell culture and somatic cell genetics of plants* **6**, (ed. J. Schell and I.K. Vasil), pp. 2–23. Academic Press, San Diego.

Komro, C.T., Di Rita, V.J., Gelvin, S.B. and Kemp, J.D. (1985). Site-specific mutagenesis in the TR-DNA region of octopine-type Ti plasmids. *Plant Molecular Biology*, **4**, 253–63.

Lewin, B. (1990). *Genes IV*. Oxford University Press, Oxford.

Loesch-Fries, L.S., Merlo, D., Zinnen, T., Burhop, L., Hill, K., Krahn, K. *et al.* (1988). Expression of alfalfa mosaic virus RNA 4 in transgenic plants confers virus resistance. *EMBO Journal*, **6**, 1845–51.

Mariani, C., De Beuckeleer, M., Truettner, J., Leemans, J., and Goldberg, R.B., (1990). Induction of male sterility in plants by a chimaeric ribonuclease gene. *Nature*, **347**, 737–41.

Matthyssee, G.A. (1984). Interaction of *Agrobacterium tumefaciens* with the plant cell surface. In *Genes involved in microbe–plant interactions*, (ed. D.P.S. Verma and T. Hohn), pp. 33–54. Springer-Verlag, Vienna.

Maurel, C., Brevet, J., Barbier-Brygoo, H., Guern, J., and Tempé, J. (1990). Auxin regulates the promoter of the root-inducing *rol B* gene of *Agrobac-*

terium rhizogenes on transgenic tobacco. *Molecular and General Genetics*, **223**, 58–64.

Mayerhofer, R., Koncz-Kalman, Z., Nawrath, C., Bakkeren, G., Crameri, A., Angelis, K. *et al.* (1991). T-DNA integration: a mode of illegitimate recombination in plants. *EMBO Journal*, **10**, 697–704.

Matzke, M.A., Primig, M., Trnovsky, J., and Matzke, A.J.M. (1989). Reversible methylation and inactivation of marker genes in sequentially transformed tobacco plants. *EMBO Journal*, **8**, 643–8.

McCormick, S., Niedermaeyer, J., Fry, J., Barnason, A., Horsch, R., and Fraley, R. (1986). Leaf disc transformation of cultivated tomato (*Lycopersicon esculentum*) using *Agrobacterium tumefaciens*. *Plant Cell Reports*, **5**, 81–4.

McDonnell, R.E., Clark, R.D., Smith, W.A., and Hinchee, M.A. (1987). A simplified method for the detection of neomycin phosphotransferase II activity in transformed plant tissues. *Plant Molecular Biology Reporter*, **5**, 380–386.

McHughen, A., Jordan, M., and McSheffrey, S. (1990). Two years of transgenic flax yield tests: What do they tell us? In *Progress in plant cellular and molecular biology—current plant science and biotechnology in agriculture* Vol. 9, *Proceedings VIIth International Congress on Plant Tissue and Cell Culture,* (ed. H.J.J. Nijkamp, L.H.W. Van der Plas, and J. Van Aartijk), pp. 207–12. Kluwer Academic Publishers, Dordrecht.

McInnes, A.J., Morgan, B.J., Mulligan, B.J., and Davey, M.R. (1991). Phenotypic effects of isolated pRiA4 TL-DNA *rol* genes in the presence of intact TR-DNA in transgenic plants of *Solanum dulcamara* L. *Journal of Experimental Botany*, **42**, 1279–86.

Melchers, L.S. and Hooykaas, P.J.J. (1987). Virulence of *Agrobacterium*. *Oxford Surveys of Plant and Cell Biology*, **4**, 167–220.

Melchers, L.S., Maroney, M.J., Den Dulk-Ras, A., Thompson, P.V., Van Vuuren, A.J., Schilperoort, R.A. *et al.* (1990). Octopine and nopaline strains of *Agrobacterium tumefaciens* differ in virulence; molecular characterization of the *vir*F locus. *Plant Molecular Biology*, **14**, 249–59.

Messens, E., Lenaerts, A., Van Montagu, M., and Hedges, R.W. (1985). Genetic basis for opine secretion from gall cells. *Molecular and General Genetics*, **199**, 344–8.

Meyer, P., Heidemann, I., Forkmann, G., and Saedler, H. (1987). A new petunia flower colour generated by transformation of a mutant with a mazie gene. *Nature*, **330**, 677–8.

Mol, J.N.M. (1991). Genetic engineering of flower colour and development. In *Horticultural exploitation of recent biological developments,* (ed. K.H. Goulding), pp. 11–15. Institute of Horticulture and the North West Branch of the Institute of Biology, London.

Moore, L., Warren, G., and Stobel, G. (1979). Involvement of a plasmid in the hairy root disease of plants caused by *Agrobacterium rhizogenes*. *Plasmid*, **2**, 617–62.

Moore, L.W., Kado, C.I., and Bouzar, H. (1990). II: Gram-negative bacteria. In *Laboratory guide for identification of plant pathogenic bacteria,* (2nd edn) (ed. N.W. Schaad), pp. 16–34. APS Press, Minnesota.

Morgan, A.J., Cox, P.N., Turner, D.A., Peel, E., Davey, M.R. Gartland, K.M.A., and Mulligan, B.J. (1987). Transformation of tomato using an Ri plasmid vector. *Plant Science*, **47**, 37–49.

Mullineaux, P.M., Lewis, D.M., Guerineaux, F., Davies, A., Kular, B., and Watts, J.W. (1987). Use of a herbicide and its resistance gene as a selectable marker for plant transformation. *Annual Report of AFRC Institute of Plant Science Research and John Innes Institute*, Norwich, UK, pp. 19–20. John Catt Ltd, Gt. Glemham.

Offringa, I.A., Melchers, L.S., Regensburg-Tuink, A.J.G., Costantino, P., Schilperoort, R.A., and Hooykaas, P.J.J. (1986). Complementation of *Agrobacterium tumefaciens* tumour inducing *aux* mutants by genes from the T_R-region of the Ri plasmid of *Agrobacterium rhizogenes*. *Proceedings of the National Academy of Sciences USA*, **83**, 6935–9.

Ooms, G. (1992). Genetic engineering of plants and cultures. In *Plant biotechnology*, Comprehensive Biotechnology Second Supplement, (ed. M.W. Fowler, G.S. Warren, and M. Moo-Young), pp. 223–7. Pergamon Press, Oxford.

Ooms, G., Twell, D., Bossen, M.E., Hoge, J.H.C., and Burrell, M.M. (1986). Development regulation of Ri Ti DNA gene expression in roots, shoots and tubers of transformed potato (*Solanum tuberosum* cv. Desiree). *Plant Molecular Biology*, **6**, 321–30.

Pazour, G.J. and Das, A. (1990). *VirG*, an *Agrobacterium tumefaciens* transcriptional activator, initiates translation at a UUG codon and is a sequence-specific DNA-binding protein. *Journal of Bacteriology*, **172**, 1241–9.

Peacock, J. (1990). Ways to pollen sterility. *Nature*, **347**, 714–15.

Peerbolte, R., Leenhouts, K., Hookyaas-Van Slogteren, G.M.S., Hoge, J.H.C., Wullems, G.J., and Schilperoort, R.A. (1986a). Clones from a shooty tobacco crown gall tumor I: deletions, rearrangements and amplifications resulting in irregular T-DNA structures and organizations. *Plant Molecular Biology*, **7**, 265–84.

Peerbolte, R., Leenhouts, K., Hookyaas-Van Slogteren, G.M.S., Hoge, J.H.C., Wullems, G.J., and Schilperoort, R.A. (1986b). Clones from a shooty tobacco crown gall tumor II: irregular T-DNA structures and organization, T-DNA methylation and conditional expression of opine genes. *Plant Molecular Biology*, **7**, 285–99.

Powell, G.K., Hommes, N.G., Castle, L.A., and Morris, M.P. (1988). Inducible expression of cytokinin biosynthesis in *Agrobacterium tumefaciens* by plant phenolics. *Molecular Plant–Microbe Interactions*, **1**, 235–42.

Powell, P.A., Stark, D.M., Sanders, P.R., and Beachy, R.N. (1989). Protection against tobacco mosaic virus in transgenic plants that express tobacco mosaic virus antisense RNA. *Proceedings of the National Academy of Sciences USA*, **86**, 6949–52.

Rech, E.L., Golds, T.J., Hammatt, N., Mulligan, B.J., and Davey, M.R. (1988). *Agrobacterium rhizogenes* mediated transformation of the wild soybeans *Glycine canescens* and *G. clandestina*: production of transgenic plants of *G. canescens*. *Journal of Experimental Botany*, **39**, 1275–85.

Riker, A.J. (1930). Studies on infectious hairy root of nursery apple trees. *Journal of Agricultural Research*, **41**, 507–40.

Ronson, C.W., Nixon, B.T., and Ausubel, F.M. (1987). Conserved domains in bacterial regulatory proteins that respond to environmental stimuli. *Cell*, **49**, 579–81.

Rubin, R.A. (1986). Genetic studies on the role of octopine T-DNA border regions in crown gall tumour formation. *Molecular and General Genetics*, **202**, 312–20.

Saito, K., Yamazaki, M., and Murakoshi, I. (1992). Transgenic medicinal plants: *Agrobacterium*-mediated foreign gene transfer and production of secondary metabolites. *Journal of Natural Products*, **2**, 149–62.

Scott, R.J. and Draper, J. (1987). Transformation of carrot tissues derived from proembryogenic suspension cells: a useful model system for gene expression studies in plants. *Plant Molecular Biology*, **8**, 265–74.

Schilperoort, R.A., Meijs, W.H., Pippel, G.M.W., and Veldstra, H. (1969). *Agrobacterium tumefaciens* cross-reacting antigens in sterile crown-gall tumors. *FEBS Letters*, **3**, 173–6.

Shah, D.M., Horsch, R.B., Klee, H.J., Kishore, G.M., Winter, J.A., Tumer, N.E. *et al.* (1986). Engineering herbicide tolerance in transgenic plants. *Science*, **233**, 478–81.

Shahin, E.A., Sukhapinda, K., Simpson, R.B., and Spivey, R. (1986). Transformation of cultivated tomato by a binary vector in *Agrobacterium rhizogenes*: transgenic plants with normal phenotypes harbour binary vector T-DNA, but no Ri plasmid T-DNA. *Theoretical and Applied Genetics*, **72**, 770–7.

Shaw, C.H., Ashby, A.M., Brown, A., Royal, C., Loake, G.J., and Shaw, C.H. (1988). *Vir*A and *Vir*G are the Ti plasmid functions required for chemotaxis of *Agrobacterium tumefaciens* towards acetosyringone. *Molecular Microbiology*, **2**, 413–18.

Shen, W.H., Petit, A, Guer, J., and Tempé, J. (1988). Hairy roots are more sensitive to auxin than normal roots. *Proceedings of the National Academy of Sciences USA*, **85**, 3417–21.

Simoens, C., Spielmann, A., Margossian, L., and McKnight, T.D. (1986). A disarmed binary vector from *Agrobacterium tumefaciens* functions in *A. rhizogenes*. *Plant Molecular Biology*, **6**, 403–15.

Simpson, R.B., Spielmann, A., Margossian, L., and McKnight, T.D. (1986). A disarmed binary vector for *Agrobacterium tumefaciens* functions in *Agrobacterium rhizogenes*. *Plant Molecular Biology*, **6**, 403–15.

Sinkar, V.P., White, F.F., and Gordon, M.P. (1987). Molecular biology of Ri-plasmid: a review. *Journal of Bioscience*, **11**, 47–57.

Sinkar, V.P., Pythoud, F., White, F.F., Nester, E.W., and Gordon, M.P. (1988). *Rol*A locus of the Ri plasmid directs developmental abnormalities in transgenic tobacco plants. *Genes and Development*, **2**, 688–97.

Smith, E.F. (1917). Embryogenesis in plants (produced by bacterial inoculations). *Bulletin of the Johns Hopkins Hospital*, **28**, 277–94.

Smith, E.F. and Townsend, C.O. (1907). A plant tumour of bacterial origin. *Science*, **25**, 671–3.

Smith, C.J.S., Watson, C., Ray, J., Bird, C.R., Morris, P.C., Schuch, W. *et al.* (1988). Antisense RNA inhibition of polygalacturonase gene expression in transgenic tomatoes. *Nature*, **334**, 724–6.

Stachel, S.E., Messens, E., Van Montagu, M., and Zambryski, P. (1985). Identification of the signal molecules produced by wounded plant cells that activate T-DNA transfer in *Agrobacterium tumefaciens. Nature,* **318,** 625–9.

Stachel, S. and Nester, E. (1986). The genetic and transcriptional organization of the *vir* region of *Agrobacterium tumefaciens. EMBO Journal,* **5,** 1445–54.

Stachel, S.E., Timmermann, B., and Zambryski, P. (1987). Activtion of *Agrobacterium tumefaciens vir* gene expression generates multiple single stranded T-strand molecules from the pTiA6 T-region: requirements for 5' *vir*D gene products. *EMBO Journal,* **6,** 857–63.

Stalker, D.M. (1991). Developing herbicide resistance in crops by gene transfer technology. In *Plant genetic engineering,* (ed. D. Grierson), pp. 82–104. Blackie, Glasgow, London.

Stalker, D.M., McBurke, K.E., and Malyj, L.D. (1988). Herbicide resistance in transgenic plants expressing a bacterial detoxification gene. *Science,* **242,** 419–23.

Streber, W.S., Timmis, K.N., and Zenk, M.H. (1987). Analysis, cloning, and high-level expressing of 2,4-dichlorophenoxyacetate monooxygenase gene *tfdA* of *Alcaligenes entrophus* JMP134. *Journal of Bacteriology,* **169,** 2950–5.

Streber, W.R. and Willmitzer, L. (1989). Transgenic tobacco plants expressing a bacterial detoxifying enzyme are reistant to 2,4-D. *Bio/Technology,* **7,** 811–16.

Sun, L.Y., Tourand, G., Charbonnier, C., and Tepfer, D. (1991). Modification of phenotype in Belgium endive (*Chicorium intybus*) through genetic transformation by *Agrobacterium rhizogenes*: conversion from biennial to annual flowering. *Transgenic Research,* **1,** 14–22.

Tempé, J. and Goldmann, A. (1982). Occurrence and biosynthesis of opines. In *Molecular biology of plant tumours,* (ed. G. Kahl and J.S. Schell), pp. 427–49. Academic Press, New York.

Tempé, J. and Casse-Delbart, F. (1989). Plant vectors and plant transformation: *Agrobacterium rhizogenes* Ri plasmids. In *Cell culture and somatic cell genetics of plants* **6,** (ed. J. Schell and I.K. Vasil), pp. 26–49. Academic Press, San Diego.

Tepfer, D. (1983). The potential uses of *Agrobacterium rhizogenes* in the genetic engineering of higher plants: nature got there first. In *Genetic engineering in eukaryotes* **61,** (NATO AS1 Series. Series A. Life Sciences) (ed. P.F. Lurquin and A. Kleinhofs), pp. 153–64. Plenum Press, New York.

Tepfer, D. (1984). Transformation of several species of higher plants by *Agrobacterium rhizogenes*: sexual transmission of the transformed genotype and phenotype. *Cell,* **37,** 959–67.

Tepfer, D. (1990). Genetic transformation using *Agrobacterium rhizogenes. Physiologia Plantarum,* **79,** 140–6.

Thomashow, M.F., Karlinsey, J.E., Marks, J.R., and Hubert, R.E. (1987). Identification of new virulence locus in *Agrobacterium tumefaciens* that affects polysaccharide composition and plant-cell attachment. *Journal of Bacteriology,* **169,** 3209–16.

Thompson, D.V., Melchers, L.S., Idler, K.B., Schilperoort, R.A., and Hooykaas, P.J.J. (1988). Analysis of the complete nucleotide sequence of the *Agrobacterium tumefaciens vir*B operon. *Nucleic Acids Research,* **16,** 4621–36.

Trinh, T.H., Mante, S., Pua, E.C. and Chua, N.H. (1987). Rapid production of transgenic flowering shoots and F1 progeny from *Nicotiana plumbaginifolia* epidermal peels. *Bio/Technology,* **5,** 1081–4.

Valvekens, D., Van Montagu, M., and Van Lijsebettens, M. (1988). *Agrobacterium tumefaciens*-mediated transformation of *Arabidopsis thaliana* root explants by using kanamycin selection. *Proceedings of the National Academy of Sciences USA*, **85,** 5536–40.

Usame, S., Morikawa, S., Takebe, I., and Machida, Y. (1987). Absence in monocotyledonous plants of the diffusible plant factors inducing T-DNA circularization and *vir* gene-expression in *Agrobacterium*. *Molecular and General Genetics*, **209,** 221–6.

Van Haaren, M.J.J., Sedee, N.J.A., Schilperoort, R.A., and Hooykaas, P.J.J. (1987). Overdrive is a T-region transfer enhancer which stimulates T-strand production in *Agrobacterium tumefaciens*. *Nucleic Acids Research*, **15,** 8983–97.

Van Larabeke, N., Engler, N.G., Holsters, M., Van der Elsacker, S., Zaenen, I., Schilperoort, R.A., and Schell, J. (1975). Large plasmids in *Agrobacterium tumefaciens* essential for crown gall inducing ability. *Nature*, **252,** 169–70.

Velten, J., Velten, L., Hains, R., and Schell, J. (1984). Isolation of a dual plant promoter fragment from the Ti plasmid of *Agrobacterium tumefaciens*. *EMBO Journal*, **3,** 2723–30.

Vilaine, F. and Casse-Delbart, F. (1987*a*). A new vector derived from *Agrobacterium rhizogenes* plasmids: a micro-Ri plasmid and its use to construct a mini-Ri plasmid. *Gene*, **55,** 105–14.

Vilaine, F. and Casse-Delbart, F. (1987*b*). Independent induction of transformed roots by the T_L and T_R regions of the Ri plasmid of agropine type *Agrobacterium rhizogenes*. *Molecular and General Genetics*, **206,** 17–23.

Walker, J.C. (1969). *Plant pathology*, (3rd edn). McGraw-Hill, New York.

Wallroth, M., Gerates, A.G.M., Rogers, S.G., Fraley, R.T., and Horsch, R. (1986). Chromosomal localization of foreign genes in *Petunia hybrida*. *Molecular and General Genetics*, **202,** 6–15.

Weising, K., Schell, J., and Kahl, G. (1988). Foreign genes in plants: transfer, structure, expression and applications. *Annual Review of Genetics*, **22,** 421–77.

White, F.F. and Nester, E.W. (1980). Hairy root: plasmid encodes virulence traits in *Agrobacterium rhizogenes*. *Journal of Bacteriology*, **141,** 1134–41.

White, F.F., Taylor, B.H., Huffman, G.A., Gordon, M.P., and Nester, E. (1985). Molecular and genetic analysis of the transferred DNA regions of the root-inducing plasmid of *Agrobacterium rhizogenes*. *Journal of Bacteriology*, **164,** 33–44.

Willmitzer, L., Sanchez-Serrano, J., Buschfeld, E., and Schell, J. (1982). DNA from *Agrobacterium rhizogenes* is transferred and expressed in axenic hairy root plant tissues. *Molecular and General Genetics*, **186,** 16–22.

Winans, S.C., Allanza, P., Stachel, S.E., McBride, K.E., and Nester, E.W. (1987). Characterization of the *vir*E operon of the *Agrobacterium* Ti plasmid pTiA6. *Nucleic Acids Research*, **15,** 825–37.

Winans, S.C., Randall, A., Kerstetter, R., and Nester, E.W. (1988). Transcripted regulation of the *VirA* and *VirG* genes of *Agrobacterium tumefaciens*. *Journal of Bacteriology*, **170,** 4047–54.

Yanofsky, M., Porter, S., Young, C., Albright, L., Gordon, M., and Nester, E. (1986). The *VirD* operon of *Agrobacterium tumefaciens* encodes a site specific endonuclease. *Cell*, **47,** 471–7.

Zaenen, I., Van Larabeke, N., Teuchy, N., Van Montagu, M., and Schell, J.

(1974). Supercoiled circular DNA in crown gall inducing *Agrobacterium* strains. *Journal of Molecular Biology*, **86,** 109–27.

Zaitlin, M. and Hull, R. (1987). Plant virus–host interactions. *Annual Review of Plant Physiology*, **38,** 291–315.

Zambryski, P., Joos, H., Genetello, C., Leemans, J., Van Montagu, M., and Schell, J. (1983). Ti plasmid vector for the introduction of DNA into plant cells without alteration of their normal regeneration capacity. *EMBO Journal*, **2**(12), 2143–50.

Zorreguieta, A., Geremia, R.A., Cavaignac, S., Cangelosi, G.A., Nester, E., and Ugalde, R.A. (1988). Identification of the product of an *Agrobacterium tumefaciens* chromosomal virulence gene. *Molecular Plant–Microbe Interaction*, **1,** 121–7.

3. British gall-causing rust fungi

T.F. PREECE* and A.J. HICK†

*Kinton, Turners Lane, Lynclys Hill, Near Oswestry, Shropshire, UK
† Department of Pure and Applied Biology, University of Leeds, Leeds, UK

Abstract

Although some fungal galls on cultivated plants have been the subject of much research (most of which has been aimed at preventing their occurrence) fungal galls on wild plants remain little studied. The literature on galls often includes long lists of rust fungi (Basidiomycotina: Uredinales) but some of these fungi do not produce abnormal growth visible to the naked eye. It is suggested that visible swelling or thickening of plant organs is used to distinguish gall-causing rust fungi. After outlining the life history of a long-cycled rust the authors provide a list of principal British gall-causing rust fungi and their hosts.

Fungi as gall causers

Swelling, thickening, or distortion of plant organs, loosely called galls, have a variety of causes other than the animals with which much of this book is concerned. Other gall causers include fungi, bacteria, viruses and mycoplasma-like and rickettsia-like organisms. Chemicals such as weed-killers may produce galls on plants which are almost indistinguishable from *Agrobacterium* infections (2,4-dichlorophenoxyacetic acid on Brassicas, for example). Though more commonly seen by plant breeders, genetically produced galls on hybrid plants also look strikingly like galls produced by bacteria or fungi.

Most fungal infections of plants do not induce galls, but result in a range of other symptoms such as necrosis, wilting, and so on. In contrast to overgrowth, stunting of whole plants may be seen, in which whole plants are smaller than usual. Dwarf bunt disease of wheat and many grasses in the USA caused by the smut fungus, *Tilletia contraversa* Kühn is an example. The reverse of this whole plant effect is seen in 'Foolish Seedling Disease' of rice caused by the ascomycete fungus *Gibberella*

Plant Galls (ed. Michèle A. J. Williams), Systematics Association Special Volume No. 49, pp. 57–66. Clarendon Press, Oxford, 1994. © The Systematics Association, 1994.

fujikuori (Saw.) Wollenw. (*Fusarium moniliforme* J. Sheld.) from which the growth promoters known as gibberellic acids have been developed as well as greatly advancing our knowledge of green plant physiology. Overgrowth of one part of a plant, including gall formation, may result in underdevelopment of another part, as in club root disease of *Brassica* caused by the myxomycete fungus *Plasmodiophora brassicae* Woronin. Infected plants are smaller above ground than non-infected ones.

Little is known of the physiology, biochemistry, or molecular biology of plant galls in general. Fungal galls of plants are no exception to this, but the involvement of plant hormones—auxins, gibberellins, abscisins, ethylenes, and cytokinins—in their formation would seem to be likely as indicated by work on other plant diseases. The involvement of cytokinins in sawfly-induced galls of willows is discussed by Leitch (Chapter 17, this volume).

We known more about the general biology of gall-forming fungi affecting crop plants such as *P. brassicae* than we know about any of the fungal galls on British wild plants, all of which await investigation.

Club root was very troublesome in Russia during the last century and was intensively studied by Woronin in the 1870s. He first used the term hypertrophy to describe the plant tissues in galls. Since then the notion (usually untested) that many types of gall 'feed' their causers has often been taken for granted. In club root, not only is there an increase in the size of host cells but also an increase in their number leading to the observed swelling. Electron microscopy revealed the unexpected mechanism by which resting spores of *P. brassicae* develop a cavity inside themselves (a 'rohr') within which a rod (a 'stachel') develops which penetrates the host plant and admits the fungus into the host cytoplasm. When infected *Brassica* root cells divide, the fungus divides too, giving rise to a mass of infected cells. Spores round off in infected tissue, which eventually disintegrates into an unpleasant smelling mess, the spores remaining alive in the soil for long periods. Here the galled tissue is of prime importance to farmers as a means of survival and dispersal of the fungus. Real control is by the use of resistant cultivars, in the production of which the fact that the fungus exists in many separate strains of races is of critical importance. The existence of races of other gall formers affecting wild plants is likely.

This kind of build-up of spores in the soil, is also the agricultural problem with the large potato tuber galls caused by the chytrid fungus, *Synchytrium endobioticum* (Schilb.) Percival The very thick-walled resting spores remain in the soil as the galled tubers rot. More details of the life history of these and other gall-forming fungi which affect crop plants is to be found in Alexopoulos and Mimms (1979) and more practical information in Parry (1990).

Gall-forming fungi in Britain

Fred Stubbs, a pioneer of field cecidology in Britain, included in his 'Check List' (Stubbs 1986) 63 galls caused by fungi and his compilation gives us a useful overview of gall-forming fungi in Britain. The fungal genera listed are *Plasmodiophora*, *Synchytrium* (see above), *Urocystis*, *Ustilago*, *Entyloma* (smut fungi), *Exobasidium* (see Ing, Chapter 4, this volume), *Epichloe*, *Protomyces*, *Taphrina* (ascomycetes), *Peronospora*, *Plasmopara* and *Albugo* (oomycetes). Approximately three-quarters of the fungi listed are rusts in the 10 genera *Puccinia*, *Uromyces*, *Melampsora*, *Melampsorella*, *Tranzschelia*, *Ochropsora*, *Triphragmium*, *Gymnosporangium*, *Phragmidium*, and *Pucciniastrum*. It is immediately clear that the gall-forming genera cut across all the major taxonomic divisions of the Kingdom Fungi. It is also clear that the conidial fungi which cause the majority of leaf spots on green plants do not cause galls.

Rust Fungi: understanding the stages in their life history

The spores and sori of a British gall-forming rust on a rose leaf was the first fungus ever observed with the aid of a compound microscope (Hooke 1665, in Preece and Hick 1990). The drawings are of the black spores (teliospores) of *Phragmidium tuberculatum*. They are seen on small raised areas of the leaf surface, probably the epidermis, with other plant cells forced upwards and outwards by the emerging mass of black spores.

Rust spores range in colour from near white, through pale yellow, rusty brown to black and represent different spore stages. Not all the various spore stages of rusts produce galls and in order to understand which stage of a rust is doing so and to speculate why a particular stage is involved it is necessary to understand these fungi in more detail.

There are still differences between workers as to what to call the various stages in the life history of rusts (Hawksworth *et al.* 1983). Those devised by Hiratsuka and Sato (1982) are used in this chapter. It is now accepted that, amongst all living things, the rust fungi have the most complex known series of events in their life cycles and their nuclear arrangements (Petersen 1974).

For almost all its life a rust carries haploid nuclei in pairs in its hyphae (dikaryotic mycelium). Five types of rust spores occur and these are denoted by numbers [that is, 0, I, II, III, and IV.] *Teliospores* (III), which were the first seen by Hooke, are usually black and are produced in *telia*. Fusion of the two haploid spores of a dikaryon occurs in them followed immediately by meiosis to give haploid spores again. Thus, sex occurs briefly in teliospores and the diploid state of the cells is transitory. The morphological characters of teliospores are the best means of identifying a rust if the plant host has been correctly identified. *Basidiospores* (IV) are borne on *basidia* which emerge from the teliospores. If two plant hosts are known for the particular rust (these two hosts always being in widely different groups of plants such as a nettle and a sedge) basidiospores infect one of the hosts. *Spermogonia* then develop and produce bacteria-sized spores termed *spermatia* (0). Spermogonia often have means of attracting insects which inadvertently transmit spermatia from one sper-mogonium to another. The spermatia fuse in pairs. Each has a single haploid nucleus and this process of dikaryotization produces dikaryotic mycelium which grows in the plant host (usually a leaf) and produces *aecia* bearing *aeciospores* (I) which have two haploid nuclei. Aeciospores are often yellow or white. In the rust *Puccinia caricina* one aecial host is the nettle (*Urtica dioica*). Aeciospores infect the alternate host, if there is one. This is a sedge, *Carex acutiformis*, for example, in the case of *Puccinia caricina*. On this second host *uredia* containing *urediniospores* (IV) are produced and they spread the rust to more leaves of the same plant and to more plants of the same host. Uredia are the orange coloured pustules from which the rusts get their common name. On this same host telia appear, often later in the season, producing teliospores (III). There are a considerable number of variations on the outline presented here; more details about the stages of individual rusts producing galls in Britain can be found in Wilson and Henderson (1966), Ellis and Ellis (1985), and Preece and Hick (1990). Not all rust fungi have five types of spore. Some have only one plant host, not two. *Puccinia menthae*, mint rust has all five stages (0–IV) on the mint plant, but only the aecia are associated with enlargement and distortion of the plant. In this chapter only *visibly* swollen or thickened rust-infected plants are considered to be galled.

A checklist of principal British rusts causing plant galls

Table 3.1 lists the principal British rusts which cause galls on plants at some stage in their life history, based on the visible swelling of a plant organ. This list includes approximately one-quarter of British rust fungi and approximately 80 per cent of those which have been recorded as gall causers. Usually only one stage in the life history of the rust is associated with galling, this is most often the aecial stage (I), producing aeciospores. Apart from gall-causing species listed from personal obser-vations by the authors the list is compiled from the previously published reports of Plowright (1889), Connold (1909), Swanton (1912), Grove (1913), Buhr (1964), Wilson and Henderson (1966), Darlington (1968) and Stubbs (1986). Scanning electron micrographs which illustrate diagnostic characters of almost all of these rusts are presented in Preece and Hick (1990).

Space does not allow detailed comment on all these gall formers, and, thus, just some species, chosen because they illustrate general points or

Table 3.1. Principal British rusts causing galls on plants at some stage in their life history

Rust	Host and rust life-cycle stage which causes galls
Chrysomyxa pirolata Winter	*Pyrola* spp. (II)
Coleosporium tussilaginis (Pers.) Lév.	*Pinus sylvestris* L. (I)
Cronartium flaccidum (Alb. & Schwein) Winter	*Pinus sylvestris* l. (I)
Cronartium ribicola Fischer	*Pinus* spp. (I)
Cumminsiella mirabilissima (Peck) Nannf.	*Mahonia* spp. (I)
Endophyllum euphorbiae-silvaticae (DC) Winter	*Euphorbia amygdoloides* L. (III)
Endophyllum sempervivi de Bary	*Sempervivum* spp. (III)
Frommea obtusa (Str.) Arth.	*Potentilla erecta* (L.) Räusch, *P. reptans* L.
Gymnosporangium clavariiforme (Pers.) DC	*Crataegus* (I), *Juniperus* (III)
Gymnosporangium confusum Plowright	*Crataegus, Cydonia, Mespilus* (I)
Gymnosporangium cornutum Kern	*Sorbus* (I), *Juniperus* (III)
Gymnosporangium fuscum DC	*Pyrus communis* L. (I), *Juniperus sabina* L. (III)
Melampsora epitea Thümen	*Dactylorchis* spp. *Gymnadenia, Listera* (I)
Melampsorella caryophllacearum Scroet.	*Abies* spp. (I)
Nyssopsora echinata (Lév) Arth.	*Meum athamanticum* Jacq. (III)
Ochrospora ariae (Fckl.) Ramsb.	*Anemone nemorosa* L. (I)
Phragmidium mucronatum (Pers.) Schlecht.	*Rosa* spp. (I)
Phragmidium sanguisorbae (DC) Schroet.	*Poterium* spp. (I)
Phragmidium tuberculatum J. Müller	*Rosa* spp. (I)
Puccinia adoxae DC	*Adoxa moschatellina* L. (III)
Puccinia aegopodii (Str.) Röhl	*Aegopodium podagraria* L. (III)
Puccinia albescens Plowr.	*Adoxa moschatellina* L. (I)
Puccinia angelicae (Schum.) Fckl.	*Angelica, Peucedanum, Selinum, Silaum* (II)
Puccinia annularis (Str.) Röhl	*Teucrium scorodonia* L. (III)
Puccinia apii Desm.	*Apium graveolens* L. (I)
Puccinia arenariae (Schum.) Winter	*Sagina* spp. *Moehringia* (III)
Puccinia argentata (Schultz) Winter	*Adoxa moschatellina* (I)
Puccinia bistortae DC	*Angelica, Conopodium* (I)
Puccinia bulbocastani Fckl.	*Bunium, Carum* (I)
Puccinia bupleuri Rudolphi	*Bupleurum falcatum* L. (I)
Puccinia buxi DC	*Buxus sempervirens* L. (III)
Puccinia calthae Link	*Caltha palustris* L. (I)
Puccinia calthicola Schroet.	*Caltha palustris* L. (I)
Puccinia caricina DC	*Urtica* spp., *Ribes* spp. (I)
Puccinia chaerophylli Purton	*Anthriscus, Chareophyllum, Myrrhis* (I)
Puccinia circaeae Pers.	*Circaea* spp. (III)
Puccinia cnici-oleracei Pers.	*Achillea millefolium* L. (III)
Puccinia conii Lagh.	*Conium maculatum* L. (II)
Puccinia coronata Corda	*Frangula, Rhamnus* (I)
Puccinia difformis Kunze	*Asperula, Galium aparine* L. (I & III)
Puccinia fergussonii Berk. & Br.	*Viola palustris* L. (III)
Puccinia galii-verni Ces.	*Galium* spp. (III)

Table 3.1. *contd.*

Rust	Host and rust life-cycle stage which causes galls
Puccinia glechomatis DC	*Glechoma hederacea* L. (III)
Puccinia glomerata Grev.	*Senecio* spp. (III)
Puccinia graminis Pers.	*Berberis, Mahonia* (I)
Puccinia heraclei Grev.	*Heracleum sphondylium* L. (I)
Puccinia hieracii Mart.	Various Compositae (I)
Puccinia lagenophorae Cooke	*Senecio* spp. (II)
Puccinia lapsanae Fckl.	*Lapsana communis* L. (I)
Puccinia libanotidis Lindr.	*Seseli libonois* (L.) Koch (II)
Puccinia liliacearum Duby	*Ornithogalum* spp. (III)
Puccinia maculosa (Str.) Röhl	*Mycelis muralis* (L.) Dumort. (I)
Puccinia major Diet.	*Crepis paludosa* (L.) Moench (III)
Puccinia malvacearum Mont.	*Althaea, Lavatera, Malva* (IIII)
Puccinia menthae Pers.	*Mentha* spp. and various Labiatae
Puccinia nemoralis Juel	*Melampyrum pratense* L. (I)
Puccinia pimpinellae (Str.) Röhl	*Pimpinella* spp. (I)
Puccinia phragmitis (Schum.) Körnicke	*Rheum, Rumex* spp. (I)
Puccinia poae-nemoralis Otth	*Berberis vulgaris* L. (I)
Puccinia poarum Niels.	*Tussilago farfara* L. (I)
Puccinia polemonii Diet. & Holw.	*Polemonium careuleum* L. (III)
Puccinia punctiformis (Str.) Röhl	*Cirsium arvense* (l.) Scop. (II)
Puccinia recondita Rob. & Desm.	Various Ranunculaceae & Boraginaceae (I)
Puccinia saniculae Grev.	*Sanicula europaea* L. (I)
Puccinia septentrionalis Juel	*Thalictrum alpinum* L. (I)
Puccinia smyrnii Biv.-Bernh.	*Smyrnium olusatrum* L. (I)
Puccinia tumida Grev.	*Conopodium majus* (Gouan) Loret (III)
Puccinia veronicae Schroet.	*Veronica* spp. (III)
Puccinia violae DC	*Viola* spp. (I)
Pucciniastrum goeppertianum (Kühn.) Kleb.	*Vaccinium* spp. (III)
Tranzschelia anemones (Pers.) Nannfeldt	*Anemone nemorosa* L. (III)
Tranzschelia discolor (Fckl.) Tranz. & Litv.	*Anemone* spp. (I)
Triphragmium filipendulae Pass.	*Filipendula vulgaris* Moench (I & III)
Triphragmium ulmariae (DC) Link	*Filipendula ulmaria* (L.) Maxim (I & II)
Uromyces betae (Pers.) Tul.	*Beta vulgaris* L. (I)
Uromyces ficariae (Alb. & Schwein) Lév.	*Ranunculus ficaria* L. (III)
Uromyces geranii (DC) Lév.	*Geranium* spp. (I)
Uromyces limonii (DC) Lév.	*Limonium* spp. (I)
Uromyces lineolatus (Desm.) Schroet.	*Oenanthe, Betula, Glaux* (I)
Uromyces minor Schroet.	*Trifolium* spp. (I)
Uromyces muscari (Duby) Lév.	*Endymion* spp. (III)
Uromyces nerviphilus (Grognot) Hotson	*Trifolium* spp. (III)
Uromyces pisi (DC) Otth	*Euphorbia cyparissias* L. (I)
Uromyces scrophulariae Fckl.	*Scrophularia* spp. (I & III)
Uromyces trifolii (DC) Lév.	*Trifolium* spp. (I)
Uromyces valerianae (DC) Lév.	*Valeriana* spp. (I)
Zaghouania phillyreae Pat.	*Phillyrea latifolia* L. (I)

Rust stages are represented by I, Aecidia; II, Uredia; III, Telia.

are of particular interest, are discussed below. It should be emphasized that all these galls need detailed microscopic analysis (most authors hitherto have described details of the spore production, not the histology of the gall tissue) and experimental work relating to the function of the gall tissue in natural conditions of every kind.

A rust fungus has to grow through a great deal of plant tissue in the stem or leaf to produce the aecia and aeciospores (I). This takes a varying length of time. Days to years may be needed. In one example the spermogonia with spermatia and aecia with aeciospores take 2–4 years to appear after basidiospores infect five-needled pines after they have arrived from the teliospores which are produced on *Ribes* leaves. As is always the case in rusts the number of aeciospores produced is finite as distinct from the urediniospores and teliospores which are produced in indefinite and usually vast numbers. In this sense in all rusts the success of the aeciospores in reaching another host if there are two hosts and to other parts of the same host if there is only one host, is vital in order that the life history of the rust can be completed. The story of *Puccinia caricina* has already been used above to illustrate the full-cycled life history of a typical rust. The aecial stage of this rust on nettle stems is one of the largest rust galls seen in Britain, whereas when developed on leaves the galls are almost flat and less striking when seen in the field. In one of the very few documented studies of the possible function of gall tissue in rusts, Savile (1954) showed that aeciospores of *P. caricina* were discharged further from the swollen aecia on stems of the nettle than from relatively flat aecia on leaves. Another aspect of the biology of this rust gall needs study and is perhaps related to Savile's observations. Nearly a century ago, an observant naturalist (Vize 1894) noted that where the two hosts of this rust are found close together, the aecia are commonly on the leaves rather than on the stems.

Another effect of the swelling and distortion of leaves by rust aecia is seen in aecial infections by *Triphragmium ulmariae* on meadow sweet, *Filipendula ulmaria*, where the whole leaf and petiole is so twisted that the lower (abaxial) surface of the leaf is presented uppermost—which may aid the dispersal of aeciospores.

There does seem to be a general rule that stages of rusts and particularly aecia on petioles, veins, and stems are more distorting than those on the laminas of leaves. Most aecia of *Melampsora populnea* on dogs mercury, *Mercurialis perennis*, in the spring are on the distal tip of the leaf, but occasionally, aecia develop on the petiole and cause considerable swelling and distortion.

Exceptions to the general rule in British rusts that galls are found associated with the aecial stage must be of interest. *Uromyces nerviphilus* is a good example of distortion when sori occur on or near leaf veins. These galls are caused by the telia, on white clover, *Trifolium repens*. Other examples of galls not associated with aecia and in which aecia are unknown are *Puccinia tumida* on *Conopodium*, *P. adoxae* on *Adoxa moschatellina*, and *P. aegopodii* on *Aegopodium podograria*. Again these have galled telial stages. Three other British rusts have swollen telia and aecidia, but none have galled uredinial stages. Finally, no monocotyledonous plant is galled here and there is general agreement that the most severe galls occur on rusted plants in the families Umbelliferae, Urticaceae, and Cupressaceae. No rust on ferns produces galls (see Hick and Preece 1990).

Structure and possible functions of rust galls

Galls caused by rusts do not have the distinct form seen in many galls caused by animals like the cynipid galls on oak fruits and leaves which resemble new plant organs on the plant host. In rust galls the principal histological change is an increase in cell size, which is largest near the developing rust sorus. This increase in cell size is accompanied by an increase, which is often very substantial, in the volume occupied by air spaces between the parenchymatous tissues of the leaf or other affected organ. Details may differ, but these two features seem central ones. An increase in cell numbers (seen in bacterial galls, for example) is unusual. One case where it is seen is in tissues of *Abies alba* affected by the telia of *Gymnosporangium cornutum* (Wornle 1894) where needle tissue maybe about four times as thick as normal, with layers of cork cells produced each year 'peeled back' to allow the teliospores to emerge. On the other hand no such cork cell multiplication was reported in aecial galls on *Picea abies* needles caused by *Melampsorella caryophyllacearum* (White and Merrill 1969). In their beautifully illustrated book about the histology of galls of all kinds, *Anatomie des galls*, Meyer and Maresquelle (1983) give examples of the increased air spaces in rust infected leaves with galls of *Uromyces pisi*, *Pucciniastrum goeppertianum*, *Puccinia buxi*, and *P. coronata*. All these examples have increased cell size and the gradient of cell size outwards from the aecia of *P. coronata* on *Rhamnus frangula* is shown clearly in a drawing of a transverse section of a leaf. We are at present working on the aecia of *Puccinia lapsanae* on *Lapsana communis* and of *Puccinia lagenophorae* on *Senecio vulgaris*, which show similar histological changes: increased cell size and increased size of air spaces in a gradient towards the aecia. Aecia of *P. caricina* on nettle cause galls that are so large that Plowright (1889) notes that they were being used in 1887 by native peoples in the Himalayas for food because they are very rich in starch!

As well as the histological changes in host tissues a gall also contains fungal hyphae in the swollen areas. Luttrell (1981) has drawn attention to a group of plant diseases, including rusts which he terms 'tissue replacement diseases' in which host tissue is replaced by fungal tissue. In all the five rust stages, relative to the size of individual plant cells a massive body of fungal tissue becomes embedded in the leaf or stem. Some of this must contribute to the swelling, as Hooke observed over 300 years ago in rose rust! Luttrell (1981) describes another part of the process by which the host cells around an aecium, for example, are crushed and flattened by mechanical means. Opposite in effect to this must be the volume occupied by the swollen hyphal endings (paraphyses) common in rust sori and the process of spore development with many spores with their surface ornamentation pushing sideways and outwards.

We must face a final question. What use are rust galls? Apart from Savile's (1954) work indicating the possible usefulness of swollen aecia in spore discharge we are ignorant. In future work, like Dawkins (1990) we should consider whether the 'extended phenotype' of rust genes represented by this swollen plant tissue might confer some benefit to the host or have done so in the coevolution of rusts and their green host plants. Pirozynski (1988) asks us to consider whether genetic material of some kind may be transferred by fungi to plants here, as is the case in *Agrobacterium* galls. Not only might we give our attention to these matters, but also Burdon (1987) asks for work on how populations of plants are affected by rusts with galls. There is much to be done if we are to know and understand gall-forming rust fungi—of which the British species are only a fraction of the world rust flora.

References

Alexopoulos, C.J. and Mimms, C.W. (1979). *Introductory mycology*, (3rd edn). John Wiley, New York.

Buhr, H, (1964–65). *Bestimmungstabellen der gallen (Zoo- und Phytocecidien) an Pflanzen Mittel- und Nordeuropas*, Vols, 1 and 2. Gustav Fischer Verlag, Jena.

Burdon, J.J. (1987). *Diseases and plant population biology.* Cambridge University Press, Cambridge.

Connold, E.T. (1909). *Plant galls of Great Britain*. Adlard and Sons, London.

Darlington, A. (1968). *The pocket encyclopedia of plant galls in colour.* Blandford Press, London.

Dawkins, R. (1990). *The extended phenotype—the long reach of the gene* Oxford University Press, Oxford.

Ellis, M.B. and Ellis, J.P. (1985). *Microfungi on land plants—an identification handbook.* Croom Helm, London.

Grove, W.B. (1913). *British rust fungi.* Cambridge University Press, Cambridge.

Hawksworth, D.L. Sutton, B.C., and Ainsworth, G.C. (1983). *Ainsworth and Bisby's dictionary of the fungi*, (7th edn). Commonwealth Mycological Institute, Kew.

Hick, A.J. and Preece, T.F. (1990). Scanning electron microscopy of the sori and spores of six species of rust fungi (Uredinales) found on ferns in Britain. *Fern Gazette*, **13** (6), 321–8.

Hiratsuka, Y. and Sato, S. (1982). Morphology and taxonomy of rust fungi *In The rust fungi*, (ed. K.J. Scott), pp. 1–36. Academic Press, London.

Hooke, R. (1665). *Micrographia*. The Royal Society, London.

Johnson, A. and Booth, C. (ed.) 1983 *Plant pathologists pocketbook*, (2nd edn.) Commonwealth Mycological Institute, Kew.

Luttrell, E.S. (1981). Tissue replacement diseases caused by fungi. *Annual Review of Phytopathology*, **19**, 373–89.

Meyer, J. and Maresquelle, H.J. (1983). *Anatomie des Galles*. Gebrüder Borntraeger, Berlin.

Parry, D.W. (1990). *Plant pathology in agriculture*. Cambridge University Press, Cambridge.

Petersen, R.H. (1974). The rust fungus life cycle. *Botanical Review*, **40**, 453–51

Pirozynski, K.A., (1988). Co-evolution by horizontal gene transfer; a speculation on the role of fungi. In *Co-evolution of fungi with plants and animals*, (ed. K.A. Pyrozynski and D.L. Hawksworth), pp. 247–68. Academic Press, London.

Plowright, W.B. (1989). *A monograph of the British Uredineae and Ustilagineae*. Kegan Paul, Trench and Co., London.

Preece, T.F. and Hick, A.J. (1990). *An introductory scanning electron microscope atlas of rust fungi*. Farrand Press, London.

Savile, D.B.O. (1954). Cellular mechanics, taxonomy and evolution in the Uredinales and Ustilaginales. *Mycologia*, **46**, 736–61.

Stubbs, F.B. (1986). *Provisional keys to British plant galls*. The British Plant Gall Society.

Swanton, E.W. (1912). *British plant galls*. Methuen, London.

Vize, J.E. (1894). Fungi in the Powys-land district, to be found in our gardens. *Collections Historical and Archaeological relating to Montgomeryshire*, **10**, 306–24.

White, D. and Merrill, W. (1969). Pathological anatomy of *Abies balsamea* infected with *Melampsorella caryophyllacearum*. *Phytopathology*, **59**, 1238–42.

Wilson, M. and Henderson, D.M. (1966). *British rust fungi*. Cambridge University Press, Cambridge.

Woernle, P. (1894). Anatomische Untersuchung der durch Gymnosporangi-Arten hervorgerufenen Mitsbildungen. *Forstlich Naturwissenschaftliche Zeitschrift*, **2** (3), 68–84; (4), 129–73.

4. European Exobasidiales and their galls

BRUCE ING

Department of Biology, Chester College of Higher Education, Cheyney Road, Chester, UK

Abstract

Exobasidiales are obligate parasites on leaves and shoots of Ericaceae, Lauraceae, Saxifragaceae, and Theaceae. They are heterobasidiomycetous fungi with no vegetative structures of their own, utilizing the somatic tissues of their hosts to support either systemic or annual mycelia and spore producing organs. The European genera are reviewed including observations on morphology, host species, and distribution. The extent of gall formation varies, from thickened, discoloured leaf spots in *Arcticomyces* and some *Exobasidium* species, through well-developed leaf pouch galls caused by *Exobasidium* to the clavarioid stem galls of *Laurobasidium*. A key to European species, based on host damage, is provided.

Introduction

The Exobasidiales are basidiomycetous fungi which are obligate parasites on flowering plants, notably Commelinaceae, Ericaceae, Lauraceae, and Theaceae. At least one species, *Exobasidium vexans* on tea, is of major economic importance. The Exobasidiales are among the simplest of fungi having an internal mycelium and not producing fruit bodies. Instead, the basidia are developed on the surface of the host tissue or emerge through the stomata. This lack of organized fungal tissue is paralleled in the ascomycete genus *Taphrina*, in which most species also gall their hosts. Although there are few morphological characters available to discriminate species, Exobasidiales show a marked degree of host specificity and this, together with the shape, size, and degree of septation of the basidiospores, allows for a fairly precise taxonomy.

The order Exobasidiales comprises one family of seven genera, in total fewer than 70 species. *Arcticomyces* (one species) is found in the leaves of arctic and alpine saxifrages; *Exobasidiellum* comprises two species which

Plant Galls (ed. Michèle A. J. Williams), Systematics Association Special Volume No. 49, pp. 67–76. Clarendon Press, Oxford, 1994. © The Systematics Association, 1994.

are parasites of grass stems and has been variously classified in Exobasidiales or in Corticiaceae, as *Limonomyces* (Jülich 1984); *Exobasidium* (with more than 50 species) is the most widespread genus, attacking Epacridaceae, Ericaceae, and *Camellia* species in temperate regions; *Kordyana* (five species) occurs on tropical Commelinaceae; *Laurobasidium* (one species) galls trunks of *Laurus* in Mediterranean Europe and Macaronesia; *Microstroma* (two species) occurs on temperate trees and was previously referred to the Hyphomycetes; lastly, *Murobasidiospora* comprises three species in India.

Within the Ericaceae, some plant species are host to more than one species of *Exobasidium* but the symptoms these cause are usually distinctive and the microscopical characters of the fungi are a reliable distinguishing guide. The species may also be geographically or ecologically separated, usually as a species pair where one is restricted to high latitudes or high altitudes, for example, *Exobasidium myrtilli* and *E. aequale*. The common evicaceous genera *Calluna* and *Erica* are not attacked.

Types of galls and other damage caused by Exobasidiales

The amount of damage caused to the host varies markedly within the different genera. Of the European genera, species of *Arcticomyces*, *Exobasidiellum* and *Microstroma* produce no obvious hypertrophy or discolouration of the host tissue and cannot be said to be gall causers. *Exobasidium* species produce symptoms ranging from coloured leaf spots with no thickening of tissue, through typical leaf blister galls and apple-galls to deformed shoots with misshapen and discoloured leaves. The spore-producing surface (hymenium) is usually found on the underside of affected leaves but may cover the whole shoot. In some species, infection is localized to individual leaves and must be continued annually from dispersed spores. In others, the mycelium is clearly systemic and perennial, moving from one stem to another via the rootstock. In between are other species in which whole branches are attacked, but not via the rootstock. These strategies confirm the biological identity of parasites infecting the same host in the same locality. The most elaborate galls are those of *Laurobasidium lauri* which resemble clavarioid fungi in shape and colour. The different types of damage are summarized in Table 4.1.

European Exobasidiales which induce a visible host response

Exobasidium aequale on *Vaccinium myrtillus* produces slightly elongated shoots with thickened, slightly enlarged leaves, which become red and then white from the hymenial surface below. The spores are large and variable,

Table 4.1. European Exobasidiales—types of galling

1. No discernible hypertrophy	*Arcticomyces*
	Exobasidiellum
	Microstroma
2. Leaf spot only	*Exobasidium*
3. Leaf pouch gall	*Exobasidium*
4. Whole shoot deformity	*Exobasidium*
5. Apple gall	*Exobasidium*
6. Antler gall	*Laurobasidium*

up to 28 μm long, aseptate. Known from montane Fennoscandia and the Alps.

Exobasidium angustisporum produces similar symptoms (to those of *E. aequate*) on *Arctostaphylos alpinus*: the spores are narrow, up to 16 μm long and 3.5 μm wide, usually aseptate. Montane Fennoscandia and the Alps.

Exobasidium arescens induces small, scarcely thickened, yellow to red leaf spots on *V. myrtillus*; spores banana-shaped, up to 14 μm long, one to three septate. Known throughout Europe but always rare.

Exobasidium camelliae causes conspicuous, fleshy apple-galls on leaves, peduncles, sepals, and petals of *Camellia japonica*. This species is not uncommon in gardens in the south of Britain but is apparently unrecorded in continental Europe. *Exobasidium cassandrae* produces a range of symptoms on *Chamaedaphne calyculata*, from scarcely thickened leaf spots to typical blister galls up to 10 μmm in diameter; spores banana-shaped, up to 15 μm long. Known from 60° latitude and northwards.

Exobasidium cassiopes infects shoots of *Cassiope tetragona* which become pale and have swollen leaves. Spores banana-shaped, to 16 μm long, usually septate; known from Arctic Fennoscandia.

Exobasidium caucasicum produces its spores on the underside of young leaves in slightly deformed shoots. This species is doubtfully European as its host, *Rhododendron caucasicum* does not reach Europe in the wild.

Exobasidium dubium causes small, yellowish leaf spots on *Rhododendron luteum*, in Poland. The spores are large, more or less cylindrical, to 24 μm long, often septate.

Exobasidium expansum infects branches and stems, but not the whole plant, of *Vaccinium uliginosum*, resulting in slightly reduced or swollen, bright red leaves; spores septate, curved, to 15 μmm; known from Scandinavia and central Europe.

Exobasidium horvathianum is found on *R. luteum* where it produces large fleshy galls up to 7 cm in diameter on the leaves. Known in the wild from Poland; the host is reported as affected in England (Rea 1922) but no specimens or data are available.

Exobasidium hypogenum infects shoots of *C. tetragona* inducing considerable enlargement of the leaves; spores more or less straight, to 15 μm long, rarely septate. Known from Fennoscandia and Russia.

Exobasidium japonicum occurs widely on Asiatic evergreen azaleas in cultivation, forming fleshy, pale galls up to 3 cm across on the leaves.

Exobasidium juelianum attacks *Vaccinium vitis-idaea*, which becomes dwarf and unusually branches. Leaves are reduced and the whole plant becomes covered with pale pink hymenium; spores banana-shaped, to 14 μm, usually septate. Known from Fennoscandia, the Alps, and mountainous areas of Wales and Scotland.

Exobasidium karstenii is common in Europe on shoots of *Andromeda polifolia* which bear much enlarged leaves which redden and become purplish-black before falling off. It has curved, septate spores up to 18 μm long.

Exobasidium ledi forms small yellowish blister galls or leaf spots on *Ledum palustre* in Arctic Europe, where it is rare.

Exobasidium myrtilli is widespread and common on *V. myrtillus* where it causes whole shoots to stand out in appearance from the rest of the plant, becoming thickened, reddened, and eventually covered with white hymenium; spores slightly curved, to 16 μm long, usually septate.

Exobasidium oxycocci attacks *Vaccinium oxycoccus* and its allies where it causes the production of upright, rather than creeping, terminal shoots which are swollen and pink; spores banana-shaped, to 15 μm long, one to three septate. Scattered across Europe but common nowhere.

Exobasidium pachysporum forms small, slightly thickened, yellow-margined red leaf spots on *V. uliginosum*. The spores are broadly banana-shaped, up to 16 μm long, septate. Widely distributed in Europe, noticeably less boreal or montane than *E. vaccinii-uliginosi*.

Exobasidium rhododendri produces conspicuous red apple-galls on the leaves of *Rhododendron ferrugineum*, *hirsutum*, and their hybrids. It is widespread and common in the Alps and Pyrenees, is known from the Nordic countries and is rare in gardens in lowland Europe.

Exobasidium savilei causes annual shoots of *C. calyculata* to carry somewhat enlarged, discoloured leaves. Spores slightly curved, to 13 μm long. Rare, except in Finland.

Exobasidium splendidum affects whole shoots of *V. vitis-idaea*, where the leaves are enlarged, often lengthened, and the whole shoot becomes shining red; spores ellipsoid, to 27 μm long, septate. High latitude or altitude in Scandinavia and the mountains of central Europe.

Exobasidium sundstroemii is found rarely on *A. polifolia* where it causes broadening of the leaves which only become lilac-pink; spores almost straight, to 13 μm long, septate. Known from Scandinavia, Estonia, and Germany.

Exobasidium sydowianum forms small orange–red, rather waxy, leaf spots on *Arctostaphylos uva-ursi*. The spores are broadly banana-shaped, to 17 μm long. Known in Europe only from Sweden, Finland, Austria, Switzerland, and Scotland and rare throughout.

Exobasidium unedonis is a rare Mediterranean species on *Arbutus* leaves, but also occurs on cultivated plants in temperate localities. The young shoots are clustered and open earlier than healthy ones; the leaves are small, yellow then red.

Exobasidium uvae-ursi causes the host, *A. uva-ursi*, to develop upward-growing clusters of deformed shoots. The leaves become dark reddish-purple. The spores are banana-shaped, to $24\,\mu$m long, septate. This species is rare, but known from Scandinavia, Estonia, Russia, the Alps, the Pyrenees, and the Balkan mountains.

Exobasidium vaccinii is a common and conspicuous parasite, forming red leaf pouches on *V. vitis-idaea*. Occasionally, shoot tips are affected but never the whole shoot as with *E. juelianum*. Spores slender, curved, to $19\,\mu$m long, septate.

Exobasidium vaccinii-uliginosi occurs on *V. uliginosum* where it attacks scattered shoots, producing enlarged and discoloured leaves. Spores oblong, to $23\,\mu$m long, aseptate. A high latitude or high altitude species known from Fennoscandia, the Alps, the Pyrenees, and the Caucasus.

Laurobasidium lauri produces bizarre antler-like galls on the trunks of *Laurus* in the south-west Europe and the Atlantic islands. The galls resemble the fruit bodies of *Macrotyphula (Clavaria) fistulosa* var. *contorta* in shape and colour but tend to be larger and more elastic.

Detailed descriptions, distribution data, and some physiological data on the European species of *Exobasidium* on Ericaceae are given by Nannfeldt (1981) and of *Laurobasidium* by Jülich (1984).

Key to Exobasidiales found in Europe, based on host damage

Andromeda polifolia
Type 4 damage—shoots slightly enlarged, leaves much broader than normal, structure reddening, becoming bluish-purple, widespread.

E. karstenii*

Type 4—branches, rarely whole shoots, not abnormally enlarged but leaves broadened, pink, often only mottled, rare.

E. sundstroemii

Arbutus andrachne, A. xandrachnoides, and *A. unedo*
Type 4—young shoots clustered, premature, leaves reduced, yellow then red, rare.

E. unedonis*

Arctostaphylos alpina
Type 4—shoots clustered, leaves enlarged, from near white to purple, widespread but nowhere common.

E. angustisporum

Arctostaphylos uva-ursi
Type 2—small, thickened reddish-orange leaf spots, rare, northern and montane.

E. sydowianum*

Type 4—shoots densely clustered, upright, swollen, bright red, rare, northern and montane.

E. uvae-ursi

Camellia japonica
Type 5—large pink, fleshy galls on leaves and flowers, in southern gardens.

E. camelliae*

Cassiope tetragona
Type 3/4—leaves very swollen but bases distinct from stem, hymenium on underside of leaf, boreal.

E. hypogenum

Type 3/4—leaves swollen, not clearly separated from stem, hymenium all over, boreal.

E. cassiopes

Chamaedaphne calyculata
Type 2/3—thick leaf spots or small concave leaf galls, reddish above, boreal.

E. cassandrae

Type 4—shoots clustered, enlarged, red or white, boreal.

E. savilei

Dactylis glomerata
Type 1—pink film on previous year's stem, rare.

El. culmigenum*

Festuca rubra
Type 1—pink to cream film on stems, rare.

El. roseipelle*

Juglans regia
Type 1—cream patches of hymenium on underside of leaves, frequent.

M. juglandis*

Laurus azorica and *L. nobilis*
Type 6—antler-like outgrowths on the trunk.

L. lauri

Ledum palustre
Type 2—small, pale, hardly thickened leaf spots, rare, boreal.

E. ledi

Lolium perenne
Type 1—pink to cream film on stems, rare.

El. roseipelle*

Quercus cerris and *Q. robur*
Type 1—white patches of hymenium on underside of leaves.

M. album*

Rhododendron amoenum, R. indicum, R. kiusianum, and hybrids
Type 5—fleshy white galls on leaves, widespread, in gardens.
E. japonicum*

Rhododendron caucasicum
Type 4—shoots slightly deformed, pale yellow, not convincingly recorded
from Europe.
E. caucasicum

Rhododendron ferrugineum, R. hirsutum and *R. x wilsonii*
Type 5—large pink apple galls on leaves, alpine, or in gardens.
E. rhododendri*

Rhododendron luteum
Type 2—scarcely thickened, yellow leaf spots, eastern.
E. dubium

Type 5—fleshy orange-red apple galls on leaves, eastern or in gardens,
where it is rare.
E. horvathianum(*)

Saxifraga aizoides, S. aizoon, S. bryoides and *S. oppositifolia*
Type 1—greyish-white patches of powdery hymenium on undersides of
leaves on side shoots, northern and montane.
A. warmingi*

Vaccinium macrocarpum, V. microcarpum and *V. oxycoccus*
Type 2—small, red, hardly thickened leaf spots.
E. rostrupii*

Type 4—shoots elongated, deformed, leaves swollen, pink.
E. oxycocci*

Vaccinium myrtillus
Type 2—small, yellow–red leaf spots, rare.
E. arescens*

Type 4—shoots taller, swollen, leaves larger, red, common.
E. myrtilli*

Type 4—shoots slightly deformed, leaves somewhat thickened, white or
red, northern and montane.
E. aequale

Vaccinium uliginosum
Type 2—thickened leaf spots, red with yellow margins.
E. pachysporum*

Type 4—stems and branches slightly deformed, leaves red but not
deformed.
E. expansum

Type 4—shoots slightly deformed, leaves markedly thickened, northern
and montane.
E. vaccinii-uliginosi

Vaccinium vitis-idaea

Type 3—red blister galls on leaves, occasionally affecting shoot tip, but never whole shoot, common.

E. vaccinii*

Type 4—whole shoot distorted, stunted, pink, leaves reduced, widespread but uncommon.

E. juelianum*

Type 4—whole shoot enlarged, leaves swollen, red, northern and montane.

E. splendidum

Key: A, Arcticomyces; E, Exobasidium; El, Exobasidiellum; L, Laurobasidium; M, Microstroma; *reported from the British Isles.

Collection and preservation of Exobasidiales

Infected plants of moorland and montane hosts are often very conspicuous, with abnormal growths standing out distinctly from the healthy population. Individual shoots which are attacked are also quite distinctive. Where only leaf spots occur it is often a matter of careful search for infected material, this is especially true of *Exobasidium arescens* which is probably not as rare as it seems. The inconspicuous leaf spots on cranberries are common enough but as they develop the diagnostic hypophyllous hymenium only late in the season, they may not always suggest an *Exobasidium*. Many of the branch infections involve only young shoots and so must be looked for in early summer. Other species cause infected leaves, which would otherwise be evergreen, to drop off in early autumn. It is helpful to have a reasonable knowledge of the location of suitable host plants when searching for exobasidia and then systematically visit as frequently as possible to seek material at all stages. In this way it is possible to recognize early symptoms of disease before there is any fungal material visible on the outside of the leaf or stem. As with other groups of gall causers a good working knowledge of the taxonomy of the host plants is essential.

Most species can be readily collected with a knife, scissors, or secateurs and transported in paper or polythene envelopes or bags. It is essential to keep each sample separate as transfer of spores could seriously hinder correct identification. Most species can be pressed with their hosts and will retain sufficient characteristics for later checking, especially microscopic details of basidia and spores. The larger blister and apple-galls may also be pressed but will need more frequent changes of paper. They may also be freeze-dried or stored in boxes of silica gel. Colour is not well preserved after drying.

As more and more members of the Ericaceae are brought into

cultivation it is likely that they will be infected either by pathogens introduced with them, which in spite of plant quarantine, is of relatively frequent occurrence, or they will succumb to native species which have a reasonably broad host tolerance. It is likely that additions to the European list (Appendix 1) will be made in the near future, especially with North American and Asiatic species which occur on European hosts or their close relatives. The effect of climatic changes on the distribution of these organisms needs to be monitored as many of the rare boreal taxa could face extinction.

References

Jülich, W. (1984). *Die Nichtblätterpilze, Gallertpiltze und Bauchpilze.* Gustav Fischer Verlag, Stuttgart.

Nannfeldt, J.A. (1981). *Exobasidium*, a taxonomic reassessment applied to the European species. *Symbolae Botanicae Upsalienses*, **23**, 1–72.

Rea, C. (1922). *British Basidiomycetae.* Cambridge University Press, Cambridge.

Appendix: checklist of European Exobasidiales

1. Exobasidiaceae

Arcticomyces Savile
  warmingii (Rostrup)Savile on *Saxifraga aizoides* L., *S. aizoon* L., *S. bryoides* L., and *S. oppositifolia* L.
Exobasidiellum Donk
  culmigenum Webster & Reid on Poaceae
  roseipelle (Stalper & Loerakker) on Poaceae
Exobasidium Woronin
  aequale Sacc. on *Vaccinium myrtillus* L.
  angustisporum Linder on *Arctostaphylos alpina* (L.) Sprengel
  arescens Nannf. on *V. myrtillus*
  ***camelliae** Shirai on *Camellia japonica* L.
  cassandrae Peck on *Chamaedaphne calyculata* (L.) Moench
  cassiopes Peck on *Cassiope tetragona* (L.) D. Don
  caucasicum Woronich. on *Rhododendron caucasicum* Pallas
  dubium Rac. on *Rhododendron luteum* Sweet
  expansum Nannf. on *Vaccinium uliginosum* L.
  horvathianum (Thomas)Nannf. on *R. luteum*
  hypogenum Nannf. on *C. tetragona*
  ***japonicum** Shirai on *Rhododendron* (evergreen azaleas)
  juelianum Nannf. on *Vaccinium vitis-idaea* L.

karstenii Sacc. & Trott.in Sacc. on *Andromeda polifolia* L.

ledi Karst. on *Ledum palustre* L.

myrtilli Siegm. on *V. myrtillus*

oxycocci Rostrup ex Shear on *Vaccinium oxycoccus* L. and *V. microcarpum* (Turcz. ex Rupr.) Schmalh.

pachysporum Nannf. on *V. uliginosum*

rhododendri (Fuckel)Cram. in Geyler on *Rhododendron ferrugineum* L., *R. hirsutum* L., and *R. x wilsonii*

rostrupii Nannf. on *V. oxycoccus, V. macrocarpum*, and *V. microcarpum*

savilei Nannf. on *C. calyculata*

splendidum Nannf. on *V. vitis-idaea*

sundstroemii Nannf. on *A. polifolia*

sydowianum Nannf. on *Arctostaphylos uva-ursi* (L.) Sprengel

unedonis R.Maire on *Arbutus andrachne* L., A. unedo L. and *A. x andrachnoides* Link

uvae-ursi (R.Maire)Juel on *A. uva-ursi*

vaccinii (Fuckel)Woronin on *V. vitis-idaea*

vaccinii-uliginosi Boud. apud Boud. & E. Fisch. on *V. uliginosum*

Laurobasidium Jülich

lauri (Geyler) Jülich on *Laurus azorica* (Seub.) Franco and *L. nobilis* L.

Microstroma Niessl

album (Desm.)Sacc. on *Quercus cerris* L. and *Q. robur* L.

juglandis (Bereng.)Sacc. on *Juglans regia* L.

*On introduced hosts.

5. The lichen thallus: a symbiotic phenotype of nutritionally specialized fungi and its response to gall producers

DAVID L. HAWKSWORTH* and ROSMARIE HONEGGER†

* International Mycological Institute, Bakeham Lane, Egham, Surrey, UK †
Institut für Pflanzenbiologie, Universität Zürich, Zollikerstrasse 107,
CH-8008 Zürich, Switzerland

Abstract

The view that lichen thalli should be regarded as analagous to insect and fungal galls produced on plants is critically analysed. Ten similarities and 13 differences are recognized and discussed. In view of the fundamental differences identified, the analogy cannot be upheld. Lichen thalli are regarded as a particular biological phenomenon with no exact analogue. An expanded definition of 'lichen' is provided. Coevolutionary and systematic aspects of the lichen mutualism are considered, stressing the need to emphasize non-coevolved structures in supraspecific taxonomies. Attention is also drawn to the galls produced on established lichen thalli by invasive fungi and other organisms, including nematodes and arthropods.

Introduction

The view that lichen thalli should be regarded as galls induced by an alga in a fungus originated with Moreau and Moreau (1918). The alga was considered to be an infective agent inducing a response in a fungal host and was later compared to fungal galls on vascular plants in particular (Moreau 1921, 1927). This view was strongly defended against criticisms, such as those of des Abbayes (1951), by Moreau (1956), who by that date did, however, appear to acknowledge that some modification might be necessary.

Plant Galls (ed. Michèle A. J. Williams), Systematics Association Special Volume No. 49, pp. 77–98. Clarendon Press, Oxford, 1994. © The Systematics Association, 1994.

This hypothesis was largely based on studies of cephalodia, discrete and often morphologically distinct outgrowths from lichen thalli including a photosynthetic partner ('photobiont') different from that of the main thallus (Fig. 5.3). Cephalodia had already been compared to galls by various investigators (see Kaule 1932). These opinions were mentioned by James and Henssen (1976) in their classic study of cephalodia, but no critical comparison of insect galls and lichen thalli has hitherto been attempted. Such an objective analysis is necessary to determine:

(1) if such an analogy can be sustained in the light of current knowledge of gall and lichen morphogenesis and structure;

(2) if knowledge of lichen morphogenesis can be expected to contribute to the understanding of insect galls and *vice versa*.

In addition, cecidiologists rarely consider the galls produced on lichen thalli by lichenicolous fungi and other organisms (Figs 5.18–5.23). These are scarcely mentioned by Mani (1964), although a considerable literature exists (Grummann 1960).

In this contribution we consider first the current definition of lichen, identify the differences and similarities between lichen thalli and insect galls, including gall-like responses of lichens to other biotic agents and, second, the validity of the gall analogy and coevolutionary and systematic aspects of the organisms involved.

Definition

Attempts to provide a succinct definition of 'lichen' are frustrated by the breadth of the fungal–algal and/or cyanobacterial relationships and the diverse morphological forms to be embraced. The various definitions proposed were analysed by Hawksworth (1988*a*) who concluded that the association between the fungal partner (mycobiont) and the photobiont must be stable and that the mycobiont should form the outer tissue-like structures (that is, be the 'exhabitant') with the photobiont included within (that is, being the 'inhabitant'). This definition excludes from 'lichen' associations in which the photobiont is the exhabitant (that is, mycophycobioses) and also fungal parasites of algae, fungi growing on lichens (that is, lichenicolous fungi) as parasites, commensals, or saprobes, and initially lichenicolous stages of finally independent lichens. However, further modification is required to exclude *Geosiphon*, in which a cyano-bacterial photobiont occurs intracellularly in a zygomycete (Mollenhauer 1992), by stressing the extracellular location of the photobiont (Honegger 1991*b*).

Our more comprehensive definition of 'lichen' is 'an ecologically obligate, stable mutualism between an exhabitant fungal partner (the mycobiont) and an inhabitant population of extracellularly located unicellular or filamentous algal or cyanobacterial cells (the photobiont)'.

A comparison of lichen thalli and insect galls

In order to objectively analyse the hypothesis that lichen thalli are analogous to insect galls, a prerequisite is an enumeration of the similarities and differences between the two biological phenomena. We have been able to identify the 10 similarities and 13 differences detailed in turn below.

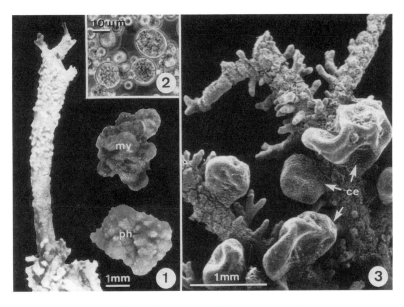

Figs 5.1–5.3. The symbiotic phenotype of lichen-forming fungi, as expressed in association with compatible photobiont(s). **Figs 5.1 and 5.2.** *Cladonia macrophylla* (Schaerer) Stenhammar (*Lecanorales, Ascomycotina*) and its unicellular green algal photobiont, *Trebouxia* sp. (*Pleurastrales, Chlorophyta*; detail in Fig. 5.2). Symbiotic phenotype (**left**) and the isolated partners in axenic culture on agar media (**right**). my, mycobiont; ph, photobiont. **Fig. 5.3.** *Stereocaulon ramulosum* (Sw.) Räusch. (*Lecanorales, Ascomycotina*) forms slim, ecorticate ramules in association with *Pseudochlorella* sp. (*Chlorococcales, Chlorophyta*) and houses nitrogen-fixing cyanobacteria (*Scytonema* sp., *Nostocales*) in bag-like, corticate structures, the so-called cephalodia (ce).

1. Similarities

1. In both cases a compatible inhabitant induces the expression of a peculiar symbiotic phenotype in the quantitatively predominant exhabitant. In lichen thalli the photobiont, by definition (see above) is the inhabitant, while in the case of insect galls the insect eggs or juvenile stages are the inhabitant and the host plant the exhabitant. While it is unusual outside lichens for an alga to be involved in gall-like deformations, the siphonaceous alga *Phyllosiphon* has such an effect on leaves (Pirozynski, personal communication).

2. Moderate to high specificity between the bionts is shown. Most mycobiont species associate with a particular photobiont genus and most gall-forming insect species produce galls on a single or closely related host species. More precise information is scarce for most lichens as the taxonomy of lichen photobionts at the species level is currently inadequate and precludes more definite statements. Experimental data are largely missing since the symbiotic phenotype is not routinely expressed in resynthesis experiments under axenic conditions (see Honegger 1990). In some instances a mycobiont can, however, form morphologically very similar crustose thalli with different photobiont genera, for example, *Chaenotheca carthusiae* (Harm.) Lettau with *Stichococcus* and *Trebouxia* (Tibell 1982).

3. A gradation of complexity occurs in both cases, from symbioses

---

Figs 5.4–5.10. Taxonomic, morphological, and anatomical diversity in lichens with either non-stratified or stratified thalli. **Fig. 5.4.** *Dimerella lutea* (Dickson) Trevisan (*Gyalectales, Ascomycotina*) forms a crustose thallus between leaf cells (lc, some of which are penetrated by the fungus) and cuticle (c) of the liverwort *Frullania dilatata* (L.) Dum. (*Jungermaniales, Bryophyta*) where it meets its photobiont (ph), probably· a *Trentepohlia* sp. (*Trentepohliales, Chlorophyta*). **Fig. 5.5.** Cross-section of the bark of *Ilex aquifolium* L. with *Graphis elegans* (Borrer ex Sm.) Ach. within it (*Graphidales, Ascomycotina*) and its photobiont (*Trentepohlia* sp., *Chlorophyta*) which reveals coccoid growth in symbiosis and forms filaments in pure culture. **Figs 5.6–5.8.** Microfilamentous thallus and ascoma (ac) of a *Coenogonium subvirescens* Nyl. (*Gyalectales, Ascomycotina*), each microfilament consisting of a filament of *Trentepohlia* (*Chlorophyta*) and peripheral fungal sheath (large arrows); small arrows point to bacterial epibionts; the fungal partner does not penetrate the cell wall of its photobiont. **Figs 5.9 and 5.10.** Foliose, dorsiventrally organized and internally stratified thalli of *Parmeliaceae* (*Lecanorales, Ascomycotina*) with unicellular photobionts of the genus *Trebouxia* (*Pleurastrales, Chlorophyta*). **Fig. 5.9.** Marginal lobe of *Parmelia sulcata* Taylor with rhizines (rh) arising from the strongly melanized lower cortex. **Fig. 5.10.** Vertical cross-section of *P. borreri* (Sm.) Turner with conglutinate upper (uc) and lower (lc) cortical layers around a gas-filled medulla (m) built up by a system of aerial hyphae; the photobiont cell population is housed and maintained underneath the upper cortex at the periphery of the gas-filled thalline interior. Scale bars equal $5\,\mu$m unless otherwise stated.

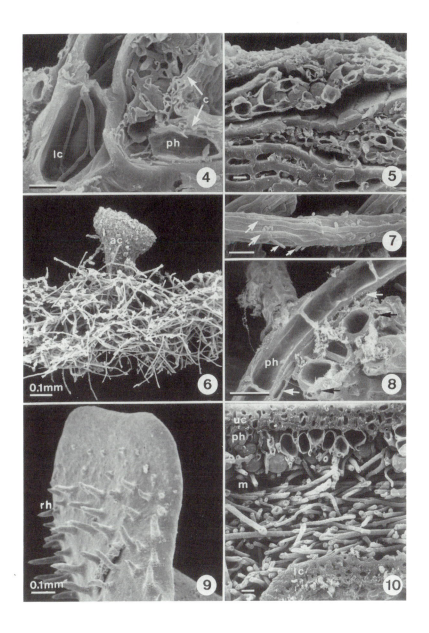

with little or no apparent morphogenetic modification, through to complex phenotypes formed in response to a compatible symbiont which are quite distinct from the individuals when grown in isolation (Fig. 5.1). Examples of less complex types are kataplasmic galls and little differentiated leprose, crustose (Figs 5.4–5.5), and filamentous lichens (Figs 5.6–5.8). More complex types include prosmoplasmic galls and the complex layered (or 'stratified') structures with clearly differentiated cell types found in foliose and fruticose lichens (Figs 5.1, 5.3, 5.9, 5.10, 5.16 and 5.17).

In the case of lichen thalli, when the isolated bionts are grown in pure culture both are almost invariably quite different in form compared with that they assume in the lichen (Fig. 5.1). The mycobionts generally form cartilaginous, undifferentiated colonies on solid agar media, while the photobiont cells produce mounded gelatinous colonies, in some cases forming filaments rather than the discrete coccoid cells found in the intact lichen (see Tschermak-Woess 1988).

The stimuli which cause the loose non-corticate associations or 'pre-thalli' which initially form in nature or in synthetic cultures between a mycobiont and a compatible photobiont, to be transformed into a structured corticate lichen thallus are unknown.

4. Growth of the bionts is harmonized so that an association between a particular exhabitant and inhabitant pair results in a constant growth form with a predictable structure. This enables both lichens and insect galls to be recognized in a similar way as independent taxonomic species. Further, the resultant growth form yields characters of taxonomic value used in the identification of one of the bionts: the exhabitant fungus in lichens and the inhabitant insect in insect galls.

5. In both cases there are examples of different inhabitants being able to produce distinctive phenotypes with the same exhabitant species. For example, *Quercus petraea* (Mattuschka) Liebl. and *Q. robur* L. form approximately 100 morphologically distinctive galls with some 40 cynipid wasp species (Shirley 1992). This situation is paralleled by lichen pho-tobiont morphotypes or ('photosymbiodemes') where a single mycobiont evidently forms at least two morphologically distinct lichens, that is, expresses different symbiotic phenotypes with a cyanobacterial as opposed to an algal photobiont; this phenomenon, which is particularly frequent in *Peltigerales*, has been well-documented (James and Henssen 1976; Jahns 1988a). Due to the possible genetic heterogeneity of lichen mycobionts involved in the formation of individual lichen thalli one cannot assume with complete certainty that one fungal genotype is capable of producing distinctly different morphotypes in association with different photobionts in all such cases. However, in one instance molecular techniques suggest that a single mycobiont is indeed involved (Armaleo and Clerc 1991).

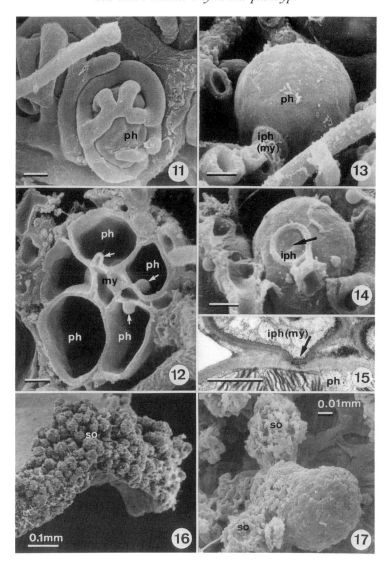

Figs 5.11–5.15. Complex types of mycobiont–photobiont interactions and dispersal by means of vegetative symbiotic propagules in *Lecanorales* (*Ascomycotina*) with stratified thallus and unicellular photobionts of the genus *Trebouxia* s. lat. (*Pleurastrales, Chlorophyta*). **Figs 5.11 and 5.12.** Mycobiont–photobiont interface in *Cladonia arbuscula*; arrows point to a peculiar type of haustorium which is ensheathed by the algal cell wall. **Figs 5.13–5.15.** Intraparietal haustoria (iph) in *Parmeliaceae*, as observed in *Parmelia borreri* (Figs 5.13 and 5.14) and *P. tiliacea* (Hoffm.) Ach. (Fig. 5.15). **Figs 5.16 and 5.17.** Propagation of the symbiotic state by means of soredia (so): minute groups of photobiont cells with a fungal sheath which form, upon landing on a suitable substratum, an undifferentiated pre-thallus and subsequently primordia of the symbiotic phenotype. Scale bars equal $2\,\mu$m unless otherwise stated.

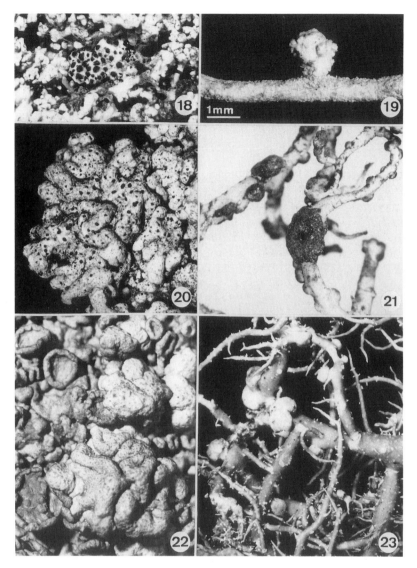

Figs 5.18–5.23. Galls on ascomycetous lichens, induced by fungi and nematodes. **Fig. 5.18.** *Polycoccum trypethelioides* (Th.Fr.) R. Sant. (*Dothideales, Ascomycotina*) forming galls on *Stereocaulon alpinum* Laurer (*Lecanorales*). **Fig. 5.19.** Gall on *Cladonia glauca* Flörke (*Lecanorales*) containing a mixture of nematodes, the causal one most probably being a *Nothanguina* species. **Fig. 5.20.** *Polycoccum galligenum* Vězda forming galls on *Physcia caesia* (Hoffm.) Fürnrohr (*Lecanorales*). **Fig. 5.21.** *Phacopsis vulpina* Tul. (*Lecanorales, Ascomycotina*) forming swollen gall-like areas below its fruit bodies on *Letharia vulpina* (L.) Hue (*Lecanorales*). **Fig. 5.22.** *Guignardia olivieri* (Vouaux) Sacc. (*Dothideales, Ascomycotina*) forming galls on *Xanthoria parietina* (L.) Th.Fr. (*Teloschistales*). **Fig. 5.23.** *Biatoropsis usnearum* Räsänen (*Tremellales, Basidiomycotina*), forming galls on *Usnea cornuta* Körber (*Lecanorales*). Same scale bar applies throughout.

6. The heterotrophic partner can exert a rejuvenating effect on the photoautotroph. Cells of the photoautotroph which had previously differentiated can start to grow once again, forming new and distinctive structures. This phenomenon has been extensively studied in insect galls and is paralleled in, for example, layered foliose and fruticose lichen thalli, in which asexual symbiotic propagules (for example, isidia, soredia) and thalline exciples, develop on areas of the thallus where there is otherwise little or no cell turnover (Fig. 5.16). The factors leading to the initiation of this process on already developed lichen thalli await elucidation (Honegger 1987*a*, *b*).

7. Almost no new cell types are produced, apart from the complex haustoria in certain lichenized groups, especially *Lecanorales* and *Teloschistales* (Figs 5.11–5.15), although these can vary in distribution and abundance. However, there are species-specific novelties evident at the organ level in both galls on plants and in the lichen phenotype. Lichen thalli have very complex growth patterns, with polarized, marginal/apical, and/or intercalary laminal growth. Conglutinate cell layers develop, usually at the periphery and a system of aerial hyphae arises in the gas-filled thalline interior (Fig. 5.10). In contrast, when mycobionts are grown on solid agar media, hydrophobic aerial hyphae are produced over the whole surface and conglutinate zones in the central part of the colonies, demonstrating centrifugal growth.

8. Other organisms may be associated with lichen thalli or insect galls. For example, insect galls can include a vast array of parasitoids, inquilines, predators and successori, and, in the case of 'ambrosia' gall midge larvae, also a mutualistic fungus (Bissett and Borkent 1988). In a comparable manner, lichen thalli can support a vast array of obligately lichenicolous fungi (Clauzade *et al.* 1989; including some gall formers, see below) and can also include fortuitously additional algae and bacteria (Fig. 5.7).

9. In both cases, at least in some examples, the inhabitant can have an independent existence remote from the exhibitant for some part of its life cycle. This applies to cyanobacterial and various algal photobionts of lichens that can also occur in the free-living state (Table 5.2) and to the free-flying and independently feeding adults of gall-forming insects.

10. In both cases the symbiotic state is not inheritable and has to be re-established in each sexual reproductive cycle.

In addition to these 10 biological similarities between insect galls and lichens, the study of the biogeography, cytology, morphogenesis, physiology, taxonomy, and ultrastructure of both lichen thalli and insect galls has been sadly neglected. Experimental work has been minimal and restricted to very few of the vast array of morphological structures involved, also critical biosystematic studies involving a full range of modern molecular, biochemical, and population biology approaches are generally lacking. The study of both phenomena has tended to be marginalized and regarded as somewhat esoteric, despite the very considerable numbers of species involved, their importance in certain ecosystems, and the scientific challenges they afford both in morphogenesis and coevolution. Further, both attract a considerable amateur following which merits encouragement to pursue more critical investigation.

2. Differences

1. The carbon-fixing photoautotroph is the inhabitant in lichen thalli (that is, the algal or cyanobacterial partner), but the exhabitant in insect galls (that is, the plant partner). The converse also applies, the carbon heterotroph being the exhabitant and predominant (to 80 per cent of the biomass) in lichens and the inhabitant constituting a small proportion of the bulk in insect galls.

2. In lichen thalli the carbon heterotroph (that is, the mycobiont) obtains fixed carbon by the absorption of actively or passively released photosynthates, whereas in insect galls the carbon heterotroph (that is, the insect larva) usually acquires fixed carbon by feeding on and ingesting the carbon-containing tissues of the photoautotroph, although in some insect galls secreted enzymes may have a role. While this difference holds for insect galls, it does not for fungal galls of plants where enzyme secretion by the gall-inducing fungus is a widespread phenomenon.

3. Specialized interconnecting haustorial cells are produced in lichen thalli at the interface of the bionts (see Honegger 1991*b*, 1992; Figs 5.11–5.15) to facilitate the bidirectional transfer of photosynthates and other compounds (including water, mineral ions, and precursors of mycobiont-derived secondary metabolites). The stylet of many gall-inducing insects has a somewhat analogous function in solute transfer.

4. Secondary metabolites not generally or more sparingly produced by either biont when grown alone are produced in the vast majority of lichen thalli and can constitute up to 10 per cent of the dry weight of the thallus. Special compounds are not usually found in insect galls. Neither in insect galls nor in lichens has a case of novel secondary compounds being formed by horizontal gene transfer been documented, such as that of the *Agrobacterium* crown gall system where a plasmid transmits genetic information (Ream 1989). The vast array of secondary compounds produced in lichen thalli are essentially derived from the mycobiont (Culberson and Elix 1989) and this circumstance was hypothesized to result from esterification resulting from inhibitors produced

by the photobiont interacting with the biochemical pathways present in the mycobiont (Culberson and Ahmadjian 1980). However, subsequent work has shown that at least some characteristically 'lichen' depsidones can also be formed in cultures of the mycobiont (Leuckert *et al.* 1990; Culberson and Armaleo 1992). The resultant compounds are generally deposited as crystals in a common hydrophobic cell wall surface layer enveloping the bionts (Honegger 1986). In many cases, different compounds accumulate in different parts of the lichen (for example, cortex *vs* medulla *vs* exciple) so some interplay is involved at least in determining which substances are to be accumulated.

5. Insect galls are generally short-lived and usually restricted to periods of a few weeks or months during a particular phase of the growing season of the exhabitant. In contrast, lichen thalli can continue to exist for extremely long periods, crustose species being the most long-lived and well-documented as being able to persist for several centuries (Winchester 1988), if not millenia.

6. There are fundamental differences in physiological parameters, especially water relations. Insect galls develop on a homoiohydrous exhabitant with controlled water relations which normally remains metabolically active throughout its lifetime. In contrast, lichen thalli are characteristically poikilohydrous, that is, adapted to regularly occurring wetting and drying cycles which can include extended periods of drought stress (see Kappen 1988). Under dry conditions, many lichen thalli survive in a metabolically inactive and effectively dormant state, whereas galls generally remain metabolically active throughout their life span.

7. In lichen thalli both the carbon photoautotroph and heterotroph are frequently genetically heterogeneous. Individuals of lichens have been shown to not infrequently develop from a mixture of independently germinating asexual propagules (Schuster *et al.* 1985). Further, originally independent thalli can coalesce to form a single unit and, during the lifetime of a lichen, additional propagules of the same species may even alight on a thallus and be incorporated into it (Larson and Carey 1986). This situation may well be the norm in many of the structurally more complex macrolichens.

8. In lichen thalli the greatest degree of specificity is shown by the exhabitant, the mycobiont, whereas in the case of insect galls the partner with the most specificity is the inhabitant, the insect. This is evident from the statistic that there are approximately 530 genera and 13 500 species of lichen-forming fungi (Hawksworth *et al.* 1983), but only approximately 23 genera and approximately 100 species of green algal photobionts, and 15 genera and an uncertain but probably much lower number of species

David L. Hawksworth and Rosmarie Honegger

Table 5.1. Orders of *Ascomycotina* and *Basidiomycotina* which include lichenized taxa

Order (Family)	Nutritional strategies	Thallus anatomy in lichenized taxa
Ascomycotina[a,b,c]		
(approximately 98 per cent of lichen-forming fungi)		
Arthoniales	**l**, nl (*lf, sap*)	**ns**
Caliciales	**l**, nl (*f, lf, sap*)	**ns**, s
Dothideales	**nl** (*sap, lf, pp*), l	**ns**
Graphidales	**l**, nl (*lf*)	**ns**
Gyalectales	**l**	**ns**
Lecanorales	**l**, nl (*sap, lf*)	**ns, s**
Leotiales	**nl** (*sap, pp, lf*), l	**ns**, s
Lichinales	**l**	**ns**, s
Opegraphales	**l**, nl (*lp, sap*)	**ns**, s
Ostropales	**nl** (*sap, pp, lf*), l	**ns**
Patellariales	**nl** (*sap, lf*), l	**ns**
Peltigerales	**l**	**s**, ns
Pertusariales	**l**	**ns**
Pyrenulales	**l**, nl (*sap*)	**ns**
Teloschistales	**l**, nl (*lf*)	**ns**, s
Verrucariales	**l**, nl (*sap, lp*)	**ns**, s
Basidiomycotina[d,e]		
(approximately 0.4 per cent of lichen-forming fungi)		
Aphyllophorales	**nl** (*sap, myc, pp, f*), l	**ns**, s
Agaricales	**nl** (*sap, myc, pp*), l	**ns, s**

[a] According to Eriksson and Hawksworth (1991).

[b] Sixteen out of 46 orders of *Ascomycotina* include lichenized taxa.

[c] Approximately 46 per cent of *Ascomycotina* are lichenized (approximately 13 250 spp.).

[d] According to Hawksworth *et al.* (1983) and Oberwinkler (1970).

[e] Only 0.3 per cent of *Basidiomycotina* are lichenized (approximately 50 spp.).

Abbreviations: f, fungicolous; l, lichenized; lf, lichenicolous; myc, mycorrhizal; nl, non-lichenized; ns, non-stratified; pp, plant pathogens; s, internally stratified; sap, saprotrophic.

Note: Bold type, predominant strategy or anatomy.

Approximately 55 per cent of lichen-forming fungi form non-stratified (crustose, microfilamentous, etc.) thalli, 20 per cent form either squamulose or placodioid thalli, and 22 per cent form either foliose or fruticose internally stratified thalli.

of cyanobacterial ones (Tschermak-Woess 1988). Further, the numbers of families and orders to which the mycobionts belong are in marked contrast to those of the photobionts (Tables 5.1 and 5.2). However, in less than 2 per cent of all lichen species has the photobiont ever been identified at the species level and the range of compatible photobiont species per mycobiont is virtually unknown (see Tschermak-Woess 1988; Honegger 1990).

9. In some groups of insect galls, notably the cynipid wasps, the galls produced by the sexual and asexual generations on the same plant host are strikingly different — so much so that they have on occasion been incorrectly described as separate species (Rohfritsch and Shorthouse 1982). At least in the *Cladoniaceae* the converse applies in most species, that is, the initiation of the sexual process in the mycobiont induces a morphogenetically distinct lichenized structure, a lichenized stipe or 'podetium' (Jahns 1970; Ahti 1982).

10. There are an increasing number of cases being documented of one exhabitant replacing another in an already existing lichen thallus to produce a morphologically distinct lichen with the same inhabitant. In effect, a sexual ascospore from one mycobiont alights on an existing lichen thallus, outcompetes or otherwise parasitizes the original mycobiont so that it dies, and then subsumes its inhabitant. Genera known to include species exhibiting this phenomenon include *Acarospora*, *Diploschistes*, *Rhizocarpon*, and *Toninia* (see Hawksworth 1988a; Timdal 1991). A parallel may also occur with inquilines on insect galls able to modify the structure and physiology of a gall and outlive the primary gall maker, but more information is required.

11. Numerous lichen-forming fungi disperse very efficiently by means of symbiotic vegetative propagules (containing both partners) such as soredia (Figs 5.16 and 5.17), isidia, schizidia, etc. or simply by thallus fragmentation.

12. Lichens are generally regarded as mutualistic symbioses, that is, the biological fitness of both partners is increased in the symbiotic state, although both mycobiont and photobiont invest in this symbiotic relationship. The photobiont releases substantial amounts of photosynthates, and the fungal partner, by producing morphologically complex, three-dimensional thalline structures in order to house the photobiont cell population and have it adequately illuminated, is subjected to distinctly more dramatic microclimatic stresses than any fungus which lives within a dead substratum or in close association with a life, homoiohydrous host (for example, plant pathogens, mycorrhizas). Insect galls, on the other hand, are mildly parasitic systems which usually do not cause irreparable damage to the plant host, but are unlikely to significantly increase its biological fitness. Exceptions are found amongst galls which attract militant ants by secreting honey-dew; these ants effectively protect the plant host (Pirozynski 1991).

13. Morphologically complex lichen thalli can differentiate galls in response to a variety of third parties and behave in a comparable manner to a vascular plant. Amongst insect galls, inquilines may have somewhat

Table 5.2. Genera of lichen photobionts (after Tschermak-Woess 1988; Honegger 1990)

Abundance[a]/genus	Occurrence[b]	Primary/secondary photobiont[c]	Abundance[a]/genus	Occurrence[b]	Cell wall type[d]
Cyanobacteria			**Chlorophyta**		
Chroococcales			Charophyceae		
Anacystis†	ph + ns	p	Klebsormidiales		
*Chroococcus**	ph(as) + ns	p	*Stichococcus*‡	ph(as) + ns	
Gloeocapsa‡	ph(as) + ns	p, s			
*Hormatonema**	ph(as) + ns	p	Chlorophyceae		
Entophysalis-like*	ph + ns	p	Chlorococcales		
Chroococcidiopsis†	ph + ns	p	*Asterochloris**	ph	
			Chlorella†	ph(as) + ns	
Pleurocapsales			*Chlorosarcinopsis**	ph(as) + ns	
*Hyella**	ph(as) + ns	p	*Coccomyxa*‡	ph(as) + ns	sporopollenin
			Dictyochloropsis‡	ph(as) + ns	intermediate
Stigonematales			*Gloeocystis*†	ph + ns	
Stigonema‡	ph(as) + ns	p, s	*Elliptochloris*†	ph	sporopollenin
Hyphomorpha†	ph + ns	p	*Myrmecia*‡	ph(as) + ns	intermediate
			Pseudochlorella†	ph + ns	intermediate
Nostocales			*Trochiscia*†	ph + ns	
Calothrix†	ph(as) + ns	p, s			
Dichothrix†	ph(as) + ns	p	Ulvophyceae		
Anabaena†	ph + ns	p, s	Cladophorales		
Nostoc‡	ph(as) + ns	p, s	*Cladophora**	ph(as) + ns	(cellulose)

Taxon	Occurrence	Primary/secondary	Cell wall type
Scytonema‡	ph(as) + ns	p, s	
*Tolypothrix**	ph + ns	p	
Ulotrichales			
Apatococcus†	ph + ns		
*Blidingia**	ph(as) + ns		
*Coccobotrys**	ph		
Pseudendoclonium†	ph(as) + ns		
Prasiolales			
Prasiola†	ph(as) + ns		(xylo-mannan)
Pleurastrales			
Pleurastrum†	ph(as) + ns		
Trebouxia¶	ph(as**) + ns		cellulose or cellulose-like
Trentepohliales			
Cephaleuros†	ph(as) + ns		
Phycopeltis‡	ph(as) + ns		
Trentepohlia§	ph(as) + ns		cellulose
Incertae sedis			
*Nannochloris**	ph + ns		
Xanthophyta			
Tribonematales			
Heterococcus	ph(as) + ns		
Phaeophyta			
Ectocarpales			
*Petroderma**	ph(as) + ns		

[a] Abundance: *so far found as photobiont of a single lichen species; †so far found in less than 10 lichen spp.; ‡so far found in less than 100 lichen spp.; §very widespread in a large number of lichen spp.; ¶the most common group of lichen photobionts.

[b] Occurrence: genera of cyanobacteria and algae with lichen photobiont(s), ph; lichen photobionts some (or all) of which have been found in the aposymbiotic (free-living) state in nature, (as); non-symbiotic algae, ns.

[c] Primary/secondary photobiont: p, cyanobacteria which have been found as primary/only photobiont of lichen thalli; s, cyanobacteria which have been found as secondary (mainly nitrogen-fixing) photobiont in cephalodia.

[d] Cell wall type: (fine structure and composition) as described in the literature. See Honegger (1990, 1991b).

** *Trebouxia* spp. *do* occur in the aposymbiotic state in nature but they have so far not been found in all habitats which support growth of lichen spp. with these particular photobionts. See Tschermak-Woess (1988).

similar effects, reacting individually or syngergistically with the primary gall maker to produce gall modifications.

Gall-like structures on lichen thalli have received very little attention to date. The first person to endeavour to examine certain of these in depth was Bachmann (see bibliography in Grummann 1960) who conducted exhaustive studies in particular on galls of *Cladoniaceae* (Bachmann 1927–1928). He also provided detailed anatomical accounts of a variety of other galls on diverse lichens he had examined (Bachmann 1929).

In a broader survey, Grummann (1960) studied a remarkable 23 471 individual galls formed on 101 lichen species, including many not previously documented. These he grouped into kataplasmic fungal galls, witches' brooms, kataplasmic animal galls, false or shiny galls, and cephalodia (see above). The gall-forming agents, where known, are predominantly fungi and gall formation is one of the main strategies in lichenicolous fungi (that is, fungi growing on lichens; Hawksworth 1982). Descriptions of these are therefore included in the literature on lichen-icolous fungi (for example, Keissler 1930; Hawksworth 1983; Clauzade *et al.* 1989; Triebel 1989; Alstrup and Hawksworth 1990). Most of the gall-forming fungi involved are ascomycetes, especially of the family *Pyrenidiaceae* (for example, *Dacampia, Polycoccum, Pyrenidium*), but also include representatives from other families (for example, *Abrothallus, Cecidonia, Guignardia, Nesolechia, Phacopsis, Thamnogalla*), basidiomycetes (for example, *Biatoropsis, Tremella*), coelomycetes (for example, *Bachmanniomyces*), and hyphomycetes (for example, *Refractohilum*). Some examples are shown in Figs 5.18 and 5.20–5.23.

In some cases the galls on lichens may be attributable to agents other than fungi, notably nematodes of which several lichenicolous species are now known (Siddiqi and Hawksworth 1982; Fig. 5.19) and also eriophyid mites (Mani 1964). However, the causal agent of many of the galls catalogued by Grummann (1960) remains unknown and further galls of unknown aetiology are readily found by the critical observer. The possibility that at least some of those with no known cause might conceivably be due to yet unidentified bacterial or viral agents should not be discounted.

No gall occurring on a lichen has yet been studied critically from a morphogenetic, ultrastructural, or physiological viewpoint. In the absence of such data, it is not possible to state categorically that these structures are fully morphogenetically comparable to insect galls, as hitherto assumed. This is a fascinating and challenging topic, ripe for original research, and awaiting a champion.

Validity of the lichen–gall analogy

On the basis of the foregoing enumeration of the similarities and differences between lichen thalli and insect galls, the validity of the proposed analogy between these two phenomena can now be assessed. From the data analysed here, it is evident that despite certain not insignificant similarities, the proposed analogy between these phenomena cannot be sustained in view of the substantive and often fundamental differences identified.

Lichen thalli also differ significantly from the types of interactions found in mycorrhizas and plant–pathogen associations, although as in the case of insect galls some similarities have been identified (Honegger 1991*a*). We conclude that lichen thalli must be recognized as a particular biological phenomenon with no close exact analogue. Morphologically complex lichen thalli can be regarded as adaptations of nutritionally specialized fungi to the cohabitation with a population of minute photobiont cells. It is the fungal partner in these systems which competes for space above ground, facilitates gas exchange, provides, although passively, water and dissolved mineral nutrients, and secures adequate illumination of the photoautotroph inhabitant. This is a unique situation in the fungal kingdom. However, lichenization is a fairly widespread nutritional strategy since it is used by approximately 21 per cent of all fungi. Stable mutualistic symbioses of C-heterotrophs with algal or cyanobacterial inhabitants are widespread among protoctista and, additionally, occur in numerous invertebrates of freshwater and marine habitats (see Reisser 1992). However, none of these associations occur in animals structurally more complex than cnidarians, molluscs, or platyhelminths. The reasons for this phenomenon are summarized and discussed by Smith (1991).

It is also appropriate to draw attention briefly to two other interesting suggestions as to the morphogenetic interpretation of lichen thalli that have been highlighted in some recent publications.

First, Hawksworth (1988*b*) suggested that lichen thalli might be regarded as lichenized stromata. The stroma of some lichen progenitors could have incorporated photobiont cells which coevolved into lichen thalli through the development of harmonized growth mechanisms (see above). As noted by Jahns (1988*b*), this hypothesis has the attraction that it is based on a postulated process in fossil fungi which occurs in a similar manner in certain contemporary species. In order to test this hypothesis it would be pertinent to determine whether the morphogenetic processes triggered by a compatible photobiont were identical to those responsible for stroma development in non-lichenized ascomycetes. Examples of non-lichenized genera whose stromata recall the different thallus types of lichens to some extent are the fruticose stromata of *Thamnomyces*, lobed stromata of *Hypocreopsis*, crustose stromata of certain groups of *Hypoxylon* and *Ustulina*, and also the remarkable *Lichenothelia* (Henssen 1987). Interestingly, and for quite other reasons, Corner (1964) argued that stromata were an ancestral feature in ascomycetes as a whole.

Second, it has been proposed that algal cells incorporated into the exciples of ascomata led to their proliferation into expansive areas of photosynthetic tissue, that is, that lichen thalli can be viewed as precocious excipular structures (Poelt and Wunder 1967; Moser-Rohrhofer 1969; Poelt 1991). Such a phenomenon could conceivably have been an early evolutionary event in the coevolution of certain lichenized groups, prior to the development of mutually beneficial morphogenetic processes. Some support for this view may be derived from the occurrence of photobiont cells in the ascomata of some essentially non-lichenized ascomycetes, such as *Orbilia* (Benny *et al.* 1978). However, we are not aware of any massively proliferating exciples in any non-lichenized ascomycete which suggests that novel morphogenetic processes must have been coevolved for this hypothesis to be historically correct.

Coevolution and systematics

A variety of structures, characters, and biological strategies found in lichens which can be interpreted as coevolved have been identified (Ahmadjian 1987; Hawksworth 1988*b*): vegetative morphology, dual propagules (Fig. 5.16), specialized haustoria (Figs 5.11–5.15), eco-physiological adaptations (including special structures for gaseous exchange, 'cyphellae', and 'pseudocyphellae';), extracellular secondary metabolite production and localization (see above; the compounds are most probably primarily antifeedents and antimicrobials), slow growth rates, sexual strategies involving the suppression of sex and dependence on dual propagules for dispersal and establishment, expanded ranges of ecological niches available for establishment compared with the isolated bionts, and, in some cases, interactions with other organisms (for example, adaptations to grazing).

Traditionally, many of the presumptive coevolved structures have been extensively used in lichen taxonomy. A major thrust in the fundamental systematics of lichen-forming fungi in the last 15 years particularly has been an increased emphasis on the sexual fungal structures in the delimitation of genera and their placement in families and orders. Progress has been substantial in the case of crustose lichens (for example, Hafellner 1984) and in the near future major progress can be expected on parallel lines in foliose and fruticose groups — as evidenced by recent studies on alectorioid and cetrarioid lichens (Kärnefelt and Thell 1992; Kärnefelt *et al.* 1992). Lichenized genera are now incorporated into overall systems for the fungi involved (Eriksson and Hawksworth 1991).

Coevolved structures remain of paramount importance at the specific level, but this does not appear to result in any particular problems except where different morphologies appear to arise from a single mycobiont interacting with disparate photobionts (Jørgensen 1991). Taxonomists should, however, be cautious in their approaches and critically assess the status of the characters they employ and also the variations in lifestyles, biology, and structure that can be evidenced through the life span of certain species (see above; Ott 1987).

Molecular approaches are only now being initiated in lichen systematics and these can be expected to shed new light on many basal aspects of our understanding of both the specificity and systematic position of the bionts involved. The prospects for the future are exciting!

Acknowledgements

Our sincere thanks are due to Dr K.A. Pirozynski for stimulating discussions, to Mrs V. Kutasi and Miss G. Godwin for photographic support, and Ms M.S. Rainbow for assistance in preparing the manuscript.

References

Ahmadjian, V. (1987). Coevolution in lichens. *Annals of the New York Academy of Sciences*, **503,** 307–15.

Ahti, T. (1982). The morphological interpretation of cladoniiform thalli in lichens. *Lichenologist*, **14,** 105–13.

Alstrup, V. and Hawksworth, D.L. (1990). The lichenicolous fungi of Greenland. *Meddelser om Grønland, Bioscience*, **31,** 1–90.

Armaleo, D. and Clerc, Ph. (1991). Lichen chimeras: DNA analysis suggests that one fungus forms two morphotypes. *Experimental Mycology*, **15,** 1–10.

Bachmann, E. (1927–1928). Die Pilzgallen einiger Cladonien. [I-] IV. *Archiv für Protistenkunde*, **57,** 58–84; **59,** 373–416; **62,** 261–306; **64,** 109–51.

Bachmann, E. (1929). Pilze-, Tier- und Scheingallen auf Flechten. *Archiv für Protistenkunde*, **66,** 459–514.

Benny, G.L., Samuelson, D.A., and Kimbrough, J.W. (1978). Ultrastructural studies on *Orbilia luteorubella* (discomycetes). *Canadian Journal of Botany*, **56,** 2006–12.

Bissett, J. and Borkent, A. (1988). Ambrosia galls: the significance of fungal nutrition in the evolution of the *Cecidomyiidae (Diptera)*. In *Coevolution of fungi with plants and animals*, (ed. K.A. Pirozynski and D.L. Hawksworth), pp. 204–25. Academic Press, London.

Clauzade, G., Roux, C., and Diederich, P. (1989). Nelikenigintaj fungoj likenlogaj illustrita determinlibro. *Bulletin de la Société Linnéenne de Provence, numéro spécial,* **1,** 1–142.

Corner, E.J.H. (1964). *The life of plants.* Weidenfeld and Nicolson, London.

Culberson, C.F. and Ahmadjian, V. (1980). Artificial reestablishment of lichens. II. Secondary products of resynthesized *Cladonia cristatella* and *Lecanora chrysoleuca. Mycologia,* **72,** 90–109.

Culberson, C.F. and Armaleo, D. (1992). Induction of a complete secondary-product pathway in a cultured lichen fungus. *Experimental Mycology,* **16,** 52–63.

Culberson, C.F. and Elix, J.A. (1989). Lichen substances. *Methods in Plant Biochemistry,* **1,** 509–35.

des Abbayes, H.A. (1951). Traité de lichénologie. *Encyclopédie Biologique,* **41,** 1–217.

Eriksson, O.E. and Hawksworth, D.L. (1991). Outline of the ascomycetes — 1990. *Systema Ascomycetum,* **9,** 39–271.

Grummann, V.J. (1960). Die Cecidien auf Lichenen. *Botanische Jahrbücher für Systematik, Pflanzengeschichte und Pflanzengeographie,* **80,** 101–44.

Hafellner, J. (1984). Studien in Richtung einer natürlicheren Gliederung der Sammelfamilien *Lecanoraceae* und *Lecideaceae. Beihefte zur Nova Hedwigia,* **79,** 241–371.

Hawksworth, D.L. (1982). Secondary fungi in lichen symbioses: parasites, saprophytes and parasymbionts. *Journal of the Hattori Botanical Laboratory,* **52,** 357–66.

Hawksworth, D.L. (1983). A key to the lichen-forming, parasitic, parasymbiotic and saprophytic fungi occurring on lichens in the British Isles. *Lichenologist,* **15,** 1–44.

Hawksworth, D.L. (1988a). The variety of fungal–algal symbioses, their evolutionary significance, and the nature of lichens. *Botanical Journal of the Linnean Society,* **96,** 3–20.

Hawksworth, D.L. (1988b). Coevolution of fungi with algae and cyanobacteria in lichen symbioses. In *Coevolution of fungi with plants and animals* (ed. K.A. Pirozynski and D.L. Hawksworth), pp. 125–48. Academic Press, London.

Hawksworth, D.L., Sutton, B.C., and Ainsworth, G.C. (1983). *Ainsworth & Bisby's dictionary of the fungi,* (7th edn). Commonwealth Mycological Institute, Kew.

Henssen, A. (1987). *Lichenothelia,* a genus of microfungi on rocks. *Bibliotheca Lichenologica,* **25,** 257–93.

Honegger, R. (1986). Ultrastructural studies in lichens. II. Mycobiont and photobiont cell wall surface layers and adhering crystalline lichen products in four *Parmeliaceae. New Phytologist,* **103,,** 797–808.

Honegger, R. (1987a). Questions about pattern formation in the algal layer of lichens with stratified (heteromerous) thalli. *Bibliotheca Lichenologica,* **25,** 59–71.

Honegger, R. (1987b). Isidium formation and the development of juvenile thalli in *Parmelia pastillifera* (*Lecanorales,* lichenized ascomycetes). *Botanica Helvetica,* **97,** 147–52.

Honegger, R. (1990). Surface interactions in lichens. In *Experimental phycology 1: Cell walls and surfaces, reproduction, photosynthesis,* (ed. W. Wiessner, D.G. Robinson, and R.C. Starr), pp. 40–54. Springer Verlag, Berlin.

Honegger, R. (1991*a*). Haustoria-like structures and hydrophobic cell wall surface layers in lichens. In *Electron microscopy of plant pathogens*, (ed. K. Mendgen and D.-E. Lesemann), pp. 277–90. Springer Verlag, Berlin.

Honegger, R. (1991*b*). Functional aspects of the lichen symbiosis. *Annual Review of Plant Physiology and Plant Molecular Biology*, **42**, 553–78.

Honegger, R. (1992). Lichens: mycobiont–photobiont relationships. In *Algae and symbioses. Plants, animals, fungi, viruses interactions explored*, (ed. W. Reisser), pp. 255–75. Biopress, Bristol.

Jahns, H.M. (1970). Untersuchungen zur Entwicklungsgeschichte der Cladoniaceen unter besonderer Berücksichtigung des Podetium-Problems. *Nova Hedwigia*, **20**, 1–177.

Jahns, H.M. (1988*a*). The establishment, individuality and growth of lichen thalli. *Botanical Journal of the Linnean Society*, **96**, 21–9.

Jahns, H.M. (1988*b*). The lichen thallus. In *CRC handbook of lichenology*, (ed. M. Galun), Vol. 1, pp. 95–143. CRC Press, Boca Raton, FL.

James, P.W. and Henssen, A. (1976). The morphological and taxonomic significance of cephalodia. In *Lichenology: progress and problems*, (ed. D.H. Brown, D.L. Hawksworth, and R.H. Bailey), pp. 27–77. Academic Press, London.

Jørgensen, P.M. (1991). Difficulties in lichen nomenclature. *Mycotaxon*, **40**, 497–501.

Kappen, L. (1988). Ecophysiological relationships in different climatic regions. In *Handbook of lichenology*, (ed. M. Galun), Vol. 2, pp. 37–100. CRC Press, Boca Raton, FL.

Kärnefelt, I., Mattsson, J.-E., and Thell, A. (1992). Evolution and phylogeny of cetrarioid lichens. *Plant Systematics and Evolution*, **183**, 113–60.

Kärnefelt, I. and Thell, A. (1992). The evaluation of characters in lichenized families, exemplified with the *Alectoriaceae* and some genera in the *Parmeliaceae*. *Plant Systematics and Evolution*, **180**, 181–204.

Kaule, A. (1932). Die Cephalodien der Flechten. *Flora*, **126**, 1–44.

Keissler, K.A. (1930). Die Flechtenparasiten. *Rabenhorst's Kryptogamen-Flora von Deutschland, Österreich und der Schweiz*, **9**, i–xi, 1–712.

Larson, D.W. and Carey, C.K. (1986). Phenotypic variation within 'individual' lichen thalli. *American Journal of Botany*, **73**, 214–23.

Leuckert, C., Ahmadjian, V., Culberson, C.F., and Johnson, A. (1990). Xanthones and depsidones of the lichen *Lecanora dispersa* in nature and of its mycobiont in culture. *Mycologia*, **82**, 370–8.

Mani, M.S. (1964). *The ecology of plant galls*. Junk, The Hague.

Mollenhauer, D. (1992). *Geosiphon pyriforme*. In *Algae and symbioses. Plants, animals, fungi, viruses interactions explored*, (ed. W. Reisser), pp. 339–51. Biopress, Bristol.

Moreau, F. (1921). Recherches sur les lichens de la famille des Stictacées. *Annales des Sciences Naturelles, Botanique, série 10*, **10**, 297–376.

Moreau, F. (1927). Les lichens: morphologie, biologie, systématique. *Encyclopédie Biologique*, **2**, 1–144.

Moreau, F. (1956). Sur la théorie biomorphogénique des lichens. *Revue Bryologique et Lichénologique*, **25**, 183–6.

Moreau, F. and Moreau, V. (1918). La biomorphogénèse chez les lichens. *Bulletin de la Société Mycologique de France*, **34**, 84–5.

Moser-Rohrhofer, M. (1969). Der vegetative Flechtenthallus — ein Derivat des

98 *David L. Hawksworth and Rosmarie Honegger*

Ascophors? *Anzeiger, Österreichische Akademie der Wissenschaften, Mathematisch-Naturwissenschaftliche Klasse,* **6,** 109–120.

Oberwinkler, F. (1970). Die Gattungen der Basidiolichenen. *Deutsche Botanische Gesellschaft Neue Folge,* **4,** 139–69.

Ott, S. (1987). Reproductive strategies in lichens. *Bibliotheca Lichenologica,* **25,** 81–93.

Pirozynski, K.A. (1991). Galls, flowers, fruits, and fungi. In *Symbiosis as a source of evolutionary innovation,* (ed. L. Margulis and R. Fester), pp. 364–80. Massachusetts Institute of Technology Press, Cambridge, MA.

Poelt, J. (1991). Homologies and analogies in the evolution of lichens. In *Frontiers in mycology,* (ed. D.L. Hawksworth), pp. 85–97. CAB International, Wallingford.

Poelt, J. and Wunder, H. (1967). Über biatorinische und lecanorische Berandung von Flechtenapothecien, untersucht am Bespiel der *Caloplaca ferruginea*-Gruppe. *Botanische Jahrbücher für Systematik, Pflanzengeschichte und Pflanzengeographie,* **80,** 101–44.

Ream, W. (1989). *Agrobacterium tumefaciens* and interkingdom genetic exchange. *Annual Review of Phytopathology,* **27,**, 583–618.

Reisser, W. (ed.) (1992). *Algae and symbioses. Plants, animals, fungi, viruses interactions explored.* Biopress, Bristol.

Rohfritsch, O. and Shorthouse, J.D. (1982). Insect galls. In *Molecular biology of plant tumors,* (ed. G. Kahl and J.S. Schell), pp. 121–52. Academic Press, New York.

Schuster, G., Ott, S., and Jahns, H.M. (1985). Artificial cultures of lichens in the natural environment. *Lichenologist,* **17,** 247–53.

Shirley, P. (1992). Oak galls. *British Wildlife,* **3,** 162–6.

Siddiqi, M.R. and Hawksworth, D.L. (1982). Nematodes associated with galls on *Cladonia glauca,* including two new species. *Lichenologist,* **14,** 175–84.

Smith, D.C. (1991). Why do so few animals form endosymbiotic associations with photosynthetic microbes? *Philosophical Transactions of the Royal Society of London, Biological Sciences,* **333,** 225–30.

Tibell, L. (1982). *Caliciales* of Costa Rica. *Lichenologist,* **14,** 219–54.

Timdal, E. (1991). A monograph of the genus *Toninia* (*Lecideaceae,* ascomycetes). *Opera Botanica,* **110,** 1–137.

Triebel, D. (1989). Lecideicole Ascomyceten. Eine Revision der obligat lichenicolen Ascomyceten auf lecideoiden Flechten. *Bibliotheca Lichenologica,* **35,** 1–278.

Tschermak-Woess, E. (1988). The algal partner. In *CRC handbook of lichenology,* (ed. M. Galun), Vol. 1, pp. 39–92. CRC Press, Boca Raton, FL.

Winchester, V. (1988). An assessment of lichenometry as a method for dating recent stone movements in two stone circles in Cumbria and Oxfordshire. *Botanical Journal of the Linnean Society,* **96,** 57–68.

6. *Proales werneckii*: a gall-causing rotifer

B.M. SPOONER

31 Balmoral Crescent, West Molesey, Surrey, UK

Abstract

Proales werneckii (Ehrenberg) is unique amongst known rotifers in being a gall causer. It is parasitic in species of *Vaucheria* and *Dichotomosiphon*, causing characteristic lateral galls on the filaments. It appears also to be unique in being the only non-fungal causer of galls in freshwater algae. The species has other unusual features, notably the presence of males which possess fully developed jaws and a functional mastax, a situation which is of very rare occurrence amongst rotifers. Galls on *Vaucheria* have been known since early in the nineteenth century and the causer was described in 1834. It has proved to be widely distributed in temperate regions, occurring in both northern and southern hemispheres and is now known to parasitize at least 16 host species. However, *P. werneckii* is still infrequently recorded and the process of gall formation remains little studied. A review of our current knowledge is presented here.

Introduction

Proales werneckii (Ehrenberg) is parasitic on various species of *Vaucheria* DC and is of interest to cecidologists for its ability to induce galls on the algal filaments. These galls are an essential part of the life history of the rotifer. They have been known since early in the nineteenth century, although their true nature was initially misunderstood and the species itself was not described until 1834. It has since proved to be widely distributed in temperate regions and is now known from both northern and southern hemispheres. However, *P. werneckii* has received little recent study; many aspects of its biology remain inadequately known and it is still infrequently recorded. The present review, which has arisen from work currently being carried out for a checklist of British gall-causing organisms, aims to bring the species to attention and hopes to stimulate

Plant Galls (ed. Michèle A. J. Williams), Systematics Association Special Volume No. 49, pp. 99–117. Clarendon Press, Oxford, 1994. © The Systematics Association, 1994.

further study of an organism which, in several ways, must be regarded as unique.

Proales werneckii is unique in being the only known gall-causing rotifer. It appears also to be unique in being the only known animal causer of galls in freshwater algae. Other animal gall causers, including nematodes and copepods, are known from marine algae, but only fungi, of various species belonging to Chytridiomycetes and Plasmodiophorales, are otherwise known to gall freshwater algae. Two gall-causing fungi are known from *Vaucheria* in Europe: *Zygorhizidium vaucheriae* Rieth (Chytridiomycetes), which causes 'witches' brooms' (Karling 1977) and *Woronina glomerata* (Cornu) Fischer (Plasmodiophorales) (Buhr 1965). Other species of chytrid are known to be parasitic in *Vaucheria*, but they do not cause galls.

Proales werneckii is again unusual and, perhaps, unique with regard to its reproductive biology, some aspects of which require further study. Males are also anatomically unusual in several respects, as further noted below.

Historical survey

Historical surveys of the study of *P. werneckii* have been presented by several authors, notably Balbiani (1878, 1879) and Merola (1956). That by the latter author is part of a detailed historical survey of the study of galls on algae and extends considerably that provided by Trotter (1901). An historical outline, covering the most significant work on *P. werneckii*, is presented here.

The earliest illustration of galls caused by this species appears to be that by Vahl (1787), who provided coloured figures of Danish material of *Vaucheria dilatata* (as *Ulva*) which clearly bears galls. However, the nature of these galls was misinterpreted at that time, as it has been subsequently on occasion by other authors. The galls were obviously considered by Vahl to be a normal part of the alga and there is no indication of any associated parasite. It may be noted that two varieties of *C. dilatata*, var. *clavata* Roth and var. *bursata* Roth, were later described based on these galled filaments (Roth 1800, 1806).

The earliest recognition of the existence of galls on 'lower plants' can be attributed to Vaucher (1803) (see Balbiani 1878, 1879). Vaucher described galls, now known to be caused by *P. werneckii*, on a genus of filamentous algae which was later named after him as *Vaucheria* DC. Vaucher regarded these structures as true galls, although he considered that they were due to an 'insect', *Cyclops lupula* Müller. Galls on *Vaucheria* from the British Isles were first illustrated by Smith (1807), who referred to them as 'possibly caused by *C. lupula* Müller'. Galls on *V. racemosa* from Denmark were reported by Lyngbye (1819), who again referred to the causer as *C. lupula*. Galls were later observed by Unger (1828) during a developmental study of *V. clavata*. Unger at that time did not realize their true nature; he noted the presence of 'animalcules', but offered no identification of the causer. In 1834 Unger collected galled filaments of *V. caespitosa* at Kitzbühel in Austria (Balbiani 1878, 1879). This material he sent to Werneck, who studied the galls and sent them on to Ehrenberg who was working in Germany. Ehrenberg (1834) was the first to recognize that the galls were caused by a rotifer, which he described as a new species, *Notommata werneckii* Ehrenberg, although this was apparently overlooked by subsequent workers for almost 40 years. Morren (1839, 1841) mistakenly referred the causer to *Rotifer vulgaris* (= *Rotaria rotatoria* (Pallas)), a common free-living species (Holloway, personal communication), but suggested that the 'vesicles' on the filaments were indeed true galls. Oliver (1860) described galls from British material but, following Morren, again attributed them to *R. vulgaris*. Cornu collected galled material of *V. terrestris* from Bordeaux in 1874 and this became the subject of a detailed study by Balbiani, who published (Balbiani 1878), a long, somewhat verbose, account which included an historical survey, a detailed description of the rotifer, and important observations on its life cycle and gall formation. An abridged English translation of this account was published the following year (Balbiani 1879).

Magnus (1876) gave a general account of the species which he reported from Germany forming galls on *V. dichotoma* and *V. racemosa*. Wollny (1877, 1878*a, b*) described galls on several species of *Vaucheria* and emphasized differences in the shape of galls developed on different species of *Vaucheria*. Those on *V. racemosa* and *V. geminata*, for example, he described as having a thin neck and being swollen at the tip with two to six projections. In contrast, galls on *V. clavata* are obpyriform and those on *V. uncinata* are more cylindrical and rounded at the tip. He considered the possibility that more than one causer might be involved. This possibility has also been suggested by later workers, such as Docters van Leeuwen (1982), Islam (1983), Christensen (1987*a*), and Entwisle (1988), but there is as yet no confirmation that this is the case.

Further detailed study of *P. werneckii*, including gall development, was undertaken by Debray (1890) and Rothert (1896*a,b*). As noted by Gabriel (1922*a*), Debray died in 1900 before completing his studies of this species. Rothert (1896*b*) provided the first description of the male of *P. werneckii*, which had not previously been recognized, although it was independently recognized and described almost simultaneously by Rousselet (1897). The structure of the male, including fully developed jaws and mastax but poorly developed digestive tract, proves to be almost unique amongst rotifers; its large size in comparison to the female is also of rare occurrence.

There has been comparatively little study of the biology of this species this century, although important accounts, including gall formation and development of the eggs, were published by Gabriel (1922*a,b*), Weidner (1952), and Weber (1960) and the wide distribution of the species has become clear.

The similarity in some cases between galls and antheridia has led more than once to misinterpretation of the former as a normal part of the alga. The varieties of *V. dilatata* described by Roth are mentioned above. In addition, *V. sacculifera* Kützing (Kützing 1856) has been shown to be based on galled filaments of *V. geminata*. More recently, Habeeb (1965) has described a new monotypic genus, *Debsalga*, for the new species *D. gigasporangia* Habeeb, which has since been shown to have been based on galled filaments of a species of *Vaucheria* (Christensen 1987*b*).

Brief account of the genus *Proales*

The genus *Proales* Gosse [Rotifera, Monogononta, Proalidae (Lecanidae *fide* Pontin 1978)] is world-wide in distribution. It is a large genus, including soft-bodied (illoricate) rotifers (typical of the Proalidae), in which the digestive tract is fully developed and a foot is present. It belongs to the order Monogononta, in which the females possess a single ovary. The mastax, perhaps the most characteristic organ in Rotifera, is malleate in species of *Proales*. This organ, which is basically a muscular bulb on the alimentary canal, contains hard parts (trophi) as well as muscle and is used for crushing food. Its form is, therefore, related to feeding habits and also to habitat. Rotifers with a malleate mastax mostly feed on small organisms (bacteria, algae, protozoa) and *Proales* species are predominantly browsers (Galliford 1961–63).

Many species of *Proales* are free-living, amongst submerged macrophytes or aquatic moss or in mud and detritus, but some are epizooic, often on Crustacea (*Cladocera, Gammarus*; Fitter and Manuel 1986) or in the eggs of aquatic snails (Ward and Whipple 1918; Hollowday, personal communication) and others are parasitic. Perhaps the closest relatives of *P. werneckii* are *P. parasita* (Ehrenberg), which is parasitic in colonies of *Volvox* spp. (Chlorophyceae), feeding on the cells but causing no deformation and *P. uroglenae* (de Beauchamp), parasitic in colonies of *Uroglena volvox* Ehrenberg (Chrysophyceae). It may be noted that the unrelated *Ascomorphella volvocicola* (Plate) (Tricocercidae) also parasitizes species of *Volvox* and that a second, recently described, parasite of *U. volvox*, *Cephalodellaedax* Hollowday (Notommatidae), has been discovered in Britain (Hollowday 1993). Nineteen species of *Proales* have been hitherto recorded from the British Isles (Hussey 1981).

Proales werneckii

Proales werneckii (Ehrenberg) Hudson and Gosse, *Rotifera*, Suppl. 1889: 23, pl. 32
  fig. 18.   Fig. 6.1
  Synonyms:
Notommata werneckii Ehrenberg in *Abh. Akad. Wiss. Berlin* 1833: 216 (1834).
Copeus werneckii (Ehrenberg) in *Infusionsthierchen* 1838: 441.
Cyclops lupula Vaucher in *Hist. Conf. d'Eau Douce* 1803: 18, figs 8r & 11s; non *C.
  lupula* Müller.
General description (see, Hudson and Gosse 1889; Rousselet 1897; Harmer and
  Shipley 1901; Ward and Whipple 1918; Harring and Myers 1924; Hyman
  1951; Donner 1966; Koste 1978):

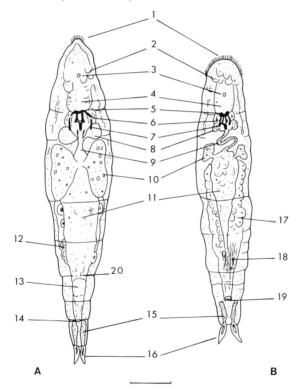

Fig. 6.1. *Proales werneckii*, (A) female, (B) male. (Redrawn from Rousselet (1897) and
Hyman (1951)). (Scale bar = 20 μm.) (1) Corona, (2) coronal glands, (3) dorsal antenna,
(4) supraoesophageal ganglion, (5) cerebral eye, (6) mastax, (7) trophi, (8) salivary glands,
(9) oesophagus, (10) gastric glands, (11) stomach, (12) yolk gland, (13) bladder, (14) anus,
(15) pedal glands, (16) toes, (17) testis, (18) sperm, (19) penis, (20) intestine.

Illoricate, without segmentation at maturity. Spindle-shaped to ver-
miform, with short, broad foot and short toes. Head and foot clearly
defined. Corona oblique; whole face of disc strongly ciliated. Mastax

malleate, modified for suction; rounded, pale, and clear, homogenous, not striate from muscle filaments (as other species are). Buccal area very simple. Eye single, usually on brain, in neck region. Fulcrum short, in line with the flattened rami.

Male: 128–150 μm long, with fully developed jaws and functional mastax; intestinal tract reduced, rudimentary. Gastric and salivary glands present; eye and dorsal antenna well developed.

Female: 140–200 μm long, with normally developed intestinal tract. Gastric and salivary glands well developed, atrophying later.

1. Life history

The most detailed studies of the life history of *P. werneckii* are those by Balbiani (1878, 1879), Rothert (1896*b*), and Weber (1960), although other authors such as Rousselet (1897), Gabriel (1922*a,b*), and Weidner (1952) have contributed important observations on various aspects of the development of this species. A summary of the available information follows.

Proales werneckii is an obligate parasite, unable to complete its life cycle outside the galls which it induces on the filaments of species of *Vaucheria*. Males occur and are entirely free-living. As with many rotifers, both sexual reproduction and parthenogenesis occur. Fertilization takes place during the free-living stage and only females enter the host filaments.

Entry of the young females into the host filaments has, apparently, never been directly observed and the method of entry has been much in dispute. It has been suggested that this is a passive process, perhaps through old galls (Magnus 1876; Balbiani 1878, 1879), but it is now generally accepted that the free-swimming female actively enters the filament through the growing point. It was suggested by Hofmeister (1867, p. 77) that penetration took place through the wall of the filament, without causing damage, but he gave no further account or explanation of the process. However, as noted by Rothert (1896*b*), the jaws in this species, as in most other rotifers, are fully developed and are evidently used to gain entry into the filament, this taking place where the membrane is thinnest, that is at the growing tip. Weber (1960) further noted, in support of this conclusion, the presence of 'scars' at the tips of freshly infected filaments. He also noted that older rotifers which have failed to invade a filament lose their ability to do so.

Inside the filament the female is initially motile, eventually settling at a point at which gall formation is induced. The stimulus for gall initiation is uncertain and requires further study. Balbiani (1878, 1879) has suggested that it is due to acid secretion from salivary glands, which, as he noted, are large in *P. werneckii* in comparison to those of other species.

Many females may enter a single filament, which can bear numerous galls as a result. The galls usually develop rapidly, reaching full size in a few days. Except in rare instances, each contains a single female, the constant presence of which is required to complete the development of the gall (Weber 1960). Should the gall be abandoned or if the female dies development ceases.

2. Gall form and development

Galls are normally developed laterally on the filament, though, rarely, they may be terminal in position (Figs 6.2 and 6.3). According to Weber (1960), 'pseudoterminal' galls, which are elongated and branch-like in form, may occur near damaged filament tips. Galls are independent structures and are not formed by modification of side branches or reproductive structures as has been suggested by some authors, notably Balbiani (1878, 1879) and Rothert (1896*b*). Evidence for this was discussed by Weber (1960).

Galls are initially low and dome-like in form, but soon develop into comparatively large structures. Their entire development may take 8–10 days. Galls are variable in form, 'multimorphic' as noted by Buhr (1965), but usually cylindric or ovate to clavate in form, 500–750 (–900) μm high, 200–300 μm in diameter. Occasionally, galls may be broader than

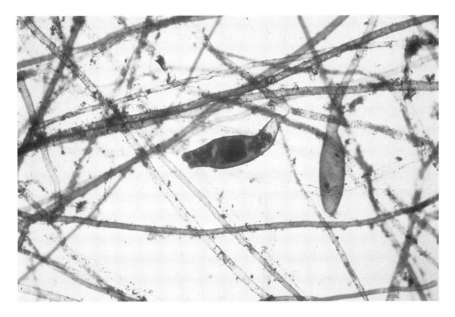

Fig. 6.2. *Proales werneckii*. Developing galls on filaments of *Vaucheria* sp. (photograph, T. Christensen).

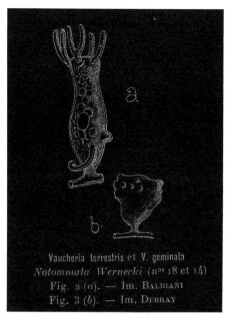

Fig. 6.3. *Proales werneckii.* Mature galls, on *Vaucheria terrestris* and *V. geminata.* Reproduced from Houard (1908).

high (Weber 1960). A gall of abnormal form was reported on *V. erythrospora* in brackish water in Australia by Christensen (1987*a*). This is saccate, with numerous short projections; it should be compared to the very similar gall figured by Debray (1890) and Houard (1908) on *V. geminata* from Europe. At maturity characteristic cylindric, horn-like projections develop, usually at the apex of the gall. These are simple or occasionally branched (Balbiani 1878, 1879) and are variable in number, apparently according to the shape of the gall. Thus, broad, transverse galls have pairs of projections at each end, whereas narrow, upright galls have just two apical projections and broad, upright galls may have 5–16 such projections and, occasionally, also a basal one (Christensen 1969). These projections contain little chlorophyll (Weber 1960) and are distinctly paler than the rest of the gall which is dark green due to accumulation of chloroplasts.

Within the gall, the young female feeds until fully mature, mainly ingesting cytoplasm exclusive of chloroplasts (Balbiani 1878, 1879), which is continually replaced from the filament outside the gall. The rotifer is motile and can normally be clearly observed within the gall, particularly when mature, 'appearing as a little black point' (Swanton 1912). At this stage, it is unsegmented and highly contractile. No faecal discharge occurs within the gall and, consequently, the body swells greatly due the accumulation of waste. As a result, the rotifer has a black, opaque centre (Weber 1960). During maturtion, the ovary develops enormously whilst the salivary and gastric glands atrophy (Balbiani 1878, 1879).

At maturity, the female lays eggs, usually 30–60 and occasionally up to 80 (Koste and Shiel 1990), per individual. She then dies, leaving the gall full of eggs. The chlorophyll within the gall then gradually loses colour and forms into irregular, discoloured masses. It soon becomes transparent and the eggs can then be clearly observed.

3. Reproduction

As noted above, both sexual reproduction and parthenogenesis occur in *P. werneckii*, as in many species of rotifer. The eggs it produces are, therefore, of more than one kind, depending at least partly on whether fertilization has taken place. However, the life cycle is complex and some aspects of the reproduction of this species are, apparently, not yet clearly understood.

The reproductive cycle of rotifers was described in detail by Ruttner-Kolisko (1974). At least three and, on rare occasions, four kinds of egg are produced, these differing with regard to morphology and/or cytology. Briefly stated, diploid, amictic females lay similarly diploid, par-thenogenetic eggs known as subitaneous eggs. These are thin-walled and normally develop immediately. They usually hatch into further diploid females but, eventually, after a variable number of such generations, a change occurs which leads to the production of haploid females from diploid eggs. Causes of this change are not fully understood, but have been linked in some species to external factors such as population density, temperature, diet, pH, and photoperiod. These females lay haploid eggs which, initially, are unfertilized and give rise to males which are then available to fertilize other mictic females. Fertilized haploid eggs differ in developing a thick, double wall. They are 'resting' eggs, often referred to as 'winter' eggs, adapted to survive long periods, of sometimes adverse conditions, before hatching. They give rise to further diploid, amictic females and the cycle is duly repeated. In a few rare cases, notably species of *Asplanchna* (Asplanchnidae) and *Conochilus* (Conochilidae) (Ruttner-Kolisko 1974), a third type of female, able to produce both haploid and diploid eggs, is known. This is termed an amphoteric female, but is not known to occur in *Proales*.

The existence in some rotifers of another form of thick-walled resting egg was reported by Ruttner-Kolisko (1946, 1974), based on a study of *Keratella quadrata*. This is a diploid egg, developed without fertilization and is termed 'pseudosexual'. It is produced in response to extreme conditions.

Proales werneckii is unique amongst rotifers in laying eggs within a gall.

Each gall can include both resting and non-resting eggs, as reported by both Balbiani (1879) and Rothert (1896a,b) amongst others, a situation which is not easy to explain based on the above cycle, as both types of egg are laid by a single female. There are possible explanations of this situation, but apparently none has been confirmed. A haploid, mictic female may, if available sperm is insufficient to fertilize all of her eggs, lay both male eggs and resting eggs. This situation was reported by Ruttner-Kolisko (1974) for *Asplanchna*, but is unlikely to be the case with *P. werneckii* due to the regular occurrence of large numbers of both egg types. However, the reported presence of both male and resting eggs by Rothert (1896b) suggests this may occur at times. Alternatively, it may be that *P. werneckii* is another example of a species in which an amphoteric female occurs, whereby both amictic (diploid) and resting eggs are produced. However, perhaps the most likely explanation is that the resting eggs are mainly pseudosexual, laid by a diploid female. The comparatively large number of resting eggs produced in the gall, up to 54 according to Rothert (1896b), suggests that they are of this nature; their number is far greater than would normally result from fertilization. However, this is unconfirmed and further study is required to clarify the situation which is certainly unusual and may prove to be unique amongst rotifers.

Balbiani (1878, 1879) described variation in the ratio of resting to non-resting eggs according to season. In spring there are few resting eggs amongst those being laid, whilst by late August the ratio is reversed. At that time the few non-resting eggs which occur are often sterile and those which develop commonly die even at an advanced stage of development. The factors which influence egg type are unclear; according to Rothert (1896b) the type is not governed by temperature, an obvious factor which has been suggested by several authors.

The eggs laid in the gall are all similar in form, mostly broadly ellipsoid to ovate. However, the non-resting or 'summer' eggs are smaller, thin-walled, and have a uniformly fine-granular, hyaline content. According to Balbiani (1878, 1879) they measure up to approximately $56 \times 42\,\mu m$, though Rothert (1896a) gives larger dimensions, $67–78 \times 44–52\,\mu m$ for female eggs and approximately $63 \times 44\,\mu m$ for males. In contrast, resting eggs measure up to $62 \times 50\,\mu m$ (Balbiani; $67–74 \times 52–56\,\mu m$ *teste* Rothert) and are brown, opaque except for a central spot and have two walls. The outermost wall is a thick, solid shell which encloses a thin inner membrane. These walls are both smooth, unlike many rotifers where the surface of the resting eggs is characteristically roughened or spiny. The non-resting eggs hatch successively within 10–15 days of being laid, whereas the resting eggs remain dormant for extended periods, usually until at least the following spring and sometimes for years.

Their subsequent development according to Weber (1960) may take approximately 3 weeks.

According to Koste and Shiel (1990), resting eggs of *P. werneckii* overwinter in the sediments. This perhaps refers to eggs which have been released from the gall or algal filament as it decays. However, as noted above, it is unclear whether true, fertilized, resting eggs are produced in the gall; it may be that fertilized females do not enter the algal filaments.

4. Development and escape of the young

The embryo of *P. werneckii* bears a distinct red spot and is motile and normally readily visible within the egg. It can develop within approximately 1 week and is especially active just prior to hatching. At this time, morphological changes occur within the gall which permit the escape of the young. These changes were described by Weber (1960) and involve the apical projections where localized dissolution of the walls creates rounded openings through which the young escape.

Young rotifers are approximately $100-120\,\mu$m long (Balbiani 1878, 1879) and actively explore the wall of the gall to find an exit. According to Balbiani (1878, 1879), they may occasionally penetrate into the main filament of the alga and then, by chance, escape via another gall unless trapped in the filament by the presence of 'false septa'.

5. Associated organisms

There are probably few species which have more than a casual association with *P. werneckii*, although information is sparse and further study would be desirable.

Known associated organisms are mostly evident in old, decaying galls after hatching and escape of the young rotifers. Bacteria and various other microorganisms are involved at this stage. According to Balbiani (1878), various species of Chytridiomycetes also invade the galls and assist in breaking down the remaining grains of chlorophyll. Rothert (1896*b*) reported various protozoa from old, open galls, including the flagellate *Peranema trichophorum* Ehrenberg, as well as unidentified small monads and ciliates and amoebae. An unidentified species of nematode was also present.

According to Thompson (1892), *Proales decipiens* (Ehrenberg) may occur in old filaments of *Vaucheria* species, usurping *P. werneckii* and laying eggs in the filaments.

6. Hosts: characters and species list

Proales werneckii occurs mainly in freshwater species of *Vaucheria*, though *V. erythrospora*, in brackish water, was reported as a host by Christensen (1987*a*). The species cannot exist in marine conditions.

Vaucheria [Tribophyceae (see Hibberd 1981), Vaucheriales, Vaucheriaceae] includes species having simple, sparingly branched, coenocytic filaments; septa are rare, formed only after injury or during reproduction. Filaments contain numerous, small, green chloroplasts and vary between 40 and 250 μm diameter according to species. Sexual reproduction is by oogonia and antheridia developed laterally on the filaments, often close together, most species being monoecious. Antheridia are usually hooked at the apex and provide important characters for delimitation of species. They may be similar in form to the galls produced by *P. werneckii*, but are usually smaller. However, there has been disagreement as to whether galls are initiated, either regularly or occasionally, from immature antheridia. The development of abnormal antheridia in North American material of *V. racemosa* was reported by Nichols (1895), though cause was not established and may have no association with *Proales*.

Most species of *Vaucheria* occur in freshwater, some being terrestrial in damp places, but avoiding acid conditions.

Recently, Del Grosso (1985) has reported galls of this species on *Dichotomosiphon tuberosus* from Italy. *Dichotomosiphon* was previously considered to be allied to *Vaucheria*, but is now placed in a separate family, Chlorophyceae. This is an interesting record and confirmation is required to exclude the possibility that a second species of *Proales* is involved.

It should be noted that *Proales* galls occur frequently on sterile filaments, which cannot be specifically determined and that mixed collections of *Vaucheria* commonly occur (Christensen, personal communication). Care is required, therefore, for host identification and identification of fertile filaments must be considered insufficient for determination of associated sterile filaments.

Host species reported for *P. werneckii* are listed below, with appropriate references given in brackets. Nomenclature, as far as possible, follows Blum (1972).

Tribophyceae: Vaucheriales
***Vaucheria appendiculata* DC** (Houard 1908)
***Vaucheria aversa* Hassall** (Lister 1884; Houard 1908; Gabriel 1922*a*; Rieth 1980; Islam 1983)
***Vaucheria bursata* Ces.** (Christensen 1969 and personal communication; Entwisle 1988)
***Vaucheria caespitosa* (Vaucher) DC** (Benkö 1882; Houard 1908)
***Vaucheria canalicularis* (L.) Christensen** (Christensen 1987)

Vaucheria clavata **Lyngbye** (Benkö 1882; Houard 1908)
Vaucheria dichotoma **(L.) Martius** (Magnus 1876; Benkö 1882; Houard 1908; Moesz 1938)
Vaucheria dillwynii **(Weber and Mohr) Agardh** (Lister 1884; Bennett 1889; Houard 1908; Swanton 1912; Christensen 1987)
Vaucheria erythrospora **Christensen** (Christensen 1987; gall of unusual shape)
Vaucheria geminata **(Vauch.) DC** (Magnus 1876; Benkö 1882; Wolle 1882; Brain 1894; Houard 1908; Moesz 1938; Weidner 1952; Buhr 1965; Meyer 1987)
Vaucheria laciniata (Benkö 1882)
Vaucheria racemosa **(Vaucher) DC** (Hassall 1845; Benkö 1882; Lister 1884; Rothert 1896a (as *V. walzi*) – most common host; Houard 1908; Koste 1978; Christensen 1987)
Vaucheria sessilis **(Vaucher) DC** (Smith 1807; Houard 1908; Moesz 1938; Islam 1983)
Vaucheria terrestris **(Vaucher) DC** (Benkö 1882; Houard 1908; Buhr 1965; Donner 1966)
Vaucheria uncinata **Kützing** (Houard 1908; Moesz 1938)
Vaucheria verticillata **Meneghini** (Islam 1983)

Chlorophyceae
Dichotomosiphon tuberosus (A. Braun) Ernst (Del Grosso 1985)

7. Distribution

Proales werneckii is widely distributed in north temperate regions and is also known from Australia. It may be locally common and the galls are sometimes abundant. However, the species remains infrequently recorded. Published records are summarized below and records of occurrence in the British Isles are also indicated in Fig. 6.4.

(a) British Isles   Cambridge (Gray 1953, no locality); Clyde (Murray 1908, Glasgow, Blantyre Moor); Devon (Bennett 1890, Buckfastleigh; Stevens 1907, Exeter Canal; Stevens 1912, Exeter Canal, Turf, Starcross); Essex (Lister 1884, Epping Forest); Hertfordshire (Palmer 1934, Hitchin); London (Rousselet 1897, Willesden; Western 1894, no locality); Mersey & Leeds Liverpool Canal (Narramore 1890); Leicestershire (Horwood 1907, no locality: Hudson & Gosse 1886, Leicester; Lowe *et al.* 1933, no locality); Lothian (Murray 1906, nr. Bavelaw Castle, Balerno); Norfolk (Brain 1894, R. Wensum); Northants (Wood 1955, Oundle); Northumberland (Oliver 1860, Prestwich); Sussex (Smith 1807; Offord 1934, St. Leonards); Warwickshire (Darlaston 1913, Birmingham); Wiltshire (Anon. 1948, W. Lavington); Yorkshire (Schroeder 1935, no. locality).

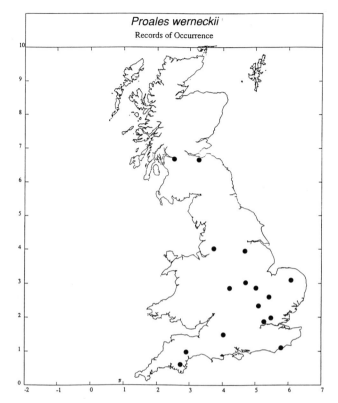

Fig. 6.4. *Proales werneckii.* Records of occurrence in Britain.

(b) Rest of Europe and Middle East **Austria** (Tyrol, Kitzbühel [type locality], see Balbiani 1878; Budde 1925)
Belgium (Morren 1839)
Denmark (Vahl 1787; Magnus 1876; Christensen 1969)
France (Bordeaux, leg. Cornu, see Balbiani 1878; Gabriel 1922*a*; Christensen 1969)
Germany (Berlin, see Magnus 1876; Zerbst, leg. Ehrenberg, see Balbiani 1878)
Iraq (Islam 1983)
Italy (Forti 1905; Del Grosso 1985)
Netherlands (Simons 1977)
Poland (leg. Wimmer, see Balbiani 1878)
Russia (Rothert 1896*a*; Hollowday, personal communication)
Switzerland (Christensen 1969)

(c) North America (Hollowday, personal communication; Wolle 1882, 1887; Nichols 1895)

(d) Australia (Christensen 1987*a*; New South Wales, Whitelegge 1889: 683; New South Wales, MacQuarie Marshes, F. Crome (unpublished; see Koste and Shiel 1990); Melbourne, Entwisle 1988)

Acknowledgements

I am grateful to several people for their generous help in preparing this account. In particular, I thank Eric Hollowday (Aylesbury) for much background information, for help with references, assistance with nomenclature, helpful discussion, and for reading the manuscript. I am also especially grateful to Professor T. Christensen (Copenhagen) for the loan of transparencies with permission for publication and for general information with regard to *Proales* and *Vaucheria*. Thanks are also due to Brian Stannard, Thomas Laessøe, and Suzy Dickersen, all of the Royal Botanic Gardens, Kew, and to Alick Henrici (London) for assistance with translation of various manuscripts, and to Georgina Godwin, International Mycological Institute, for preparation of transparencies.

References

Anon. (1948). *Dauntsey fauna list. 1931–48*, (2nd edn), pp. viii–100. Dauntsey School, West Lavington, Wiltshire.

Balbiani, G. (1878). Observations sur le Notommate de Werneck et sur son parasitisme dans le tube des Vauchéries. *Annales des Sciences Naturelles, Zoologie*, series 6, **7**, 1–40.

Balbiani, G. (1879). Observations on *Notommata werneckii*, and its parasitism in the tubes of *Vaucheria*. *Journal of the Royal Microscopical Society*, **2**, 530–44.

Benkö, G. (1882). Vaucheria-gubacsok. *Magyar Növénytani Lapok*, **6**, 146–52.

Bennett, A.W. (1889). *Vaucheria*-galls. *Annals of Botany*, **4**, 172–4.

Bennett, A.W. (1890). *Vaucheria*-galls. *Annals of Botany*, **4**, 300–1.

Blum, J.L. (1972). Vaucheriaceae. *North American Flora*, **II**(8), 1–64.

Brain, J.L. (1894). An inhabitant of Vaucheria. *Science Gossip* (ns), **1**, 201–2.

Budde, E. (1925). Die Parasitischen Rädertiere mit besonderer berücksichtigung der in der Umgegend von minden I.W. beobachten arten. *Zeitschrift für Morphologie und Oekologie der Tiere*, **3** (5), 706–84.

Buhr, H. (1965). *Bestimmungstabellen der Gallen (Zoo- und Phytocecidien) an Pflanzen Mittel- und Nordeuropas*, Vol. 2. Gustav Fischer, Jena.

Christensen, T. (1969). *Vaucheria* collections from Vaucher's Region. *Biologiske Skrifter*, **16** (4), 1–35.

Christensen, T. (1987*a*). Some collections of *Vaucheria* (Tribophyceae) from Southeastern Australia. *Australian Journal of Botany*, **35**, 617–29.

Christensen, T. (1987*b*). *Seaweeds of the British Isles*, vol. **4,** *Tribophyceae (Xanthophyceae)*. British Museum (Natural History), London.

Darlaston, H.W.H. (1913). Microscopic aquatic fauna. In *A handbook for Birmingham and the neighbourhood*, (ed. G.A. Auden), pp. 533–8. British Association for the Advancement of Science, Birmingham.

Debray, F. (1890). Sur *Notommata werneckii* Ehrb., parasite des Vauchéries. *Bulletin Scientifique de la France et de la Belgique*, **22**, 222–42.

Del Grosso, F. (1985). Sulla Formazione di galle in 'Dichotomosiphon tuberosus' (A. Br.) Ernst ('Chlorophyceae', 'Dichotomosiphonales'), Prodotte dal Rotifero 'Proales werneckii'. *Giornale Botanico Italiano*, **119**, suppl. 2, 47–8.

Docters van Leeuwen, W.M. (1982). *Gallenboek*. Thieme, Zutphen.

Donner, J. (1966). *Rotifers*. Warne, London.

Ehrenberg, C.G. (1834). Erkenntnis grosser Organization in der Richtung des Kleinsten Raumes. *Abhandlungen der Königlichen Akademie der Wissenschaften in Berlin*, **3**, 145–336.

Entwisle, T.J. (1988). A monograph of *Vaucheria* (Vaucheriaceae, Chrysophyta) in South-eastern Australia. *Australian Systematic Botany*, **1**, 1–77.

Fitter, R. and Manuel, R. (1986). *Collins field guide to freshwater life*. Collins, London.

Forti, A. (1905). I cecidi di Notommata Wernecki Ehr. in Italia. *Atti del Istituto Veneto di Scienze, Letteri ed Arti*, **64**, 1751–2.

Gabriel, C. (1922*a*). Cécidie de *Vaucheria aversa* produites par *Notommata werneckii*. *Compte Rendu des Séances de la Société de Biologie*, **86**, 453–5.

Gabriel, C. (1922*b*). La ponte de *Notommata werneckii* dans les galles de *Vaucheria aversa*. *Compte Rendu des Séances de la Société de Biologie*, **86**, 696–8.

Galliford, A.L. (1961–63). How to begin the study of rotifers. *Country-side*, n.s., **19**, 150–6, 188–94, 246–50, 291–4, 334–9, 382–8, 424–30.

Gray, E.A. (1953). The ecology of rotifers in Cambridgeshire. *Journal of Animal Ecology*, **22** (2), 208–16.

Habeeb, H. (1965). *Debsalga gigasporangia* new genus and species of Vaucheriaceae. *Leaflet of Acadian Biology*, **38**, 1–2.

Harmer, S.F. and Shipley, A.E. (ed.) (1901). *Worms, rotifers and polyzoa*. Cambridge Natural History, Macmillan, London.

Harring, H.K. and Myers, F.J. (1924). A review of the Notommatid rotifers exclusive of the Dicranophorinae. Rotifer fauna of Wisconsin—II. *Transactions of the Wisconsin Academy of Science, Arts and Letters*, **21**, 426–8.

Hassall, A.H. (1845). *A history of the British freshwater algae*. Highley, London.

Hibberd, D.J. (1981). Notes on the taxonomy and nomenclature of the algal classes Eustigmatophyceae and Tribophyceae (Synonym Xanthophyceae). *Botanical Journal of the Linnean Society*, **82**, 93–119.

Hofmeister, W. (1867). Die Lehre von der Pflanzenzelle. *Handbuch der Physiologischen Botanik*, Vol. 1. Engelmann, Leipzig.

Hollowday, E. (1993). *Cephalodella edax* sp. nov. A rotifer parasite in the motile colonial alga *Uroglena volvox* Ehrenberg. *Hydrobiologia*, **255/256**, 445–8.

Horwood, A.R. (1907). Zoolog. In *A guide to Leicester & district*, (2nd edn) (ed. G.C. Nuttall), pp. 346–62. British Association for the Advancement of Science, Leicester.

Houard, C. (1908). *Les Zoocécidies des Plantes d'Europe et du Basin de la Méditerranée*, Vol. 1. Hermann, Paris.

Hudson, C.T. and Gosse, P.H. (1886). *The rotifera or wheel animalcules*, Vol. 2, p. 134. Longmans, Green & Co., London.

Hudson, C.T. and Gosse, P.H. (1889). *The rotifera or wheel animalcules*, Supplement, pp. 23–4, pl. 32. Longmans, Green & Co., London.

Hussey, C.G. (1981). *A checklist and bibliography of records of rotifera (Rotatoria) in Britain*. British Museum (Natural History), London. (Microfiche.)

Hyman, L.H. (1951). *The invertebrates. Acanthocephala, Aschelminthes, and Entoprocta. The pseudocoelomate Bilateria*, Vol. 3. McGraw-Hill, New York.

Islam, A.K.M.N. (1983). Gall-formation in Vaucheria spp. by parasitic rotatorian members. *Bangladesh Journal of Botany*, **12,** 87–9.

Karling, J.S. (1977). *Chytridiomycetarum Iconographia*. Cramer, Vaduz.

Koste, W. (1978). *Rotatoria*, 2 Vols. Borntraeger, Berlin and Stuttgart.

Koste, W. and Shiel, R.J. (1990). Rotifera from Australian inland waters. VI. Proalidae, Lindidae. *Transactions of the Royal Society of South Australia*, **114,** 129–43.

Kützing, F.T. (1856). *Tabulae Phycologicae*, **6,** 22, pl. 63 fig. III.

Lister, A. (1884). On the parasitism of rotifers in cysts on Vaucheria. *Proceedings of the Essex Field Club*, **3,** xlv–xlviii.

Lowe, E.E., Mayes, W.E., Wagstaffe, R., and Taylor, S.O. (1933). The zoology of Leicestershire. In *A scientific survey of Leicester and district*, (ed. P.W. Bryan), pp. 33–40. British Association for the Advancement of Science, London.

Lyngbye, H.C. (1819). *Tentamen Hydrophytologiae Danicae*. Hafniae, Copenhagen.

Magnus, P. (1876). Ueber die Gallen, die ein Raederthierchen Notommata Werneckii Ehrenb. an Vaucheria-Faeden erzeugt. *Verhandlungen des Botanischen Vereins der Provinz Brandenberg*, **18,** 125–7.

Merola, A. (1956). Le galle nelle Alghe. Parte I: Storia della cecidogenesi nelle alghe. *Annali di Botanica*, **25,** 260–81.

Meyer, J. (1987). *Plant galls and gall inducers*. Borntraeger, Berlin and Stuttgart.

Moesz, G. (1938). Magyarorszáy gubacsai. (Gallen ungarn). *Botanikai közlemények*, **35,** 97–206.

Morren, C. (1839). De l'existence des infusoires dans les plantes. *Bulletin de l'Académie Royal de Bruxelles*, **6,** 298–302.

Morren, C. (1841). On the existence of infusoria in plants. *Annals and Magazine of Natural History*, **6,** 344–6.

Murray, J. (1906). Some rotifera of the Forth with description of a new species. *Annals of Scottish Natural History*, **58,** 88–93.

Murray, J. (1908). Scottish rotifers collected by the Lake Survey (Supplement). *Transactions of the Royal Society of Edinburgh*, **46,** 189–201.

Narramore, W. (1890). Vaucheria and a parasitic rotiferon. *Journal of the Liverpool Microscopical Society*, **1 (3),** 61–76.

Nichols, M.A. (1895). Abnormal fruiting of Vaucheria. *Botanical Gazette*, **20,** 269–70.

Offord, J.M. (1934). Presidential address. *Journal of the Quekett Microscopical Club*, series 3, **1,** 37–51.

Oliver, D. (1860). Note upon the occurrence of a rotiferon in Vaucheria. *Transactions of the Tyneside Naturalists Field Club*, **4,** 263–5.

Palmer, R. (1934). Rotifers (Wheel Animalcules). In *Natural history of the Hitchin region*, (ed. R.L. Hine), pp. 85–8.

116 *B.M. Spooner*

Pontin, R.M. (1978). *A key to British freshwater planktonic rotifera.* Freshwater Biological Association, Ambleside.

Rieth, A. (1980). *Süsswasserflora von Mitteleuropa,* Vol. 4, *Xanthophyceae,* part 2. Gustav Fischer, Jena.

Roth, A.G. (1800). *Catalecta botanica,* Fasc. II, p. 194. Lipsiae (Leipzig).

Roth, A.G. (1806). *Catalecta botanica.* Fasc. III, pp. 183–4. Lipsiae (Leipzig).

Rothert, W. (1896a). Über die Gallen der Rotatorie Notommata werneckii auf Vaucheria walzi n.sp. *Jahrbuch für wissenschaftliche Botanik,* **29,** 525–94.

Rothert, W. (1896b). Zur Kenntnis der in *Vaucheria*—Arten parasitirenden Rotatorie *Notommata wernecki. Zoologische Jahrbücher,* **9,** 673–713.

Rothert, W. (1906). *Vaucheria walzi* n. sp. *Nuova Notarisia,* **7,** 81.

Rousselet, C.F. (1897). On the male of *Proales wernecki. Journal of the Quekett Microscopical Club,* series 2, **6,** 415–18.

Ruttner-Kolisko, A. (1946). Über das Auftreten unbefruchteter Dauereier bein *Keratella quadrata. Hydrobiologia,* **1,** 425–68.

Ruttner-Kolisko, A. (1974). Plankton rotifers: biology and taxonomy. *Die Binnengewäffer,* **26** (1), Supplement, 1–146.

Schroeder, W.C. (1935). From a microscopist's notebook. *Naturalist, Hull,* **1935,** 2–3.

Simons, J. (1977). De Nederlandse *Vaucheria*-soorten. *Wetenschappelijke mededelingen van de koninklijke nederlandse natuurhistorische vereniging,* **120,** 1–32

Smith, J.E. (1807). *English Botany,* **25,** pl. 1765. R. Taylor, London.

Stevens, J. (1907). Rotifera of Exeter district. *Proceedings of the College Field Club and Natural History Society, Exeter,* **1,** 30–52.

Stevens, J. (1912). Some of the rotifera of Devon. *Report and Transactions Devonshire Association for the Advancement of Science,* series 3, **4,** 681–91.

Swanton, E.W. (1912). *British plant-galls. A classified textbook of cecidology.* Methuen, London.

Thompson, P.G. (1892). Notes on the parasitic tendency of rotifers of the genus *Proales,* with an account of a new species. *Science Gossip,* **28,** 219–21.

Trotter, A. (1901). Studi cecidologici I. La cecidogenesi nelle alghe. *Nuova Notarisia,* **12,** 7–24.

Unger, F. (1828). Sur les Métamorphoses et le Mouvement des corps reproducteurs de diverses Conferves, et particulièrement de l'Ectosperma clavata de Vaucher. *Annales des Sciences Naturelles,* **13,** 428–44.

Vahl. M. (ed.) (1787). *Icones Plantarum Flora Danicae,* **6,** fasc. 16.

Vaucher, J.P.E. (1803). *Histoire des conferves d'eau douce.* Paschoud, Genève.

Ward, H.B. and Whipple, G.C. (1918). *Freshwater biology.* Wiley, New York.

Weber, W. (1960). Rädertiergallen an der Schlauchalge Vaucheria. *Mikrokosmos,* **49,** 97–102.

Weidner, H. (1952). *Proales wernecki,* ein in *Vaucheria* parasitierendes Rotator. *Nachrichten des Naturwissenschaftlichen Museums der Aschaffenburg,* **35,** 39–47.

Western, G. (1894). An inhabitant of Vaucheria. *Science Gossip* (ns), **1,** 233.

Whitelegge, T. (1889). List of the marine and freshwater invertebrate fauna of Port Jackson and the neighbourhood. *Journal and Proceedings of the Royal Society of New South Wales,* **23,** 163–323.

Wolle, F. (1882). Rotifer nests. *American Monthly Microscopical Journal,* **3,** 101–2.

Wolle, F. (1887). *Fresh-water algae of the United States (exclusive of the Diatomaceae)*. Comenius Press, Bethlehem.

Wollny, R. (1877). Ueber die Gallen an Vaucheria. *Hedwigia*, **16,** 163–5.

Wollny, R. (1878*a*). Weitere beobachtungen über die Entwickelung der Notommata in einer Aussackung der Vaucheria. *Hedwigia*, **17,** 5–6.

Wollny, R. (1878*b*). Beitrag zur Kenntniss der Vaucheria-Gallen. *Hedwigia*, **17,** 97–8.

Wood, S.M. (1955). Rotifera. *Report of the Oundle School Natural History Society*, **1955,** 31–5.

7. Structure and systematics of eriophyid mites (Acari: Eriophyoidea) and their relationship to host plants

JAN BOCZEK* and D.A. GRIFFITHS†

*Agricultural University of Warsaw, 02–766 Warsaw, Poland † Bunting Biological Control Ltd, Great Horkesley, Colchester, Essex, UK.

Abstract

The eriophyid fauna is represented by 2400 species contained within six families, all of which are phytophagous and said to be monophagous to one species of plant. They possess a morphology unlike all other mites, with a minute cylindrical, worm-like or spindle-like body and two pairs of legs. Their habitat and effect on the plant ranges form leaf surface dwellers, with no detectable physical effect, through causers of leaf russeting, curling, edge-rolling, bud swelling, and brooming to producers of erinea and other complex galls. This paper discusses these phenomena and considers family ecology in relation to the gymnosperm and angiosperm hosts.

Introduction

Approximately 2400 eriophyid species belonging to 224 genera have been described (Boczek *et al.* 1989), probably representing not more than 5–10 per cent of the world fauna of these mites. This indicates the need for intensive work, both in surveying and in taxonomic studies on eriophyids. The following discussion will, therefore, necessarily be based on rather fragmentary data.

Eriophyid mites probably originated from eight-legged tenuipalpids which lost their hind legs (Jeppson *et al.* 1975). They are exclusively plant-feeding arthropods infesting both gymnosperms and angiosperms. Some 127 species have been described from 90 species of gymnosperms (Boczek and Shevtchenko, in press) and all others live on angiosperms. The majority of the species are monophagous, attacking one species of

Plant Galls (ed. Michèle A. J. Williams), Systematics Association Special Volume No. 49, pp. 119–29. Clarendon Press, Oxford, 1994. © The Systematics Association, 1994.

plant. Only a few are known to be or thought to be polyphagous although the specificity of such species should be studied further. At least some of them are probably sibling species specific to a single plant species.

Minute body size, worm-like or spindle-like shape, the presence of two instead of four pairs of legs in all stages, and a very reduced chaetotaxy differentiate these mites from other Acari. The presence of spermatophores which are left on the leaf surface, a smooth musculature, parenchyma (large mobile cells in the body cavity), and a peristaltic movement, coupled with the absence of excretion and of circulatory and respiratory organs, indicate a regressive evolution of the group, having features in common with the Tardigrada and Platyhelminthes rather than the Arthropoda (Silvere and Stein-Margolina 1976).

Eriophyid mites are obligatorily phytophagous, adapted to form plant galls or to live freely and they have few natural enemies; principally other mites. Dimorphism of females which allows overwintering on their host plants, suckers on palps, and a specialized extremity to the limbs (an empodium and an adhesive sucker-like system) enable them to stay on leaves and exploit their environment. They have three salivary glands, three segmented palpi, and stylet mouthparts for piercing the leaf surface. During the insertion of the stylets into plant tissues a telescopic movement of the pedipalpi takes place preceded by arching of the opisthosoma which, by using anal suckers, anchors the mite to the leaf surface. This is followed by rhythmic pulsations of the cheliceral pump as the anterior part of the opisthosoma becomes depressed on the leaf surface (Nuzzaci 1976*a,b*; 1979*a,b*).

Eriophyids as gall causers

All plant parts except roots are inhabited by eriophyids. Many species cause no detectable alterations of or damage to their host plants. However, others do. Eriophyids with spatulate or shovel-like shaped appendages, for example, have been suspected of mining on mango leaves. Another type of injury is russeting which occurs on leaves and on fruits. Heavily infested leaves shrivel and become bronzed. Digestive enzymes, are the main cause of russeting.

Growth modifications are initiated exclusively on embryonic plant tissues, the mite causing minute punctures which, as the plant grows, bring about discolouration, russeting, or various types of tissue deformation. It is the saliva of gall-forming species which causes the distortion of leaf tissues. Such species are usually worm-like in shape, having a dorsal shield lacking a lobe and a microtuberculate body with undifferentiated microtuberculate rings on the opisthosoma. In comparison, the body of free-living species is usually spindle-like, more or less robust, possessing a shield with the lobe overhanging the rostrum. The body rings are differentiated into a few broad, often smooth tergites dorsally, with more numerous, microtuberculate sternites ventrally. Their legs are usually longer and their chelicerae and rostrum larger. These different characters allow these mites to live on the leaf surface, suggesting a progressive, secondary evolution, with a reversion to ancestry (Shevtchenko 1982).

Different species living on the same host plant can produce strikingly different effects. For example, on pear trees, *Pyrus*, *Epitrimerus pyri* (Nal.) causes russeting and shrivelling whilst *Eriophyes pyri*, Pgst. causes the leaf to blister. *Phytoptus tiliae* Pgst. causes nail galls on linden (*Tilia*) leaves and *Vasates tiliaevagrans* Boczek causes erinea along leaf ribs. *Aceria tulipae* K. produces kernel red streak on corn, curls wheat leaves, dries onion leaves, and in store causes premature drying of garlic cloves. *Cecidophyopsis ribis* (Westwood) causes the well-known big bud galls of currants (*Ribes* spp.); this leads to serious losses, not uncommonly affecting 90 per cent of the fruit yield. Two other biotypes of this species, unrecognizable morphologically, attack white and red currants causing similar but less serious damage.

Mites of Ashieldophyidae, Nalepellidae and Eriophyidae have short oral stylets. These are principally bud mites, gall mites, and rust mites, which cause leaf rolling or erinea. These mites can only penetrate plant tissue to a depth of 10–40 μm. Longer stylets (40–70 μm) are possessed by some Nalepellidae and all Diptilomiopidae although usually they do little damage to their host plants. All these free-living suck out cell contents and the saliva they inject promotes deformation of the plant tissue giving rise to various types of growth appendages such as erinea, leaf edge rolling, brooming, bud enlargements, and other galls. Sometimes different gall types can be caused by the same species, for example, leaf edge roll and surface erinea by *Eriophyes goniothorax* (Nal.)

Erinea take the form of fine hair-like outgrowths which when aggregated appear as hairy masses sometimes covering much of the leaf surface. Felt-like erineum pads are the work of *Aceria erineus* (Nal.) (Eriophyidae) on walnut leaves and of *Colomerus vitis* (Pgst.) (Eriophyidae) on grapevine leaves and there are many other examples.

Edge rolling is a very common deformity of leaves and mite colonies develop inside these rolls. Leaves of *Fraxinus* are often rolled by *Aceria fraxinivorus* (Nal.) and leaves of *Sambucus nigra* by *Epitrimerus trilobus* (Nal.).

Brooming appears as twig elongation or bud proliferation accompanied by either stunted leaves or denuding and internode shortening. Twig and flower clusters may show this effect. In this case saliva inhibits growth, causing whole branches or flower clusters to become deformed. Such deformations are often observed on willows, caused by *Eriophyes triradiatus* Nal. Other species also living on willows cause other deformations:

Phyllocoptes magnirostris Nal. causes edge rolling and *Aceria salicinus* (Nal.) induces red galls on leaves.

Bud deformation can be serious, often causing death of the bud in spring or early summer. Mite feeding causes the interior parts of the bud to swell and become succulent. High populations of thousands of mites develop. Then, when the brood leaves, the bud dies. At least two species of such mites are serious pests in Europe: currants (*Ribes* spp.) are attacked by *Cecidophyopsis ribis* (Wes.) and hazelnut (*Corylus avellana* L.) by *Phytocoptella avellanae* (Nal.)

Galls develop from epidermal cells that are distorted through being injected with growth regulators. Many species of eriophyids produce galls on various plant parts. The shape of the galls is specific for the species of mite. Typically, overwintered females feed at the beginning of May on the undersurface of young leaves and usually they remain there for a few days. As a result, the tissue around the female becomes thicker, forming a roll and the female finds herself in a loculus. Later, the female is enclosed by the purse-like structure and she then starts to lay eggs. In *Eriophyes pyri* (Pgst.) the process is different. The saliva of the female causes the disruption of parenchymal cells, which in turn causes pressure on the epidermal cells, leading to the eruption of the epidermal layer. The female enters through such an opening to lay her eggs. The whole process of gall formation is quite rapid, taking approximately 2 weeks.

The gall provides the mites with food and ensures protection against natural enemies. Gall tissues promote a sink of assimilates to the gall from the surrounding tissues which is of specific benefit to the mite. The nutritional tissue of leaf galls of *Tilia cordata* induced by?*Eriophyes tiliae* contains cells which attain a higher degree of ploidy, $4n$'s and $8n$'s, presumably due to endomitosis (Wcislo 1977).

Leaf galls are usually exploited by the mites for approximately 2 months. Then in the second part of July they turn brown and the mites leave to look for overwintering sites. Infested leaves have lowered photosynthesis and drop earlier. Species which are free-living on leaves also affect their host plants, changing their biochemical composition and their physiological processes.

Small numbers of mites per leaf possibly stimulate gaseous exchange. Larger populations, however, inhibit photosynthesis and photorespiration and increase transpiration. Plum (*Prunus domestica* L.) leaves heavily infested by *Aculus fockeui* (Nal. & Trt.) showed a photosynthesis decrease of 70–80 per cent. Photorespiration decreased by 51 per cent while, at the same time, an increase in the compensation point of CO_2 of 69 per cent was observed. Such leaves had a 30 per cent higher dry matter content and showed increases in nitrogen (43 per cent), calcium (47 per cent), and potassium (73 per cent), while sodium and chlorophyll also increased. At higher infestation levels water transpiration also increased

Table 7.1. Eriophyid mites as vectors of plant pathogens (Boczek *et al.* 1978; Oldfield 1970; Razviazkina *et al.* 1969)

Eriophyid species	Pathogen	Name	Plant	Region
Aceria tulipae K.	Virus	Wheat streak mosaic	Grasses	North America, Europe, Asia
	Virus	Wheat spot	Grasses	Canada
	Virus	Onion mosaic	Onion	Russia
	Virus (?)	Kerner red streak	Corn	North America
Abacarus hystrix (Nal.)	Virus	Ryegrass mosaic	Grasses	North America, Europe
Eriophyes inaequalis (W. & O.)	Virus	Cherry mottle leaf	Cherry	USA
Vasates fockeui (Nal. & Tr.)	Virus	Latent virus	Plum	Europe
Calacarus citrifolii K.	Virus (?)	Concentric ring	Citrus	Africa
Cecidophyopsis ribis (Wes.)	Mycoplasma	Reversion	Black currant	North America, Europe
Colomerus vitis (Pgst.)	Virus	Panaschure	Grapevine	North America
Aceria fici Essig	Virus	Mosaic	Fig	USA
Vasates insidiosus (Keifer & Wilson)	Virus	Mosaic	Peach	USA
?	?	Cadang Cadang	Coconut palm	Philippines
Aceria cajani Chan.	Virus	Sterility	Pigeon pea	Asia
Aceria mangiferae (Sa.)	Virus (?)	Malformation	Mango	Asia
Phyllocoptes fructiphilus K. & *P. slinkardiensis* K.	Virus	Rosette	Rose	USA
Eriophyes tristriatus (Nal.)	Bacteria	*Xanthomonas juglandis*	Walnut	Europe
Phyllocoptes gracilis (Nal.)	Bacteria		Raspberry	Europe

? = suspected

by 53 per cent (Zawadzki 1975). Plants of *Geranium pratense* L. infested with *Epitrimerus geranii* (Liro) and *Geranium molle* L. infested with *Aceria geranii* (Can.) showed lower levels of carbohydrates, soluble proteins, and reducing sugars (Tomczyk and Boczek, unpublished).

Eriophyid mites as specific plant feeders are also known to be vectors of plant viruses, mycoplasmas, and bacteria. At least 11 eriphyid species transmit viruses and one species, a pathogenic mycoplasma (Table 7.1).

1. A review of the eriophyid mite families and their relationship to their host plants

At present the superfamily Eriophyoidea is divided into five families according to the structure of the dorsal shield and its setae and the structure of the gnathosoma (Table 7.2).

(a) Pentasetacidae This, the most archaic group, have five, the largest number of setae on the dorsal shield and seem to be ancestral for all other eriophyids (Shevtchenko 1991). Only one species belonging to this family is known at present, *Pentasetacus araucariae* Schliesske (1985), from *Araucaria araucana* (Molina) K. Koch (Araucariaceae) in the Chilean Andes mountains. It causes galls on leaves and on twigs. *Araucariaceae* is one of the oldest plant families first appearing in the late Permian era. *Pentasetacus araucariae* has an asymmetric, divided empodium (featherclaw), a dorsal shield with a lobe, and microtuberculate rings. The galls which it forms are relatively large, 8–15 mm in length and up to 8 mm in height.

(b) Ashieldophyidae One species, *Ashieldophyes pennadamensis* Mohan-asundaram (1984), from India, is known. This mite possesses an indistinct shield which lacks a lobe and dorsal setae. The opisthosoma bears smooth rings, with one pair of coxal setae only and with two pairs of ventral setae missing. This mite has short legs, rostrum, and chelicerae. It is found on *Casearia tomentosa (Flacortiaceae)* as an under-surface leaf vagrant, not causing any damage.

(c) Nalepellidae Seventy-one species are known belonging to six genera. These mites have one or three shield setae with or without subdorsal setae. The body is worm-like or spindle-form. All the species live on

Table 7.2. Number of eriophyid mite species found on gymnosperm families (Davis *et al.* 1982)

Plant family	Number of species
Pinaceae	83
Cupressaceae	36
Araucariaceae	1
Taxodiaceae	4
Ephedraceae	1
Podocarpaceae	1

Table 7.3.   List of eriophyid mites inhabiting coniferous plants

Family	Genus	Species of eriophyids known	
		All	On conifers
Pentasetacidae	*Pentasetacus*	1	1
Nalepellidae	*Boczekella*	2	2
	Trisetacus	48	48
	Setoptus	6	6
	Phantacrus	1	1
	Nalepella	13	13
Eriophyidae	*Litaculus*	7	1
	Platyphytoptus	9[a]	8
	Keiferella	3	3
	Cupacarus	1	1
	Cecidophyes	7	1
	Eriophyes	490	4
	Phyllocoptes	168	9
	Arectus	1	1
	Acaricalus	10	1
	Paracalacarus	1	1
	Vasates	210	1
	Epitrimerus	97	16
	Calepitrimerus	41	2
	Proartacris	1	1
	Tegonotus	69	2
Diptilomiopidae	*Asetacus*	4	1

[a] *Platyphytoptus vitalbae* Farkas does not belong to *Platyphytoptus.*

gymnosperms, usually free-living on needles or on stems. Only *Trisetacus pini* (Nal.) causes galls, these occurring on twigs of *Pinus*. Other species cause damage to the needles, buds, flowers, and fruits. *Nalepella haarlovi* Boczek has, in recent years, caused serious damage of 4-year-old seedlings of *Picea* and *Abies* in Finland. The mite sucks the needles which turn yellow, become dry and die (Löyttyniemi 1969).

Out of 126 species of eriophyid mite found on gymnosperms the majority, (83 species) were taken from plants of *Pinaceae* (see Tables 7.2, 7.3, and 7.4) and 36 from *Cupressaceae*. *Pinidae* are widely distributed, including *Ginkgoales* and they have more species and genera than any other family in *Gymnospermae*. Approximately 690 species of conifers belonging to 57 genera are known. Eriophyid mites have been found on only 86 species belonging to 24 genera. Species representing 12 genera of eriophyids have never been recorded on angiosperms.

(d) Phytoptidae   These are mites with shields having four or only two setae situated proximally. The body is worm-like or spindle-form, the shield with or without lobe over the rostrum, often with subdorsal setae

Table 7.4. List of coniferous genera inhabited by eriophyids (Krussmann 1972)

Host plant		Species		Number of eriophyid sp. found
Family	Genus	Known*	With eriophyids	
Pinaceae	*Abies*	45	2	17
	Cedrus	4	2	2
	Larix	15	3	5
	Picea	34	7	2
	Pinus	85	26	35
	Pseudotsuga	5	2	4
	Tsuga	18	4	7
Cupressaceae	*Callitris*	14	1	1
	Chamaecyparis	7	1	1
	Cupressus	15	6	6
	Juniperus	55	11	21
	Libocedrus	5	2	2
	Thuja	6	1	6
Araucariaceae	*Araucaria*	14	1	1
Taxodiaceae	*Cunninghamia*	3	1	1
	Sequoia	1	1	1
	Taxodium	3	1	2
Taxaceae	*Taxus*	8	1	2
	Torreya	6	1	1
Ephedraceae	*Ephedra*	26	1	1
Podocarpaceae	*Podocarpus*	105	1	1

present. Some 115 species are known at present in this family belonging to nine genera. The largest genus *Phytoptus* contains 100 species whilst many genera are monotypical or represented by just a few species. These mites are either leaf vagrants not causing their host plant visible damage or they may be inducers of galls on various plant parts. They infest only angiosperms.

(e) Eriophyidae This is the largest family, containing more than 2000 species which exhibit a wide diversity of shape and structure. They have nil or two setae on the dorsal shield, the rostrum and chelicerae are short, the opisthosoma has a differentiated or undifferentiated rings, and body shapes are various. All kinds of relationships to their host plants are exhibited. Some species are vagrant on leaves without causing damage, whilst others may cause all kinds of distortions and deformations, producing galls of various shapes on different parts of the plant. Only 53 species of this family (15 genera) are recorded on conifers, always living as vagrants on needles or stems. The remainder live on angiosperms.

Only mites of this family are known as vectors of mycoplasma and viruses.

Some examples of eriophyid mites which cause serious damage are listed below: *Aculops lycopersici* (Massee) is a serious pest of tomato plants causing browning and withering of whole plants. In heavily infested plants the fruits become russeted and roughened and one half of the yield may be lost. *Phyllocoptes gracilis* (Nal.) attacks fruits and leaves of raspberries and blackberries. Affected berries remain red or green and do not ripen normally. Some berries have entirely red drupelets. *Aceria phloeocoptes* (Nal.), occurring commonly in Europe on plum and almond trees, causes permanent, irregular, 1.3–1.8 mm diameter, subspherical galls around the buds on twigs. Infested trees fail to form fruit buds and lose vigour. This often results in the trees dying early.

(f) Diptilomiopidae   This family includes distinctive eriophyids having a large rostrum and strong, abruptly curved chelicerae. The shield may or may not possess a lobe and setae. The body is spindle-like and robust. Body rings are usually but not always differentiated into broader tergites and narrower sternites. All these are free-living as vagrants on leaves, only rarely causing slight discoloration or rusting of the leaves. Their chelicerae are strongly sclerotized, up to five to six times longer than those of other eriophyids. Frequently, many hundreds of these mites may occur within 1 cm^2 of their host plant leaves without causing any distinct damage.

Table 7.5.   List of eriophyid families with numbers of described genera and species

Family	Number of genera	Number of species
Pentasetacidae	1	1
Ashieldophyidae	1	1
Nalepellidae	6	71
Phytoptidae	9	115
Eriophyidae	171	2021
Diptilomiopidae	36	191
Total	224	2400

Approximately 190 species belonging to 36 genera are grouped in this family. Of these, 16 genera are monotypic. The majority of species belong to the genera *Rhyncaphytoptus* (64), *Diptacus* (26), and *Diptilomiopus* (29). Only one species has been recorded from gymnosperms.

In summarizing these families (Table 7.5) we see large differences both in the number of species described, in economic importance, and in morphological structures. A vast amount of work must be done to broaden our knowledge of the fauna of these mites across world regions. Studies of their role in plant pathogen transmission and direct damage to their host plants has barely begun. Their taxonomy remains poorly developed, not least because of an apparent lack of gross morphological differentiation. However, use of scanning electron microscopy may advance our understanding, at least at the generic level.

References

Boczek, J. and Shevtchenko, V.G. (In press) Eriophyid mites on gymnospermous plants. In *Eriophyid mites: their biology, natural enemies and control.* Elsevier Science Publishers BV, Amsterdam.

Boczek, J., Tomczyk, A., and Kropczynska, D. (1978). Injuriousness of plant feeding mites. (In Polish.) *Postepy Nauk Rolniczych, Warszawa,* **3,** 45–60.

Boczek, J., Shevtchenko, V.G., and Davis, R. (1989). *Generic key to world fauna of eriophyid mites (Acarida: Eriophyoidea).* Warsaw Agricultural University Press, Warsaw.

Davis, R., Flechtmann, C.H.W., Boczek, J., and Barke, H.E. (1982). *Catalogue of eriophyid mites (Acari: Eriophyoidea).* Warsaw Agricultural University Press, Warsaw.

Jeppson, L.R., Keifer, H.H., and Baker, E.W. (1975). *Mites injurious to economic plants,* pp. 327–614. University of California Press, Berkeley, CA.

Krüssmann, G. (1972). *Handbuch der Nadelgehölze.* P. Parey, Berlin and Hamburg.

Löyttyniemi, K. (1969). An Eriophyidae species damaging spruce seedlings in nurseries. (In Finnish.) *Silva Fennica,* **3,** 191–200.

Nuzzaci, G. (1976*a*). Contributo alla conoscenza dell anatomia degli Acari eriofidi. *Entomologica, Bari,* **12,** 21–55.

Nuzzaci, G. (1976*b*). Compartamento degli Acari eriofidi nell assunzione dell alimento. *Entomologica, Bari,* **12,** 75–80.

Nuzzaci, G. (1979*a*). Studies on structure and function of mouth parts of eriophyid mites. *Recent Advances in Acarology,* **2,** 411–15.

Nuzzaci, G. (1979*b*). Contributo alla conoscenza dello gnatosoma degli eriofidi. *Entomologica, Bari,* **15,** 73–101.

Oldfield, G.N. (1970). Mite transmission of plant viruses. *Annual Review of Entomology,* **15,** 343–60.

Razviazkina, G.M., Kapkova, E.A., and Ceremushkina, E.P. (1969). Klesci *Aceria tulipae (Eriophyoidea)*—perenosciki mosaiki luka. *Zoologiceskij Zhurnal,* **48,** 288–9.

Shevtchenko, V.G. (1982). Progresivne i regresivne preobrazovanja i ich rol w evolucii cetyrechnogich klescej (*Acariformes, Tetrapodili*). *Vestnik Leningradskovo Gosudarstvovovo Universiteta,* **9,** 13–22.

Shevtchenko, V.G. (1991). A new family of *Pentasetacidae (Acariformes: Tetrapodili)* and its role in treatment of the origin and evolution of the group. (In Russian.) *Zoologiceskij Zhurnal,* **70,** 47–53.

Silvere, A.P. and Stein-Margolina, V. (1976). *Tetrapodili—cetyrechnogije klesci. Elektronomikroskopiceskaja anatomia, problemy evoluci i wzaimoodnosenja s vosbuditelam boleznej rastenij*. Valgus, Tallin, Institut. Experimentalnoj Biologii Akademi Nauk ESSR.

Wcislo, H. (1977). Observations on leaf galls of *Tilia cordata* Mill. induced by *Eriophyes tiliae*. *Acta Biologica Cracoviensia, Botanica*, **20,** 147–52.

Zawadzki, W. (1975). The preliminary observations on injuriousness of *Aculus fockeui* (Nal. & Trt.). (In Polish.) *Zeszyty Problemowe Postepow Nauk Rolniczych, Warszawa*, **71,** 157–66.

8. Thrips and gall induction: a search for patterns

LAURENCE A. MOUND

The Natural History Museum, London, UK

Abstract

Thrips suck the contents of individual cells one at a time, and in some plants the surrounding cells react by redeveloping meristematic characteristics leading to gall production. Plant reactions vary in their degree of complexity, from simple leaf curling and distortion to complex multicellular structures of regular form. These galls, however, are wound responses; cell division ceases if the thrips are removed. This ability, to produce new cells and, hence, galls in response to feeding by thrips on young tissue, has arisen in a wide range of unrelated plants. The thrips which induce these reactions in plants come from three widely separated phyletic lineages in the major subfamily, Phlaeothripinae, although most occur in a range of genera in the leaf-feeding *Liothrips* lineage. The ecological and behavioural relationships between thrips within any one gall, both intra- and interspecific, have not been well-studied. Many leaf-roll galls are probably induced by the feeding activity of (more than one) conspecific individuals; on some plants in India, gall induction may involve feeding by two or more thrips species, while in Australia, some galls on *Acacia* phyllodes are induced by single individuals who then actively protect their investment.

Introduction

Thysanoptera, the insects we call thrips, are largely tropical and the galls they induce are almost entirely on tropical or southern hemisphere plants. Most publications about Thysanoptera have been written from temperate regions. This asymmetry causes problems, with observations reported at second-hand, plant and insect species misidentified, and reported associations repeated through the literature without further confirmation. Consequently, this paper deals only with those Phlaeothripine thrips for which there is evidence or good reason to believe that they actually

Plant Galls (ed. Michèle A. J. Williams), Systematics Association Special Volume No. 49, pp. 131–49. Clarendon Press, Oxford, 1994. © The Systematics Association, 1994.

induce plant galls. Excluded from consideration are the many species that are found in thrips gall communities as inquilines, usurpers, or merely tourists. In this way it is hoped to distinguish any biological 'signal' concerning insect/plant evolutionary relationships from the undoubted 'noise' that surrounds the subject. Conversely, Ananthakrishnan and Raman (1989) summarize all the published data concerning thrips found within plant deformations, together with their own original observations, and their extensive bibliography of 148 references is not repeated here. Plant names quoted here have been checked against Mabberley (1990), and that author's adoption of Cronquist's systematic arrangement of angiosperm families is also followed.

Some aspects of thrips biology

When considering the process of gall induction by thrips three aspects of their biology are particularly relevant: feeding, egg-laying and life cycle, and behaviour. All Thysanoptera have only one mandible—the right one never develops beyond the embryo. Larvae and adults all have a pair of maxillary stylets which are closely joined to each other by a tongue and groove mechanism when protruded from the mouth cone. This leaves a single central channel for food and saliva. They feed by using the single mandible to punch a hole through which the maxillary stylets are then introduced. The contents of plant cells are sucked out individually, just as certain species of thrips suck out the contents of pollen grains one at a time (Chisholm and Lewis 1984). (Spore feeding thrips have wider feeding stylets allowing them to take in whole fungal spores, these then being crushed by a proventricular mill.) The stylets are withdrawn and reintroduced into a new cell as each cell is sucked dry. It is this feeding activity, the killing of individual cells one at a time, which induces responses in the surrounding cells of some plants and gives rise to galls.

Most gall-inducing thrips are members of the suborder Tubulifera, family Phlaeothripidae, and deposit their eggs, suitably glued, onto their feeding substrate. Oviposition is not a stimulus for gall formation and it usually does not occur until after a gall is well-formed. When an egg hatches it gives rise to an actively feeding larva. There are two feeding larval stages, followed by two or three non-feeding pupal stages, then the adult stage (Mound and Heming 1991). Adults typically have wings, but wingless adults are common. The various pupal stages are almost immobile, but larvae are sufficiently active to move from leaf to leaf, although many probably do not do so. Adults, even wingless adults, can move between plants, and some adults move very great distances on air currents more or less passively (Mound 1983).

Thrips are markedly thigmotactic, that is, they are in the habit of crawling into dark corners with the maximum of their body surface in contact with the surroundings. This seems to be an essential component of the biology of Thysanoptera and results in thrips being found in many unlikely situations—inside polystyrene blocks, inside smoke detectors, and inside packaged tampons and surgical dressings straight from the manufacturers. Given a small cavity, such as a gall, a thrips will crawl into it.

A second important aspect of thrips behaviour that is still little studied concerns competition for resources. Recent studies have indicated that, in some species, males compete actively for females (Terry 1991). Moreover, when a food resource is sufficiently important, such as a good patch of fungus or a developing gall, then competition can become lethal. Male/male competition probably occurs in many fungus-feeding thrips (Crespi 1988), but female/female as well as male/male fighting is now known in several Australian gall thrips (Crespi 1992*a*). Unfortunately, we know almost nothing about the individual behaviour of the large number of oriental gall-inducing thrips.

Thrips taxa involved with galls

Almost 5000 species of Thysanoptera are known. Half of these have a saw-like ovipositor which is used to insert the eggs into plant tissue, and these species are placed in the suborder Terebrantia. These are almost all plant-feeding, although the group probably evolved from detritus-feeding ancestors (Mound *et al.* 1980). Seven terebrantiate families are generally recognized, but approximately 80 per cent of the species belong in the Thripidae. In this large family two subfamilies are recognized, the Panchaetothripinae and the Thripinae. The first includes approximately 110 species, often feeding on the mature leaves of plants. The Thripinae includes approximately 1500 species, including the typical flower thrips as well as many species on grass leaves and florets and many which feed on the buds and young leaves of dicotyledonous herbs and trees. Several species of Thripinae have been reported from or even as causing plant galls, but only three species from India seem to form sufficiently regular plant deformations to be considered as primary gall formers (*Amphithrips argutus* Annthakrishnan, *Aneurothrips priesneri* Bhatti and *Aneurothrips punctipennis* Karny) (Ananthakrishnan and Raman 1989).

The second suborder, the Tubulifera, includes a single family of approximately 3000 species, the Phlaeothripidae, and these have a chute-like ovipositor that deposits the eggs superficially on the feeding substrate. Two subfamilies are recognized: Idolothripinae with approximately 670 species which feed on fungal spores, and Phlaeothripinae with the

Fig. 8.1. *Kladothrips rugosus* pouch gall on *Acacia pendula* phyllode (herbarium specimen).

Fig. 8.2. *Oncothrips antennatus* pouch gall on *Acacia aneura* phyllode.

Fig. 8.3. *Kladothrips acaciae* pouch gall on *Acacia harpophylla* phyllode.

Fig. 8.4. *Hoplandrothrips coffeae* rolled leaf gall on *Coffea robusta*.

remaining 2350 species, of which approximately half feed on fungal hyphae on dead wood and in leaf litter. All of the gall-inducing species are Phlaeothripines and this subfamily also includes genera of inquiline or usurper species in galls (for example, *Androthrips* in India and *Koptothrips* in Australia), as well as the major flower-living genus *Haplothrips* and various smaller groups of species that are predators (Palmer and Mound 1991) or that feed on mosses (Mound 1990).

Unfortunately, the classification of the Phlaeothripinae is weak at all levels, specific, generic, and tribal, but three major lineages are usually recognized, the *Haplothrips, Liothrips,* and *Phlaeothrips* lineages. The primary feeding substrates of these three lineages are flowers, green leaves, and fungal hyphae, respectively.

In summary and in round numbers, half of the Thysanoptera are Thripidae and feed on plants but do not induce galls; the other half are Phlaeothripidae of which only half the species are plant-feeding but some of which induce galls. Table 8.1 lists the 57 genera of Phlaeothripinae known to include gall-inducing species and indicates the lineages to which they can be assigned. The leaf-feeding *Liothrips* lineage includes the largest number of gall-inducing species and *Liothrips* itself is one of the largest genera of Thysanoptera. The detailed taxonomy of this genus is currently inadequate to support any investigation at species group level, but gall induction is associated with widely separated species from south America, Africa, and Asia. Several of the monobasic genera listed under the *Liothrips* lineage (Table 8.1) are closely related to *Liothrips*, but three other subgroups are more distantly related—*Eugynothrips, Teuchothrips,* and *Gynaikothrips/Gigantothrips*. Within the *Haplothrips* lineage gall induction involves primarily the oriental genus *Mesothrips* and its presumed Pacific derivative *Euoplothrips*. Of greater interest is the occurrence of leaf feeding and gall induction in the primarily fungus-feeding *Phlaeothrips* lineage. Most *Hoplandrothrips* species live under bark or on dead branches, but one species group from Africa induces leaf galls (see below). The other genera listed under this lineage include a group from Australia and another group from the Neotropics, but the systematic relationships of these genera need further study.

Systematics of plants with thrips-induced galls

The term 'gall' is often used in the thrips literature for almost any plant deformation. Many thrips, by killing cells in young leaves, cause such leaves to deform as they expand. Members of the genus *Scirtothrips* typically feed on very young leaves; since their feeding kills cells such young leaves do not expand in the normal way. They often emerge distorted and discoloured and this damage has even been described as

Table 8.1. Phlaeothripine genera with species inducing galls

Liothrips lineage Thrips genus	Gall-thrips	*Haplothrips* lineage Thrips genus	Gall-thrips	*Phlaeothrips* lineage Thrips genus	Gall-thrips
Acaciothrips	1 (1)	*Dolichothrips*	1 (20)	*Amynothrips*	1 (1)
Aclystothrips	1 (2)	*Euoplothrips*	2 (6)	*Austrothrips?*	1 (4)
Adelphothrips	1 (2)	*Haplothrips*	1 (277)	*Choleothrips*	2 (2)
Aliothrips	1 (2)	*Mesothrips*	18 (35)	*Dixothrips?*	1 (1)
Ananthakrishnanothrips	1 (1)	*Senegathrips*	1 (1)	*Hoplandrothrips*	3 (83)
Arrhenothrips	3 (8)			*Iotatubothrips*	1 (1)
Brachythrips	1 (10)			*Myxothrips*	2 (2)
Byctothrips	1 (1)			*Moultonides*	1 (1)
Chaetokarnyia	1 (1)			*Neocecidothrips*	1 (1)
Coryphothrips	1 (2)			*Phrasterothrips*	1 (5)
Crotonothrips	7 (10)			*Sacothrips*	1 (7)
Dimorphothrips	2 (2)			*Vuilletia*	1 (1)
Eothrips	7 (16)				
Eugynothrips	11 (16)				
Gigantothrips	6 (22)				
Gynaikothrips	18 (55)				
Isotrichothrips	1 (2)				
Kladothrips	4 (4)				
Kochummania	1 (1)				
Leeuwenia	3 (15)				
Liothrips	73 (293)				
Lygothrips	1 (1)				
Mallothrips	1 (2)				
Megeugynothrips	1 (1)				
Mesicothrips	1 (2)				
Oncothrips	3 (4)				
Onychothrips	2 (2)				
Phaeothrips	1 (1)				
Phasmothrips	1 (1)				
Phorinothrips	2 (3)				
Ponticulothrips	1 (1)				
Schedothrips	1 (1)				
Sphingothrips	1 (1)				
Syringothrips	1 (1)				
Tetradothrips	1 (1)				
Teuchothrips	14 (29)				
Thilakothrips?	1 (1)				
Thlibothrips	2 (9)				
Xiphidothrips	1 (1)				
Zelotothrips	1 (1)				

Total species in genus in parentheses.
? = systematic position uncertain.

viral in origin (Mound and Palmer 1981). The highly polyphagous onion thrips, *Thrips tabaci*, causes leaf deformation when feeding on a few plants, and several other Thripines in Europe, such as *Anaphothrips* on *Euphorbia* and *Galium* species, and *Tmetothrips* on *Stellaria*, have been recorded as forming galls. This type of leaf distortion caused by some Thripidae will not be considered further here.

The characteristic plant deformations induced by certain Phlaeothripinae and commonly referred to as thrips-galls have been recorded from the species of 78 genera of plants from 47 families (Table 8.2). It must be emphasized that the database is inadequate. Most thrips-induced galls have been reported by very few authors, such as Docters van Leeuwen-Reijnvaan (1926) in Indonesia, Ananthakrishnan (1976) in India, and Mound (1971*a,b*) in Australia. Few thrips have been collected from galls in the neotropics, although the author has evidence from recent field work in Costa Rica that leaf galls induced by *Liothrips* and *Holopothrips* (including *Myxothrips* and *Phrasterothrips*) species are probably fairly common and widespread.

The list of host plant families (Table 8.2) shows little pattern of exploitation, beyond an absence of the major herbaceous groups in both monocotyledons and dicotyledons. Within the largest Cronquist subclasses, Hamamelidae, Dilleniidae, and Rosidae, the plant families listed represent most of the major plant orders. The Myrtales, Rubiales, Urticales, and the Fabales/Santalales which botanists now place together, have a higher proportion of thrips recorded as inducing galls, but these groups include some of the commonest tree genera in the tropics.

Equally, little pattern can be found if the families of host plants of particular genera of Phlaeothripines are considered (Tables 8.3–8.6). Certain plant genera have been exploited preferentially, such as *Ficus*, *Piper*, *Schlefflera*, and *Eugenia*, but thrips galls have been taken from many common mesophytic trees in the tropics. Genera such as *Acacia* represent commonly available resources in arid areas of Australia and Africa and these are attacked by thrips. Similarly, *Teuchothrips* has moved onto shrubby Compositae in Australia where these plants are such a major part of the flora. Host selection has thus evolved more along ecological lines than along phyletic lines. Subsequent speciation on a particular host plant genus has occurred in a few instances, such as *Crotonothrips* on *Memecylon* in India (Ananthakrishnan 1976) and *Kladothrips*, *Oncothrips*, and *Onychothrips* on phyllodinous *Acacia* trees in Australia (Mound 1971*a*).

Thrips-induced modifications to plant tissues

Gall induction by thrips is apparently possibly only on very young tissue. Various unrelated plants respond to the killing of cells in young tissues

Table 8.2.   Plant families with phlaeothripine galls

Plant family	Plant order	Cronquist subclass	Thrips genera
Lauraceae	Laurales	1 Magnoliidae	3
Monimiaceae	Laurales	1	1
Piperaceae	Piperales	1	2
Casuarinaceae	Casuarinales	2 Hamamelidae	1
Fagaceae	Fagales	2	1
Myricaceae	Myricales	2	1
Cecropiaceae	Urticales	2	3
Moraceae	Urticales	2	11
Urticaceae	Urticales	2	2
Amaranthaceae	Caryophyllales	3 Caryophyllidae	1
Chenopodiaceae	Caryophyllales	3	1
Capparidaceae	Capparidales	4 Dilleniidae	1
Ebenaceae	Ebenales	4	2
Sapotaceae	Ebenales	4	1
Lecythidaceae	Lecythidales	4	2
Elaeocarpaceae	Malvales	4	2
Tiliaceae	Malvales	4	2
Myrsinaceae	Primulales	4	3
Flacourtiaceae	Violales	4	1
Dipterocarpaceae	Theales	4	2
Theaceae	Theales	4	1
Araliaceae	Apiales	5 Rosidae	4
Celastraceae	Celastrales	5	2
Euphorbiaceae	Euphorbiales	5	5
Leguminosae	Fabales	5	6
Combretaceae	Myrtales	5	6
Melastomataceae	Myrtales	5	6
Myrtaceae	Myrtales	5	7
Rhamnaceae	Rhamnales	5	1
Vitaceae	Rhamnales	5	2
Grossulariaceae	Rosales	5	1
Pittosporaceae	Rosales	5	2
Loranthaceae	Santalales	5	3
Santalaceae	Santalales	5	2
Rutaceae	Sapindales	5	4
Zygophyllaceae	Sapindales	5	1
Compositae	Asterales	6 Asteridae	1
Apocynaceae	Gentianales	6	2
Asclepiadaceae	Gentianales	6	1
Loganiaceae	Gentianales	6	1
Rubiaceae	Rubiales	6	5
Oleaceae	Scrophulariales	6	2
Araceae	Arales	M2 Arecidae	5
Agavaceae	Liliales	M5 Liliidae	1
Hanguanaceae	Liliales	M5 Liliidae	1
Smilacaceae	Liliales	M5 Liliidae	1
Gnetaceae		Gymnospermae	2

Table 8.3. Host plant genera of gall-inducing *Liothrips* species

Plant genus	Plant family	Cronquist subclass	Thrips species
Cinnamomum	Lauraceae	1	1
Piper	Piperaceae	1	17
Poikilospermum	Cecropiaceae	2	4
Ficus	Moraceae	2	6
Elatostema	Urticaceae	2	2
Dipterocarpus	Dipterocarpaceae	4	1
Planchonia	Lecythidaceae	4	1
Ardisia	Myrsinaceae	4	2
Schefflera	Araliaceae	5	7
Salacia	Celastraceae	5	1
Guiera	Combretaceae	5	1
Macaranga	Euphorbiaceae	5	2
Mallotus	Euphorbiaceae	5	4
Polyosma	Grossulariaceae	5	1
Loranthus	Loranthaceae	5	1
Dissochaeta	Melastomataceae	5	1
Medinilla	Melastomataceae	5	1
Melastoma	Melastomataceae	5	2
Omphalopus	Melastomataceae	5	1
Eugenia	Myrtaceae	5	7
Tetrastigma	Vitaceae	5	2
Vitis	Vitaceae	5	5
Alyxia	Apocynaceae	6	1
Fagraea	Loganiaceae	6	3
Lasianthus	Rubiaceae	6	1
Pavetta	Rubiaceae	6	1
Rubia	Rubiaceae	6	1
Gnetum	Gnetaceae	G	1
Rhaphidophora	Araceae	M2	1

Total species in genus: 293.

Table 8.4. Host plant genera of gall-inducing *Gynaikothrips* species

Plant genus	Plant family	Cronquist subclass	Thrips species
Ficus	Moraceae	2	12
Casearia	Flacourtiaceae	4	1
Schefflera	Araliaceae	5	2
Melastoma	Melastomataceae	5	1
Rhodamnia	Myrtaceae	5	1
Morinda	Rubiaceae	6	1

Total species in genus: 55.

Table 8.5.   Host plant genera of gall-inducing *Mesothrips* species

Plant genus	Plant family	Cronquist subclass	Thrips species
Poikilospermum	Cecropiaceae	2	1
Ficus	Moraceae	2	5
Ardisia	Myrsinaceae	4	1
Schoutenia	Tiliaceae	4	2
Maytenus	Celastraceae	5	2
Melastoma	Melastomataceae	5	2
Memecylon	Melastomataceae	5	1
Eugenia	Myrtaceae	5	1
Santalum	Santalaceae	5	1
Vitis	Vitaceae	5	1
Pothos	Araceae	M2	1

Total species in genus: 35.

Table 8.6   Host plant genera of gall-inducing *Teuchothrips* species

Plant genus	Plant family	Cronquist subclass	Thrips species
Piper	Piperaceae	1	1
Cleistanthus	Euphorbiaceae	5	1
Callistemon	Myrtaceae	5	2
Eugenia	Myrtaceae	5	1
Melaleuca	Myrtaceae	5	4
Pittosporum	Pittosporaceae	5	1
Bursaria	Pittosporaceae	5	1
Geijera	Rutaceae	5	1
Cassinia	Compositae	6	1
Olearia	Compositae	6	1
Pavetta	Rubiaceae	6	1

Total species in genus: 29.

close to meristematic areas by becoming meristematic again. These are usually palisade or mesophyll cells. In the relatively few species which have been studied, it seems that the cells nearest to those which the thrips have killed start dividing again, the plane of division varying with the thrips species and plant species involved. Such cells show the typical large nuclei, dense cytoplasm, and reduced vacuoles and plastids of meristematic tissues. They may also show curious thickenings of the cell walls, which Ananthakrishnan and Raman (1989) suggest are related to an increase in the surface area of the cell membranes. These tissues have been termed the nutrititive zone. Surrounding this there is commonly a further band of cells in which tannins are deposited.

The plant cells thus communicate with each other and some form of gradient is formed between the damaged cells and the normal leaf tissue. It is interesting to contrast the response of gall production with the normal feeding damage to young leaves by typical Thripinae species. When large populations of *Taeniothrips inconsequens* (Uzel) feed in the leaf buds of sugar maple (*Acer* sp.) trees in eastern North America in a year with low spring rainfall, many square miles of forest may be defoliated (Kolb and Teulon 1991). The tissues of these *Acer* trees do not respond to thrips attack by increased cell division; the leaves fail to expand and are shed, setting the growth of the trees back by a whole year. A similar population of thrips attacking the leaf buds of certain *Acacia* trees in Australia, India, or Africa will induce galls, but these will not constitute a major setback for the growth of the trees. Gall formation by these plants might thus be seen as an adaptation to limit the damage caused by the thrips. From this point is becomes easy to develop the concept of mutualism which underlies so many discussions on insect galls. But the different responses by *Acer* and *Acacia* species might have a functional basis in the deciduous habit of the former, which pre-adapts them to losing their leaves when stressed, in contrast to the essentially evergreen habit of the latter, which pre-adapts them to living with conditions of stress.

The difference between leaf damage and gall induction is sometimes just one of degree, even if galls are defined as novel plant structures whose development is dependent on the activities of a second organism. The problem with calling any plant deformation a 'gall' is that these deformations are then polyphyletic analogies and any homologies that might exist are obscured. In a few situations, a gall is induced by a chemical introduced by an insect, for example some Hymenoptera, and gall development may continue even if the insect fails to survive. Usually, gall growth requires the continued input of some chemical, such as a component of saliva, and aphid galls cease to grow if the aphids are removed. Unlike aphids, thrips do not have a salivary channel in their feeding stylets and injection of saliva into plant cells surrounding those on which a thrips has actually fed remains to be demonstrated. However, since some species introduce viruses into plants, penetration of saliva into the surrounding cells is not unlikely. Thrips galls are thus unusual wound responses, but the gradient in insect–plant relationships between different groups is such that it is difficult to distinguish feeding damage categorically from the most sophisticated gall induction.

The most common type of thrips-induced gall is a rolled or distorted or folded leaf. An adult thrips, when feeding on a young leaflet, frequently lies parallel to and near the midrib and the leaf curls inwards (usually adaxially) from one or both margins; sometimes a leaf may fold along the midline. Subsequent feeding produces irregular depressions and blotches on the leaf as progressively more cells are damaged and more cells respond by increased division. A typical young leaf on a *Ficus* plant is supple and can be rolled in the fingers in either plane; soon after gall induction such a leaf becomes much more brittle and any attempt to manipulate it will result in breakage and exudation of latex.

Leaf galls usually involve single leaves, although a heavily attacked plant may bear hundreds of such independently galled leaves. Sometimes, however, gall induction may involve more than one leaf from a growing point. The largest thrips-induced gall the author has seen was from a 'vine' in the Solomon Islands (Mound 1970). It was a spherical mass of leaf tissue (one or more leaves was not apparent) 10–15 cm in diameter and contained 10 000 or more thrips.

Thrips attacks on the apical buds of *Acacia arabica* Willd. in Africa and *Acacia leucophloea* Willd. in India result in the production of terminal rosette galls. In these, the main axis of the shoot becomes highly foreshortened and the surrounding leaflets become short and thick. The insects feed on the upper surface of the leaflets, the palisade tissue becomes meristematic, and the tissues further away develop large quantities of tannins. The resultant gall encloses the thrips to some extent, but is not closed because other small organisms can enter the gall spaces.

The pouch galls on *Calycopteris floribundus* Lam. induced by *Austrothrips cochinchinensis* Karny in India develop from axillary buds. They are produced by the feeding of several thrips together around the meristem and surrounding leaf primordia. This leads to cell proliferation and the fusion of the primordia and results in the development of a pouch approximately 8 cm long and 4 cm wide. The terminal ostiole of the gall is restricted by dense hairs and the gall walls thicken progressively; calcium oxalate crystals are reported to be deposited away from the nutrititive zone. These pouch galls commonly contain 2000–5000 thrips, but Ananthakrishnan and Raman (1989, p.71) refer in a footnote to one such gall which included 108 000 adults and larvae.

Gall induction on the phyllodinous *Acacia* trees of Australia involves a different process. In these, a single thrips feeding on the upper surface of a young phyllode induces a small depression which rapidly becomes deeper to form a pouch. The lips of this pouch become closely pressed together and the thrips (commonly a male and a female) and subsequent progeny are sealed inside. These galls are thus particularly interesting because they are not open to invasion by predators or usurpers once complete; this has important implications for them as a resource to be defended by the thrips (Crespi 1992*a,b*).

Host specificity

The evidence for host specificity is essentially negative, when a thrips species has not been found on more than one species of plant. Less commonly collected species are recorded from single host species, but whether this is of statistical or biological significance is not easily determined. Evidence for host specificity only becomes significant when an insect has been collected from the same plant species on a number of occasions, preferably through both space and time. However, most species of thrips are known from single localities or single samples or even individuals.

The hypotheses we call 'species' often need testing in order to determine the range of structural variation within and between galls on the same tree, and to compare this with the variation within and between populations on different trees of the same species at the same and different sites. Without such repetitive collecting it is difficult to be sure of species concepts or to know to what extent host specificity exists. The taxonomy of the larger thrips genera from plant galls, particularly *Liothrips*, is probably unsatisfactory and conclusions on host relationships would therefore, at present, be ill-founded. But there is good evidence of host specificity in the thrips from *Acacia* trees in Australia, although one of the most common and widespread species, *Oncothrips tepperi* Karny, has been taken from galls on more than one species of tree (see below).

The literature includes many comments suggesting coevolution and synchrony between thrips and their host plants (Ananthakrishnan and Raman 1989), but the evidence is not strong. No agricultural entomologist would suggest that the biologies of onions and the onion thrips are synchronized, simply because when the seeds of the first germinate the second usually appears. Suggestions of synchrony between gall-inducing thrips and their host plants probably arise largely from lack of information on how a thrips population survives periods when its host plant is least available. Virtually no information exists on the population biology of any gall-inducing thrips, such as their vagility or the percentage of available host plants in an area which is attacked. It is thus impossible to assess the rapidity with which a diffuse population of thrips might respond to a patchily distributed host resource.

Some evolutionary patterns

Despite the apparent lack of any general pattern is host exploitation, several patterns can be detected at the generic and specific level. Further examination of these may lead to a broader understanding in due course.

1. One thrips species/one plant species

Sixty-eight species have been referred to the genus *Gynaikothrips* of which 55 are currently accepted as valid species. The low number of synonyms possibly indicates that further studies will reduce the number of valid species recognized in the genus. Approximately 18 *Gynaikothrips* species are reported to induce leaf deformation in their host plants (Table 8.1). Twelve species are recorded from galls on Moraceae, all from *Ficus*, and six from five other plant families (Table 8.4). *Gynaikothrips ficorum* (Marchal) is apparently specific to *F. microcarpa* Vahl. This tree is widely planted as an ornamental throughout the tropics and subtropics, but it has several synonyms (for example, *retusa* L. and *nitida* Thunb.) and is widely misidentified as *F. benjamina*. As a result, the published host records of *G. ficorum* are probably confused. Where *F. microcarpa* and *F. benjamina* occur together this thrips does not occur on the latter plant and recent attempts in Costa Rica to transfer specimens from one host to the other experimentally were unsuccessful. However, it may be that the thrips is capable of feeding on the non-host plant without breeding and adults have been taken from *F. elastica* Roxb. in European greenhouses. In Egypt, Tawfik (1967) reported an impressive assemblage of insects in the rolled-leaf galls caused by *G. ficorum*. Two anthocorids and a termitophylid (Heteroptera) were predators on the thrips, together with a chrysopid, also a pyemotid mite predatory on thrips eggs and a eulophid parasite of the larvae. In addition, the leaf rolls sometimes contained up to five species of coccoids and these sometimes attracted ants which killed any other arthropods present. An aleyrodid was also sometimes common in this community, also at least two species of mites, one of which was preyed on by a *Scolothrips* species.

2. One thrips species/one plant genus

The genus *Teuchothrips* is particularly well developed in Australia. The described species are currently ill-defined and several undescribed species are also known. These thrips cause leaf-roll and bud galls on a range of common Australian shrubs and trees, but no particular genus of plant seems to be favoured (Table 8.6). One species, *Teuchothrips disjunctus* (Hood), is particularly well known in New Zealand as a pest of the popular Australian garden plant *Callistemon citrinum* Linn. It causes the terminal leaves of this plant to turn pink and roll, and severe infestations can set back the growth of a shrub over a period of years. The species was described originally from a single specimen collected near the Torres Straits in northern Australia, but records from the botanic gardens in Canberra suggest that it is able to breed on other species of *Callistemon*.

3. *Several related thrips species/one plant species*

The tree *Geijera parviflora* Lindl. is common on the inner slopes of the great dividing range in northern New South Wales and southern Queensland. It has a drooping, willow-like crown which is curiously dense and shady for that part of Australia. It is a popular tree for cattle to sit under and also to graze on. It is also remarkably popular with Thysanoptera and supports considerable populations of a species swarm which is currently divided into three genera containing one, two, and seven species, respectively (Mound 1971*b*). The evolution of a series of closely related and coexisting species on this one tree species in semi-arid Australia can be contrasted with the single species *Acaciothrips ebneri* (Karny) on *Acacia nilotica* Linn. in the African sahel zone. The climatic surroundings are fairly similar, hot and usually dry, but the microhabitat offered by the host plants is quite different. The *Acacia* trees stand exposed to the prevailing weather, although the galls themselves provide some protection to the thrips. In contrast, the *Geijera* crown is dense and shady and creates a new habitat within itself, the galls themselves being open leaf curls. It is possible that the shady crowns of these trees provide conditions which the thrips do not readily leave, resulting in reduced gene flow between populations on different trees and a much increased rate of speciation. Certainly this hypothesis could be tested using biochemical methods.

4. *Several related thrips species/several related plant genera*

The genus *Hoplandrothrips* includes more than 80 species. These are found throughout the world and they are almost all associated with fungal hyphae on dead wood. In Africa, however, there is a small group of closely related species which cannot at present be distinguished from this genus, and these species are known to induce leaf roll galls on several different plants. This host shift is so unusual that the author reared members of the commonest species, *H. coffeae* Bagnall, a minor pest of coffee in eastern Africa, to ensure that the leaf rolling is induced by the thrips and not by a fungus to which the thrips might have been attracted. It is a typical thrips leaf roll, which is induced by the feeding activity of one or more adults. Very similar species have been collected in Africa from the following plant genera: *Jasminum* (Oleaceae), *Landolphia* (Apocynaceae), and *Morinda* (Rubiaceae). These genera are all members of the most advanced subclass of flowering plants, the Asteridae.

5. *Several related thrips species/ several congeneric plants*

The only unequivocal example of thrips radiation within a plant genus is the group of thrips genera associated with phyllodinous *Acacia* trees in Australia. The genera *Gynaikothrips, Liothrips,* and *Mesothrips* all have several species on the genus *Ficus* in the oriental region, but they are not restricted to that genus. Similarly, although the species of *Crotonothrips* appear to have radiated on the genus *Memecylon* in India, this thrips genus is also recorded from other plant genera. In contrast, in Australia, three closely related genera, *Kladothrips, Oncothrips,* and *Onychothrips,* are known to induce galls on various *Acacia* trees and these genera do not occur on other plants. Each of the 10 thrips species involved is probably largely specific to one host plant species although, as noted above, one of the most common and widespread has been recorded from several *Acacia* species. The published record of one species from *Hakea* is possibly an error due to the similar appearance and coexistence of some members of these genera in Western Australia (Mound 1971*a*). The behaviour of the thrips inducing these galls has been extensively studied recently and a female will actively defend a developing gall from intruders. She will pick up a conspecific female in her forelegs and drive her fore tarsi into the body, thus killing the intruder. One species, whose gall does not close tightly, even seems to have developed soldiers, with enlarged forelegs and short wings and these individuals guard the gall aperture to repel invaders (Crespi 1992*b*). Despite this, at least one further genus (*Koptothrips*) seems to have evolved from this group, in which the species specialize in taking over the incipient galls and raising their brood within (Crespi 1992*a*).

6. *Woody galls with thrips on Casuarina trees*

Galls in woody tissue are usually associated with fungal attack. However, in northern New South Wales and southern Queensland *Casuarina cristata* Miq. trees commonly have small swellings containing lobed cavities on their younger branches. These were first described many years ago as having been induced by *Thaumatothrips froggatti* Karny, an insect known from no other plant. Recent work by Dr Bernard Crespi has revealed the presence of three species of thrips in these galls and, although they are probably derived one from the other, the differences between them are such that they have been placed in three monobasic genera (Mound and Crespi 1992). Currently, there is no clear evidence as to how these galls are induced, nor on what tissues the thrips feed, but circumstantial evidence indicates that the smallest of the three species is involved in the gall initiation. This species can produce populations of several thousand individuals. The other two species invade the galls subsequently, killing the original inhabitants, and then breeding in the available space (Crespi 1992*a*).

References

Ananthakrishnan, T.N. (1976). New gall thrips of the genus *Crotonothrips* (Thysanoptera: Insecta). *Oriental Insects*, **10**, 411–19.

Ananthakrishnan, T.N. and Raman, A. (1989). *Thrips and gall dynamics*. E.J. Brill, Leiden.

Chisholm, I.F. and Lewis, T. (1984). A new look at thrips (Thysanoptera) mouthparts, their action and effects of feeding on plant tissue. *Bulletin of Entomological Research*, **74**, 663–75.

Crespi, B.J. (1988). Adaptation, compromise and constraint: the development, morphometrics and behavioural basis of a fighter-flier polymorphism in male *Hoplothrips karnyi*. *Behavioural Ecology and Sociobiology*, **23**, 93–104.

Crespi, B.J. (1992*a*). Behavioural ecology of Australian gall thrips. *Journal of Natural History*, **26**, 769–809.

Crespi, B.J. (1992*b*). Eusociality in Australian gall thrips. *Nature*, **359**, 724–6.

Docters van Leeuwen-Reijnvaan, W. and J. (1926). *The Zoocecidia of the Netherlands East Indies*. Batavia.

Kolb, T.E. and Teulon, D.A.J. (1991). Relationship between sugar maple budburst phenology and pear thrips damage. *Canadian Journal of Forestry Research*, **21**, 1043–8.

Mabberley D.J. (1990). *The plant book*. Cambridge University Press, Cambridge.

Mound, L.A. (1970). Thysanoptera from the Solomon Islands. *Bulletin of the British Museum (Natural History), Entomology*, **24**, 83–126.

Mound, L.A. (1971*a*). Gall-forming thrips and allied species (Thysanoptera: Phlaeothripinae) from *Acacia* trees in Australia. *Bulletin of the British Museum (Natural History), Entomology*, **25**, 387–466.

Mound, L.A. (1971*b*). The complex of Thysanoptera in rolled leaf galls on *Geijera*. *Journal of the Australian Entomological Society*, **10**, 83–97.

Mound, L.A. (1983). Natural and disrupted patterns of geographical distribution in Thysanoptera (Insecta). *Journal of Biogeography*, **10**, 119–33.

Mound, L.A. (1990). Systematics of thrips (Insecta: Thysanoptera) associated with mosses. *Zoological Journal of the Linnean Society*, **96**, 1–17.

Mound, L.A. and Crespi, B. (1992). The complex of phlaeothripine thrips (Insecta, Thysanoptera) in woody stem galls of *Casuarina* in Australia. *Journal of Natural History*, **26**, 395–406.

Mound, L.A. and Heming B., (1991). Thysanoptera. In *The insects of Australia*, pp. 510–15. Melbourne University Press, Carlton, Victoria.

Mound L.A. and Palmer, J.M. (1981). Identification, distribution and host-plants of the pest species of *Scirtothrips* (Thysanoptera: Thripidae). *Bulletin of Entomological Research*, **71**, 467–79.

Mound, L.A., Heming, B., and Palmer, J.M. (1980). Phylogenetic relationships between the families of recent Thysanoptera. *Zoological Journal of the Linnean Society of London*, **69**, 111–41.

Palmer, J.M. and Mound L.A. (1991). Thysanoptera. In *The armoured scale insects, their biology, natural enemies and control*, (ed. D. Rosen), Vol. B, pp. 67–76. Elsevier, Amsterdam.

Tawfik, M.F.S. (1967). Microfauna of the leaf-rolls of *Ficus nitida* Thunb.-Hort. *Bulletin of the Entomological Society of Egypt*, **51,** 483–7.

Terry, L.I. (1991). Swarming behaviour in *Frankliniella occidentalis*. *Bulletin of the Agricultural Experiment Station, University of Vermont*, **698,** 27.

9. Interactions between *Trichochermes walkeri* (Homoptera: Psylloidea) and other Homoptera on *Rhamnus catharticus*

I.F.G. McLEAN

109 Miller Way, Brampton, Huntingdon, Cambridgeshire, UK

Abstract

The psyllid *Trichochermes walkeri* Förster (Psylloidea: Triozidae) forms leaf roll galls on buckthorn, *Rhamnus catharticus* L. As part of a long-term study of the population dynamics of *T. walkeri* the interactions between this gall-forming insect and other, free-living, Homoptera associated with buckthorn have been investigated. The free-living psyllids and aphids are associated with leaves galled by *T. walkeri* to varying extents, ranging from obligate, through showing a preference for feeding on galled leaves, to showing no significant association. These different degrees of association can be related to the phenologies of the free-living species, while it is speculated that the nutritional advantage of feeding on galled leaves has been responsible for free-living Homoptera becoming more closely associated with, and eventually dependent upon, the presence of *T. walkeri* galls.

Introduction

Trichochermes walkeri Förster (Psylloidea: Triozidae) is an unusual jumping plant louse by virtue of the strongly patterned forewings (illustrated by Hodkinson and White (1979) in Figure 234) and because the nymphal development takes place in galls formed by upward rolling of the leaf margin of the host plant, buckthorn (sometimes called purging buckthorn), *Rhamnus catharticus* L. The gall is depicted in Figure 985 in Docters van Leeuwen (1982). Buckthorn is apparently the sole host in Britain (McLean 1993) despite previous publications (Hodkinson and White 1979, p. 84;

Plant Galls (ed. Michèle A. J. Williams), Systematics Association Special Volume No. 49, pp. 151–60. Clarendon Press, Oxford, 1994. © The Systematics Association, 1994.

Godwin 1943, p. 66) indicating that alder buckthorn *Frangula alnus* Mill. is an alternative host plant.

A long-term study of *T. walkeri* populations was commenced in 1982 with the objective of investigating those factors which cause population change or regulate numbers in this species. Gall-forming insects are excellent for such studies because of the ease of locating galls compared with individuals of free-living species and because the causes of mortality can often be readily discerned. During the census of *T. walkeri* galls on buckthorn study bushes it was noticed that other sap-sucking Homoptera were frequently found feeding on leaves galled by *T. walkeri*. This observation is the origin of the study which led to this paper.

Five other sap-sucking Homoptera within the Aphidoidea and Psylloidea feed on buckthorn in Britain; their biological characteristics are given in Table 9.1 and their life cycles are summarized in Fig. 9.1, together with that of *T. walkeri*.

Methods

Four study sites have been used to examine *T. walkeri* populations during this investigation; the principal site has been Chippenham Fen National Nature Reserve (NNR) Vice County, Cambridgeshire (grid reference TL 647694), with additional observations at Brampton, V.C. Huntingdonshire (TL 225709), Cavenham Heath NNR, V.C. West Suffolk

Table 9.1. The biology of Homoptera which feed on buckthorn (*Rhamnus catharticus*) in Britain

---

Psylloidea (after Hodkinson and White (1979) and personal observations)
Trichochermes walkeri Förster: first instar nymphs establish galls only on buckthorn; all five nymphal instars are spent within galls, each containing one or more nymphs
Psylla rhamnicola Scott: free-living only on buckthorn
Trioza rhamni (Schrank): free-living only on buckthorn; the nymphs settle and feed on the lower surface of leaves
Aphidoidea (after Stroyan (1984), Heie (1986), McLean (1988), and personal observations)
Aphis commensalis Stroyan: only on buckthorn. Spring and early summer generations feed on the upper surface of leaves (causing upward curling of leaf margins); from July until autumn feeding only on leaves galled by *T. walkeri*
Aphis mammulata Gimingham & Hille Ris Lambers. Spring and early summer generations not known (probably on buckthorn); from July onwards clustering on berry petioles or forming colonies under leaves of buckthorn where they are assiduously attended by the ant *L. fuliginosus* Latreille
Aphis nasturtii (Kaltenbach): a host-alternating species with spring generations on buckthorn (causing distortion of young leaves), then migrating to a wide variety of secondary hosts before returning to the primary host in late summer or autumn

---

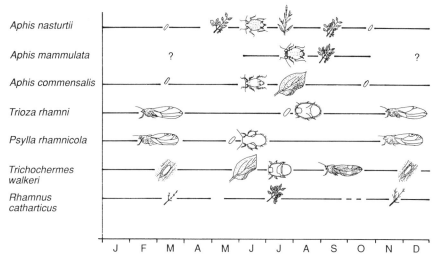

Fig. 9.1. The life cycles of Homoptera which feed on buckthorn (*Rhamnus catharticus*) in Britain.

(TL 757728), and Foulden Common, V.C. West Norfolk (TF 7600). A buckthorn bush, planted in the Brampton garden, has been used for experimental studies of aphids and free-living psyllids feeding on leaves galled by *T. walkeri*.

For growth rate experiments, aphids and free-living psyllids were confined to individual Buckthorn leaves by sleeves of nylon netting measuring approximately 10 cm × 7 cm fastened snugly around each petiole with a drawstring inside a seam. Aphids and free-living psyllids were measured using a 100 division 10 mm eyepiece graticule fitted to an Olympus SZ3 binocular microscope at ×60 magnification (×4 objective and ×15 eyepiece). The length and breadth of each individual was multiplied together to give a simple size index. In each experiment a galled leaf was paired with an ungalled 'opposite' leaf (buckthorn leaves typically grow in a sub-opposite configuration, pairs of leaves being widely separated on shoots and occurring in a more closely packed arrangement on spurs), these sub-opposite leaves being of similar size. Nymphs of aphids or free-living psyllids were carefully measured before the start of each experiment and individuals of very similar or identical size were placed on each pair of galled/ungalled leaves. The aphids and psyllids were measured again at the conclusion of each experiment and the changes in the size index calculated. The leaves used were measured (length × breadth to the nearest mm) at the beginning and end of each experiment.

When assessing the distribution of free-living psyllids on buckthorn leaves galled by *T. walkeri* compared with ungalled leaves, a random sample of leaves from the same part of the bush was chosen, at a height of 1–2 m.

Results

These are presented for two types of interaction between *T. walkeri* and five other Homoptera on buckthorn: firstly, the observed degrees of association between *T. walkeri* and free-living Homoptera on buckthorn and, secondly, a comparison of the growth rates of nymphal Homoptera on ungalled leaves of buckthorn with their growth on leaves galled by *T. walkeri*.

1. The observed degrees of association between T. walkeri *and free-living Homoptera on buckthorn*

The association between *T. walkeri* and the two free-living psyllids has been investigated by counting psyllid nymphs (which are less mobile than adult psyllids) on galled and ungalled leaves.

Table 9.2.   The distribution of *P. rhamnicola* nymphs on leaves galled by *T. walkeri* and on ungalled leaves at Chippenham NNR on 18 June 1985

	With P. rhamnicola	No P. rhamnicola	Total
Leaves with *T. walkeri* galls	5	51	56
Leaves with no *T. walkeri*	7	302	309
Total	12	353	365

These results show that the positive association between *P. rhamnicola* and leaves galled by *T. walkeri* is significant at the 5 per cent level ($\chi^2 = 6.62$).

Psylla rhamnicola has been counted on galled and ungalled leaves at Chippenham NNR (see Table 9.2). Other counts have shown a similar pattern of association to that presented here.

Trioza rhamni has been counted once on galled and ungalled leaves at Foulden Common (see Table 9.3).

The association between *T. walkeri* and the three aphids has not yet been assessed by counts on galled and ungalled leaves. However, some personal observations supplemented with the statements for these species by Stroyan (1984) and Heie (1986) are sufficient to indicate the degrees of association in general terms. *Aphis commensalis* is free-living on the upperside of buckthorn leaves from May onwards; later generations (July onwards) settle on leaves galled by *T. walkeri* and feed there before and

Table 9.3.   The distribution of *T. rhamni* nymphs on leaves galled by *T. walkeri* and on ungalled leaves at Foulden Common, 6 July 1992

	With T. rhamni	No T. rhamni	Total
Leaves with *T. walkeri* galls	28	172	200
Leaves with no *T. walkeri*	13	187	200
Total	41	359	400

These results show that the positive association between *T. rhamni* and leaves galled by *T. walkeri* is significant at the 5 per cent level ($\chi^2 = 6.11$).

after gall opening (during August and early September). *Aphis mammulata* is free-living under buckthorn leaves and on berry petioles; the population reported by McLean (1988) was abundant in late summer on the single bush where it occurred at Cavenham Heath NNR, with high numbers observed on galled and ungalled leaves alike. *Aphis nasturtii* has been frequently observed on galled leaves at Brampton by the author, but no systematic counts have yet been made to determine the proportion of colonies or individuals occurring on galled versus ungalled leaves.

2. A comparison of the growth rates of nymphal Homoptera on ungalled leaves of buckthorn with their growth on leaves galled by T. walkeri

Two species have been investigated experimentally to determine their nymphal growth rates on leaves galled by *T. walkeri* compared with ungalled leaves: *P. rhamnicola* and *A. nasturtii*. In both species faster nymphal growth occurred on galled leaves (see Tables 9.4 and 9.5).

During these growth rate experiments it was observed that galled leaves continued to increase in area (as measured by a simple length × breadth index) faster than ungalled leaves (Table 9.6). This may be an indication of the continued diversion of plant resources towards leaves galled by *T. walkeri*.

Table 9.4.   The growth rate of male *P. rhamnicola* reared singly on leaves galled by *T. walkeri* compared with those reared on ungalled leaves, 25–30 May and 31 May to 8 June 1992

Mean increase in size index on ungalled leaves	876.9 ± 414.6 SD ($n = 7$)
Mean increase in size index on galled leaves	1151.9 ± 427.6 SD ($n = 7$)

$t_{(12)} = 1.22$ not significant at the 5 per cent level, more experiments with greater replication are required to test whether the growth rate of *P. rhamnicola* is significantly increased by feeding on buckthorn leaves galled by *T. walkeri*

Table 9.5. The growth rate of *A. nasturtii* reared singly on leaves galled by *T. walkeri* compared with those reared on ungalled leaves, 3–8 June 1992

Mean increase in size index on ungalled leaves	180.5 ± 255.5 SD $(n = 4)$
Mean increase in size index on galled leaves	659.5 ± 279.6 SD $(n = 4)$
$t_{(6)} = 2.52$ significant at the 5 per cent level	

Table 9.6. Increase in leaf index during growth rate experiment with *A. nasturtii*; 3–8 June 1992

Mean increase in leaf index for ungalled leaves	5.13 per cent $(n = 4)$
Mean increase in leaf index for galled leaves	1.67 per cent $(n = 4)$

Discussion

The results of this study, combined with previous observations on the biology of aphids feeding on buckthorn (Stroyan 1984; Heie 1986) reveal a complex web of interactions between *T. walkeri* galls and other Homoptera on buckthorn in Britain. The timing of the life-cycles of the free-living Homoptera is crucial in determining aspects of the interaction of the species with the relatively long-lived galls of *T. walkeri*. Because *T. walkeri* galls remain on buckthorn from soon after the appearance of leaves in the spring until leaf fall in autumn (and are occupied by nymphs from May until August or early September), they are present throughout the periods when free-living psyllids and aphids feed on buckthorn as nymphs and adults (see Fig. 9.1).

Nymphs of the free-living psyllids and aphids on buckthorn develop more rapidly than those of *T. walkeri*. While *P. rhamnicola* and *T. rhamni* have relatively short and well-synchronized single generations (in May–June and July–August, respectively) when nymphs develop, the three aphid species have rapid individual development of nymphs combined with multiple, overlapping generations. With the exception of *A. nasturtii*, which leaves buckthorn in summer to exploit herbs, all these free-living Homoptera are apparently associated only with buckthorn in Britain and they have evolved different phenologies and strategies for exploiting their plant host through the growing season. The phenologies of Homoptera on trees and shrubs like buckthorn are likely to have evolved to match periods in their life cycles which have greater nutritional needs (for example, nymphal growth and adult reproduction) with times when better quality plant sap is available.

Galling of buckthorn leaves by *T. walkeri* provides a nutritional opportunity for five other sap-sucking Homoptera in Britain. The likely diversion of plant resources into galled leaves resulting from the feeding activities of *T. walkeri* apparently means that these leaves (typically representing 2–20 per cent of all leaves on bushes in East Anglia) offer superior quality sap for free-living sap-suckers. This has been shown to result in faster growth rates for *A. nasturtii* and *P. rhamnicola*. It is suggested here that the nutritional benefits of feeding on galled leaves are strong reasons why free-living Homoptera may feed on galled leaves longer than ungalled leaves and, hence, become positively associated with galled leaves, at least during certain parts of their life cycles.

After rapid spring growth and before leaf senescence in autumn, the summer period offers poorer quality nutrition on trees and shrubs for many aphids (see Dixon 1985, p. 69 and following). Psyllids presumably must also overcome poorer nutritional conditions on woody hosts during the summer. Galled leaves provide one means of bridging this gap and it is likely that similar patterns of association between galls and sap-sucking insects will occur on plants other than buckthorn. That four out of the five species have apparently some association with *T. walkeri* and that the species with the strongest association (*A. commensalis*) has the longest period of growth and reproduction away from the spring or autumn seasons, suggests a link between phenology of free-living species and the development and degree of association with *T. walkeri*.

The presence of a *T. walkeri* gall is in effect a signal to other Homoptera on buckthorn that good quality food is available on a leaf; this signal is probably more stable and long-lasting than where leaves have been modified by colonies of free-living sap-suckers. It is certainly a conspicuous signal in the sense that a gall is an obvious structure on a leaf and can presumably be detected quickly and easily by other searching insects.

It would be of considerable interest to discover whether plants other than buckthorn support assemblages of sap-sucking insects interacting in the ways outlined in this paper. Such patterns might be detected initially through looking for simple spatial associations between species and then by examining the nature of the interactions in more detail. More thorough investigations of the consequences for both gall formers and free-living species will be more time-consuming, but at the same time offer the prospect of greater ecological understanding by unravelling the costs and benefits for the participants in these associations.

The nutritional advantages of feeding on galled leaves are speculated here to be the principal reason for free-living insects becoming at first loosely and then more closely associated with galled leaves (and potentially other galled plant structures). The evolutionary impetus for the development of the galling lifestyle has been the subject of competing alternative hypotheses and experimental investigations (Price *et al.* 1987; see also papers and references in Shorthouse and Rohfritsch 1992). Avoiding natural enemies, physical protection or shelter against adverse weather (or similar physical factors), and improved nutrition are the three most plausible reasons which have been advanced to account for the evolution of the gall-forming habit. It is suggested here that the third of these alternatives, improved nutrition, is likely to have been most frequently responsible for the evolution of closer associations between gall formers and hitherto free-living species.

The five free-living Homoptera associated with buckthorn and discussed in this paper, can be viewed as being at different points along an evolutionary sequence starting with no significant association with *T. walkeri* galls (*A. mammulata*), continuing with species which are free-living, can survive well in the absence of *T. walkeri* galls, but nevertheless can benefit from the presence of these galls (*A. nasturtii*, *P. rhamnicola*, and *T. rhamni*), and concluding with a species which appears to be dependent on the presence of *T. walkeri* galls during one part of its life-cycle (*A. commensalis*). That the rare and highly specialized *A. mammulata* reaches very high population levels where it occurs and that it is apparently dependent upon being attended by *Lasius fuliginosus* for its survival, are factors which indicate it has evolved other adaptations than associating with *T. walkeri* to enable it to exploit buckthorn successfully.

This postulated, idealized, evolutionary sequence may indicate one way in which free-living insects can become associated, at first facultatively but eventually obligatorily, with insect galls. It is most unlikely to be the only route to developing such obligatory associations, as inquilines have evolved independently in a number of insect groups from free-living or gall-forming ancestors and such inquilines may be associated with more than one gall-forming species (see papers and references in Shorthouse and Rohfritsch 1992).

The initial stages of an association between a free-living species and a gall former are unlikely to offer any protection against natural enemies of the free-living species. Some favourable changes in microclimate are likely in the vicinity of galls or within galls with an open structure allowing free-living species to enter, so the physical protection offered by galls may benefit some species which take shelter. However, this is hard to envisage for the majority of 'closed' galls with no ready access to the sheltered interior, including those formed by *T. walkeri*.

The development of, at first, commensal and then closer relationships (including possibly mutualism and then inquilines dependent upon feeding on gall tissue) between gall formers and free-living species is of equal interest in attempting to understand how these communities have evolved and are structured. The relative importance of interspecific competition, compared with commensalism or mutualism in determining the relationships between species, is also of great interest. For leaf feeders, chewers are likely to interact with each other (and with sap-suckers) most frequently as competitors, having negative effects on other chewers (both interspecifically and intraspecifically) or on sap-suckers.

Sap-suckers, on the other hand, are able to interact positively with other sap-suckers by stimulating the metabolism of their host plant and thereby modifying the quality of leaves or other plant tissues, where they feed (Dixon (1985, p. 18) cites relevant examples for aphids). Within a species this means that growth rates can be higher for individuals within colonies rather than, feeding alone (McLean, unpublished observations on *T. walkeri*). The results presented here show that feeding by *T. walkeri* facilitates the growth of *A. nasturtii* and *P. rhamnicola* in a similar way to the intraspecific effects cited above. This is akin to the interaction between *Periphyllus acericola* (Walker) and *Drepanosiphum platanoidis* (Schrank) (Dixon 1985, p. 129). These interactions are tentatively classed here as commensalism. If the *T. walkeri* also benefit from this association as well as *A. nasturtii* and *P. rhamnicola* then the relationship would be correctly described as mutualism.

Whether sap-suckers are ever able to improve food quantity or quality for chewers seems likely, but no references have been traced during the preparation of this paper. Such effects would be most easily detected by studying sap-suckers which form galls and then looking for species feeding more frequently on galled compared with ungalled plant tissues. Examples of moth larvae which are inquilines in sawfly galls (see Price 1992, p. 213) indicate that this interaction is a probable route for the establishment of at least some gall–inquiline relationships.

If the hypothesis that the evolution of commensalism (and possibly mutualism) is more likely between sap-suckers, then further examples of closely related species developing mutualistic relations should be discovered among the Homoptera. Hutchinson (1978) observed that most conspicuous mutualism (he used the term symbiosis) occurs between very diverse kinds of organisms; perhaps inconspicuous mutualism may be frequent among quite closely related Homoptera on the same host plant, particularly when one of the species is a gall former! This suggestion will require testing for a number of host plants and their associated insect assemblages.

Whether the presence of aphids or psyllids feeding on galled leaves has any consequences for *T. walkeri* has not yet been investigated and would pose some interesting technical problems in attempting to observe growth rates of insects within galls. It is possible that any effects on subsequent size of adult *T. walkeri* could be ascertained through replicated experiments with appropriate controls; the author plans to investigate this aspect of the interaction between *T. walkeri* and other Homoptera in future.

Acknowledgements

I am grateful to the Nature Conservancy Council (since 1991 the Nature Conservancy Council for England) for permission to study *T. walkeri* at Chippenham Fen NNR and Cavenham Heath NNR and to Martin Twyman-Musgrave and Malcolm Wright, successively the wardens of these reserves, for their interest in and support of entomological investigations conducted on their sites. My wife, Christine, has helped with data collection for some aspects of this study and her continued support in many ways has enabled me to continue to research the ecology of *T. walkeri* and associated species.

References

Dixon, A.F.G. (1985). *Aphid ecology*. Blackie, Glasgow and London.

Docters van Leeuwen, W.M. (Revised by Wiebes-Rijks, A.A. and Houtman, G.) (1982). *Gallenboek*. Thieme & Cie, Zutphen.

Godwin, H. (1943). Biological Flora. Rhamnaceae, *Rhamnus cathartica* L. and *Frangula alnus* Miller. *Journal of Ecology*, **31**, 66–92.

Heie, O.E. (1986). The Aphidoidea (Hemiptera) of Fennoscandia and Denmark. III Family Aphididae: subfamily Pterocommatinae & tribe Aphidini of subfamily Aphidinae. *Fauna Entomologica Scandinavica* **17**, 1–314. E.J. Brill/Scandinavian Science Press, Leiden and Copenhagen.

Hodkinson, I.D. and White, I.M. (1979). Homoptera, Psylloidea. *Handbooks for the identification of British insects*, **2**(5a), (ed. A. Watson), pp. 1–98. Royal Entomological Society of London, London.

Hutchinson, G.E. (1978). *An introduction to population ecology*. Yale University Press, New Haven and London.

McLean, I.F.G. (1988). A second British locality for *Aphis mammulata* Gimingham & Hille Ris Lambers (Homoptera: Aphididae). *British Journal of Entomology and Natural History*, **1**, 188–9.

McLean, I.F.G. (1993). The host plant association and life history of *Trichochermes walkeri* Förster (Psylloidea: Triozidae). *British Journal of Entomology and Natural History*, **6**, 13–16.

Price, P.W. (1992). Evolution and ecology of gall-inducing sawflies. In *Biology of insect-induced galls*, (ed. J.D. Shorthouse and O. Rohfritsch), pp. 208–24. Oxford University Press, New York and Oxford.

Price, P.W., Fernandes, G.W., and Waring, G.L. (1987). Adaptive nature of insect galls. *Environmental Entomology*, **16**, 15–24.

Shorthouse, J.D. and Rohfritsch, O. (ed.) (1992). *Biology of insect-induced galls*. Oxford University Press, New York and Oxford.

Stroyan, H.L.G. (1984). Aphids—Pterocommatinae and Aphidinae (Aphidini) Homoptera, Aphididae. *Handbooks for the identification of British insects*, **2**(6), (ed. M.G. Fitton), pp. 1–232. Royal Entomological Society of London, London.

10. Galls and the evolution of social behaviour in aphids

W.A. FOSTER and P.A. NORTHCOTT

Department of Zoology, University of Cambridge, Downing Street, Cambridge, UK

Abstract

In several aphid species there is a soldier caste that actively defends the aphid colony. All the aphids that produce soldiers belong to species that induce galls on their host plants. Gall formation and soldier production are restricted to two of the families within the Aphidoidea—the Hormaphididae and the Pemphigidae. Information about the behaviour of the species (approximately 30) in which soldiers have been described is summarized. The paper examines how living in a gall might affect the social organization of aphids and increase the likelihood of the production of soldiers and the evolution of cooperation. The argument is proposed that the gall is important in providing a resource that can be readily defended and kept clean. In addition, the gall is important in providing a ring-fence around the clones, increasing the likelihood that aphids are related to their close neighbours. Soldier production might be encouraged in those galls that are relatively long-lived.

Introduction

It is now well-established that several aphid species are organized into societies that are defended by a specialized soldier caste (for example, Aoki 1977*a*; Itô 1989; Foster 1990; Kurosu and Aoki 1991*a*). These aphids are of interest to evolutionary biologists because they provide a new group in which to investigate the evolution of altruistic behaviour. The most widely applicable explanation for the evolution of altruism is Hamilton's (1964) idea of kin selection: an individual will be selected to value the reproduction of another according to how closely the two individuals are related. This is usually expressed in terms of Hamilton's Rule, which gives the conditions under which a gene for altruism might

Plant Galls (ed. Michèle A. J. Williams), Systematics Association Special Volume No. 49, pp. 161–82. Clarendon Press, Oxford, 1994. © The Systematics Association, 1994.

be expected to spread in a population (Grafen 1991). Imagine a donor provides help to a beneficiary, which incurs costs (c) to the donor and benefits (b) to the beneficiary. This helping behaviour will be selected for if the following inequality is satisfied:

$$b/c > r \text{ (donor to own offspring)}/r \text{ (donor to beneficiary's offspring)} \tag{10.1}$$

where b is the increase in the number of the beneficiary's offspring as a result of the helping behaviour of the donor, c is the decrease in the number of the donor's offspring as a result of her helping behaviour, and r is relatedness.

Since aphids are parthenogenetic, an aphid colony has the potential to be a pure clone, in which all the individuals are related to each other with a value of 1. Hamilton's rule is then simple to apply: as long as b exceeds c, altruism will be selected for. Therefore, in clonal organisms the genetic predispositions for helping behaviour are high: in a pure clone, barring mutation, there is the *certainty* (rather than probability) that the relatedness between two individuals, no matter how distant genealogically, is 1.

Aphid soldiers and the gall-forming habit

Soldiers have been described in species from two aphid families: Hormaphididae and Pemphigidae. (We are following here the classification of Heie (1980); see also Ilharco and van Harten 1987). We are using here a broad, behavioural definition of 'soldier': any set of individuals that act defensively with some likely fitness loss to themselves will be considered to be soldiers. Several authors have used a more restrictive definition, with the requirement that soldiers should be sterile (for example, Aoki 1982a: Itô 1989). However, social insect castes are usually defined in terms of their behavioural roles (for example, Hölldobler and Wilson (1990); but see also Peeters and Crozier 1988): therefore, we will refer to all sets of aphids with well-defined defensive behaviour as soldiers, whether or not they have been shown to be sterile and whether or not they are morphologically distinct from non-defensive individuals. All of the aphid species that have soldiers also form galls on the primary host or are from gall-forming genera, in the case of species whose primary host is not known (Table 10.1). Indeed, true gall-forming species are almost entirely restricted to the Hormaphididae, Pemphigidae, Adelgidae, and Phylloxeridae (Wool 1984). (The last two families are generally regarded as belonging to the non-aphid superfamily, Phylloxeroidea.) The broad purpose of this contribution is to establish whether there is any biological significance in the fact that all the soldier-producing aphids are also gallformers. Does the gall-forming habit in some way predispose the evolution of soldiers and eusociality in aphids?

In this paper, we will briefly describe the different types of soldier aphids, their taxonomic distribution, and whether or not they live in galls. We will then ask how, in terms of Hamilton's Rule, the gall-forming habit may have been an enabling factor in the evolution of social behaviour in aphids.

The association between the gall-forming habit and soldier production

Table 10.1 shows the relationship between gall forming and soldier production in the major groups of aphids. At first glance, it seems to provide an indication that this relationship is firm: all the genera of aphids with soldiers are from gall-forming taxa and the vast majority of aphids that do not produce soldiers also do not form galls. However, there are at least three reasons for thinking that this association is not as secure as the table might seem to imply. First, in order to make any kind of claim about the correlation between two traits in a group of organisms, it is essential to have some idea about how often the traits have evolved independently and for this a knowledge of the phylogeny of the organisms is required (for example, Ridley 1983; Harvey and Pagel 1991). Clearly, if gall forming and soldier production arose independently in each of the soldier-producing genera, this is a much stronger case for the biological significance of the association than if the two traits evolved only once, for example, in the common ancestor of the Hormaphididae and Pemphigidae. However, as far as we know only one cladogram of the Aphidoidea has ever been attempted (Heie 1987) and there are no published accounts of the cladistic relationships within the Hormaphididae and Pemphigidae. Secondly, soldier behaviour has been looked for in only a very small number of aphid species. Undoubtedly, many more aphid species with soldiers await discovery. Finally, in some species, the soldier morphs are on the secondary host, where galls are not usually found, and it is not clear whether these cases should be cited as evidence for an association between gall forming and soldier production.

Aphid galls in relation to the life cycle

The life cycles of the two gall-producing aphid families are very similar. There is an alternation of hosts between a primary host, on which the sexuals mate and lay an egg, from which the fundatrix hatches and forms

Table 10.1. Gall-forming and social aphids

Taxon	Gall forming	Genera in which defender morphs are (a) known	(b) not known	No. of species in the taxon
Phylloxeroidea				
Adelgidae	Yes	None	All	**47**
Phylloxeridae	Yes	None	All	**69**
Aphidoidea				
Lachnidae				
Chaitophoridae				
Drepanosiphonidae[a]				
Aphididae[b]	No	None	All	**3323**
Greenideidae				
Phloeomyzidae				
Anoeciidae				
Thelaxidae				
Hormaphididae				**171**
Cerataphidinae	Yes	*Aleurodaphis* *Astegopteryx* *Cerataphis* *Ceratoglyphina* *Ceratovacuna* *Pseudoregma* *Tuberaphis*		81
Nipponaphidinae	Yes	*Nipponaphis*		82
Hormaphidinae	Yes		*Hamamelistes* *Hormaphis*	8
Pemphigidae				**266**
Eriosomatinae				63
Eriosomatini	Yes	*Eriosoma*	*Schizoneura*	
Tetraneurini	Yes	*Colophina* *Hemipodaphis*	*Tetraneura* *Kaltenbachiella*	
Pemphiginae				146
Prociphilini	Yes	*Pachypappa*	*Pachypappella* *Gootiella* *Prociphilus*	
Pemphigini	Yes	*Pemphigus*	*Thecabius*	
Fordinae				57
Fordini	Yes		*Aploneura* *Forda* *Baizongia*	
Melaphidini	Yes		*Melaphis*	

[a] One genus *Tamalia* is gall forming and is accorded sub family status by some authors (Remaudière and Stroyan 1984).
[b] Two genera *Tuberocephalus* and *Eumyzus* have one or more gall-forming species.

a gall, and the secondary host, where successive, entirely parthenogenetic, generations of females develop (Fig. 10.1a). Winged gall emigrants fly from the primary to the secondary host and winged sexuparae travel from the secondary to the primary host, transporting the sexuals inside them; for example, Foster and Benton (1992). The primary hosts of each of the major subgroups of these two families are usually woody and tend to be highly conserved: for example, the Pemphigini form galls on species of *Populus*, the Cerataphidinae on *Styrax*, the Melaphidini on *Rhus*, and the Fordini on *Pistacia*. The secondary hosts used by the species in each subgroup tend to be very diverse and for this reason it is argued that the association with the primary host is older than that with the secondary host (for example, Heie 1980). The Prociphilini provide an exception: the sexuals, egg, and fundatrix live on a range of woody angiosperms and the other parthenogenetic generations occur on the roots of Pinaceae (Moran 1992).

There are important variations on this basic life cycle in these two families. Some species remain all year on the primary host (monoecious): the winged aphids that leave the gall are sexuparae that migrate to another part of the primary host, for example, the bark, where they give birth to the sexuals (Fig. 10.1b). Other species have lost the sexual phase, and remain all year on the secondary host (anholocyclic). The genus *Pemphigus* in Britain provides clear examples: most species alternate between galls on *Populus* and the roots of herbaceous secondary hosts, but *Pemphigus spyrothecae* remains all year on the primary host and *Pemphigus saliciradicis* is anholocyclic on the roots of *Salix*. Other examples are given in Table 10.2. Some aphid genera, for example, members of the Fordini, show the usual alternation of generations in parts of the world where the primary hosts (*Pistacia*) occur, but are anholocyclic on the secondary host in areas where the primary host is not found, for example, in Northern Europe. With this background we will now consider the types of soldier that are found and the number of times that the soldier habit might have evolved.

How often has soldier behaviour evolved?

It is generally accepted that the Hormaphididae and the Pemphigidae are closely related. They share several apomorphic characters (Heie 1987), but the Anoeciidae may belong to the same monophyletic group, perhaps being more closely related to the Pemphigidae than the Hormaphididae (Heie 1987). They are both probably rather ancient groups, whose habits, for example, gall formation and their host associations and life cycles, were fixed a long time ago. Gall formation was therefore

(a)   LIFE CYCLE OF <u>PEMPHIGUS</u> <u>BURSARIUS</u>

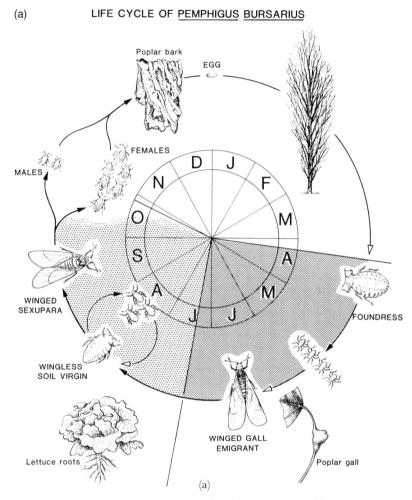

(a)

Fig. 10.1. Life cycles of Pemphigid aphids. (a.) Life cycle of *Pemphigus bursarius*, (b.) life cycle of *Pemphigus spyrothecae*. The dense stippling indicates the time spent in the gall, the lighter stippling indicates the time spent on the secondary host. Dark-headed arrows indicate 'gives birth to', light-headed arrows 'grows up into'.

probably a feature of the common ancestor of these two families and would have been a shared *derived* character if the more ancient aphids from which they evolved did not form galls. It could also be argued, in a similar way, that soldier production is a shared feature of the two families and evolved on only one occasion. This seems highly unlikely. However, given our ignorance about the cladistic relationships of the

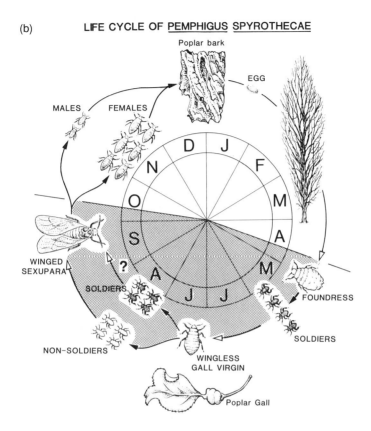

(b) **LIFE CYCLE OF <u>PEMPHIGUS</u> <u>SPYROTHECAE</u>**

taxa within the Hormaphididae and Pemphigidae and about the true extent of soldier behaviour, we should not be too dogmatic.

The best evidence that soldier behaviour has evolved more than once is provided by the morphology and behaviour of the soldiers themselves. Table 10.2 shows all the records of soldier behaviour that we have been able to extract from the literature and Table 10.3 categorizes aphid soldiers into four main types. These correspond approximately to the categories proposed by Aoki (1987) and Itô (1989).

Within the Pemphigidae, there have probably been at least two independent origins of soldier behaviour. In the *Colophina*-type, the soldiers on the secondary host are first instars and for weapons they use their stylets and their enlarged and sclerotized fore and mid-legs. In some species (for example, *C. clematis*), the soldiers are sterile and distinct in morphology from the normal first-instars ('dimorphic') and in others (for example, *C. clematicola*), all the first instars are soldiers and can develop into

Table 10.2. The chief types of aphid soldiers

	Host	Instar	Weapons	Sterility	Examples
1 Colophina type					
A	Secondary	I	Stylets, fore and mid-legs	Dimorphic[a] Sterile	Colophina clematis, arma, clematicola, monstrifica
B	Primary	I/II	Stylets, fore- and mid-legs	Di/monomorphic Not sterile	Colophina arma, clematis, Hemipodaphis persimilis
2 Pemphigus type	Primary	I	Stylets, hind-legs	Mono/dimorphic Not sterile	Pemphigus spyrothecae, monophagus, dorocola
3 Horned type	Secondary	I	Horns, fore-legs	Dimorphic Sterile	Ceratovacuna japonica; Pseudoregma alexanderi, bambucicola, panicola
4 Styrax-gall type					
A	Primary	II	Stylets	Dimorphic Sterile	Astegopteryx bambucifoliae, styraci; Ceratoglyphina bambusae; Aleurodaphis
B	Primary	I	Stylets ('outsiders')	Monomorphic Not sterile/sterile	A. bambucifoliae; Ceratoglyphina bambusae; Ceratovacuna nekoashi
C	Secondary	I	Stylets	Monomorphic Not sterile	A. bambucifoliae

[a] Dimorphic means that two morphological forms (one of which is defensive) occur in the same instar.
Monomorphic means that the defenders are not morphologically distinct from other aphids of the same instar.

reproducing adults ('monomorphic'). *Colophina*-type aphids also produce soldiers in the gall on the primary host: these are usually second instars, but they use the same weapons as the secondary host soldiers and they are not sterile. It is likely that the soldiers on the primary and secondary host are homologous and represent a single origin of soldier behaviour. The sterile, dimorphic soldiers on the secondary host are probably the most derived type of soldiers in this group.

The *Pemphigus*-type soldiers all occur in galls on the primary host: they use their stylets and hind-legs as weapons and none of them is sterile. In one species, *Pemphigus spyrothecae*, the first instars of the same generation are dimorphic (Lampel, 1968–1969), but both the soldiers and the normal 1st instars are able to moult to the next instar (Aoki and Kurosu 1986; Foster 1990). The soldiers produced towards the end of the season are probably functionally sterile, since they will not have time to develop into sexuparae before leaf fall. The soldier behaviour observed by Akimoto (1983) in *Eriosoma moriokense* may represent a third independent origin of soldier behaviour in the Pemphigidae. He observed the second and third instars, which are not sterile, behaving defensively, using their stylets.

At least two distinct types of soldier are found in the Hormaphididae. The first type, the horned soldiers, occur on the secondary host (typically a grass, such as bamboo): they are first instars with greatly thickened forelegs armed with strong claws, they have strongly sclerotized tergites, and the frontal horns are sharp and well-developed. Two genera are involved: all the species of *Pseudoregma* appear to have a morphologically distinct soldier caste, but in at least one of the *Ceratovacuna* species (*C.lanigera*) the soldiers are monomorphic and are not sterile. The second type is confined to galls on *Styrax*, the primary host and we have named them the *Styrax*-gall-type of soldier. They are typically sterile, dimorphic, second instars using their stylets as weapons (the aphids on the primary hosts do not have frontal horns). They also have stout spines on the head (four in *Astegopteryx*, two in *Cerataphis fransseni*). The legs are not especially well-sclerotized. Some species (for example, possibly *Pseudoregma koshunensis*) may produce both types of soldier during their life cycle.

Aoki (1987) argues that these two types of soldier, even if they occur in the same species, are not homologous and represent independent origins of soldier behaviour. Part of the evidence for this is the different morphology of the two types of soldiers: those on the primary host are second instars and attack predators with their stylets, those on the secondary host are first instars and pierce predators with their frontal horns. However, in some *Styrax* aphids there are soldier first instars (called 'outsiders' by Kurosu and Aoki 1988*a*), so the instar distinction is probably not fundamental. Further support for the idea that the two types of soldiers are distinct is the fact that it is possible to build up a

Table 10.3. Aphid species in which soldier behaviour has been observed

Species	Primary host	Gall	Secondary host	Primary host soldiers[a]	Secondary host soldiers[a]	References
Hormaphididae Cerataphidinae						
Aleurodaphis takenouchii (Takahashi)	*Styrax japonica*	Broccoli-shaped		II, sterile, di styl, morph	Unknown	Aoki and Usaba (1989)
Astegopteryx bambucifoliae (Takahashi)	*Styrax suberifolia*	Banana-bundle	Bamboos *Dendrocalamus Bambusa*, etc.	II, sterile, di styl, exp	I, not sterile, mono styl, obs	Aoki and Kurosu (1989*a,c,d*), Kurosu and Aoki (1991*a*)
styraci Matsumura	*Styrax obassia*	Coral-like	Monoecious	II, sterile, di styl, exp	–	Aoki and Kurosu (1989*b*, 1990)
Cerataphis fransseni (Hille Ris Lambers)	*Styrax benzoin*	'Bud gall'		II, sterile, di styl, obs		Noordam (1991)
Ceratoglyphina bumbusae[b] van der Goot	*Styrax suberifolia*	Cauliflower	Bamboo (*Pleioblastus*)	II, sterile, di styl, exp	Unknown	Aoki *et al.* (1977), Aoki (1979*a*), Aoki and Kurosu (1989*d*), Aoki *et al.* (1991), Kurosu and Aoki (1991*b,c*)
Ceratovacuna floccifera Noordam			Bamboo		I, sterile(?), di horns, fl, morph	Noordam (1991)
japonica (Takahashi)	*Styrax japonica*	Cat's-paw	Bamboos *Pleioblastus chino*	II	I, sterile, di horns, fl, obs	Aoki *et al.* (1981), Aoki (1987), Aoki and Kurosu (1989*a*), Kurosu *et al.* (1990), Aoki and Kurosu (1991*b*)

Species	Host plant	Gall	Second host	Characteristics (gall)	Characteristics (exule)	References
lanigera Zehntner	Not known		*Miscanthus sinensis*; Sugar cane		I, not sterile, mono horns, exp	Aoki *et al.* (1984), Aoki and Kurosu (1987), Arakaki (1989)
longifila (Takahashi)	Not known				Personal observation	Aoki and Kurosu (1989a)
nekoashi (Sasaki)	*Styrax japonica*	Cat's paw	*Microstegium vimineum*	I, not sterile, mono styl, exp	?	Kurosu and Aoki (1988, 1990)
Distylaphis foliorum (Van der Goot)	*Distylium stellare*	Gall	*Distylium stellare* (free, on leaves)	I, mono styl, fl, morph	No soldiers?	Noordam (1991)
Nipponaphis distyliicola Monzen	*Distylium racemosum* Siebold and Zucc.			I, styl, obs		Itô (1989)
Pseudoregma alexanderi (Takahashi)			*Dendrocalamus sp.* (bamboo)		I, sterile, di horns, fl, exp	Aoki and Miyazaki (1978), Aoki *et al.* (1981)
bambucicola (Takahashi)	*Styrax suberifolia*	Banana-bundle	*Bambusa* (Bamboos)	II, sterile, di styl, fl, exp	I, sterile, di horns, fl, exp	Ohara (1985a,b), Aoki (1987), Sakata and Itô (1991), Sakata *et al.* (1991), Noordam (1991), Aoki and Kurosu (1992)
koshunensis (Takahashi)	*Styrax suberifolia*	Banana-bundle	*Bambusa sp.* (Bamboos)	II, sterile, di styl, fl, morph	I, horns, fl, morph	Aoki *et al.* (1981), Aoki (1982b), Aoki and Kurosu (1991a)
montana (Van der Goot)			Bamboo		I, di horns, fl, morph	Noordam (1991)
paniola (Takahashi)			*Optismenus* spp. *Cyrtococcum*		I, sterile, di horns, fl, morph	Aoki *et al.* (1981), Aoki (1982b), Aoki (1987), Noordam (1991)
pendleburyi (Takahashi)			Bamboos		I, di horns, fl, morph	Noordam (1991)
sundanica (Van der Goot)			Zingiberaceae		I, di horns, fl, morph	Noordam (1991), Schütze and Maschwitz (1991)
Tuberaphis taiwana (Takahashi)	*Styrax formosana*		Loranthaceae	II, sterile styl		Aoki and Korosu (1989a)

Table 10.3—*contd.* Aphid species in which soldier behaviour has been observed

Species	Primary host	Gall	Secondary host	Primary host soldiers[a]	Secondary host soldiers[a]	References
PEMPHIGIDAE						
ERIOSOMATINAE						
Colophina arma Aoki	*Zelkova serrata* (Thunb.) Makino	Globular, loosely closed	*Clematis stans*	II, not sterile, mono styl, fl, ml, exp	I, sterile, di styl, fl, ml, exp	Aoki (1977b, 1980b)
clematicola (Shinji)			*Clematis terniflora*		I, not sterile, mono styl, legs, exp	Kurosu and Aoki (1988b)
clematis (Shinji)	*Zelkova serrata*	Globular, loosely closed	*Clematis apiifolia*	II, not sterile, mono styl, fl, ml, exp	I, sterile, di styl, fl, ml, exp	Aoki (1977a,b, 1980b)
monstrifica Aoki			*Clematis floribunda*		I, sterile, di styl, fl, ml, obs	Aoki (1983)
Eriosoma moriokense Akimoto	*Ulmus davidiana* Planchon	Twisted leaf-galls	*Sedum* spp.	II, III, not sterile, mono styl, obs		Akimoto (1983)
Hemipodaphis persimilis Akimoto	*Zelkova serrata*	Loosely closed pseudogalls		I, not sterile, mono styl, legs, exp		Aoki (1978), Akimoto (1983, 1985)
PEMPHIGINAE						
Pachypappa marsupialis Koch	*Populus maximowiczii* A. Henry	Open leaf gall		I, not sterile, di styl, obs		Aoki (1979b)
Pemphigus dorocola Matsumura	*Populus dorocola*	Pouch gall		I, not sterile, mono styl, hl, obs	?	Aoki (1978, 1980a)
monophagus Maxson	*P. angustifolia*, *P. balsamifera* L.	Pouch gall	Monoecious	I, not sterile, mono styl, hl, morph	—	Aoki and Kurosu (1988)
spyrothecae Passerini	*Populus nigra* L.	Spiral petiole gall	Monoecious	I, not sterile, di styl, hl, exp	—	Aoki and Kurosu (1986), Foster (1990), Benton and Foster (1992)

[a] This column shows, in sequence: whether the soldiers are I or II instars, sterile or not, mono or dimorphic (see Table 10.2); the weapons used (styl (stylets); fl, ml, hl fore-, mid-, or hind-legs); horns; the type of evidence available (morph: based on morphology; obs: observations on live aphids, exp: experimental manipulations).

plausible argument for the gradual evolution of the horned soldiers. The horns probably originally evolved as weapons for dislodging conspecific aphids from their feeding sites (as observed in *Astegopteryx bambucifoliae* (Aoki and Kurosu 1985) and *A. minuta* (Foster, unpublished observations); aphids with this butting behaviour then turned their attention to predators, perhaps in the first instance piercing predators' eggs (as seen in *Ceratovacuna langigera*, which has monomorphic first instars): finally species evolved with fully sterile first instars specialized to stab predators with their horns. This scenario does not involve arguments relating to the aphids on the primary host. If soldier behaviour had evolved on the secondary host as a kind of carry-over from the soldiers in the *Styrax* gall, then we would expect to find that these secondary host soldiers would use their stylets. It is interesting that Aoki and Kurosu (1989*c*) observed such behaviour in *Astegopteryx bambucifoliae* on bamboo.

As further observations are made, it is almost certain that other distinct kinds of soldier will be described. It is likely, for example, that some of the Nipponaphidini that make galls on *Distylium* produce a distinctive type of soldier. The soldier of *Distylaphis foliorum*, drawn by Noordam (1991), looks quite distinct from the other Hormaphidid soldiers, in having massive fore-legs but no frontal horns. Finally, to our knowledge no one has looked critically to see if there might be soldiers in the Fordinae: this seems a particularly fruitful group to investigate.

How might the gall-forming habit act as an enabling factor in the evolution of aphid soldiers?

All but one of the four or five independent origins of soldier behaviour in aphids have occurred in colonies that now live in galls. The exception is provided by the horned cerataphidine soldiers, which probably evolved independently on the secondary host. In what ways might living in a gall predispose the evolution of altruistic behaviour and sterile soldiers in an insect? If we are to discuss this question in the context of kin selection, we need to consider both sides of Hamilton's inequality. In what way might living in a gall

(1) enhance the benefits and reduce the costs of helping by individual soldiers;

(2) foster high values of relatedness between individuals?

1. Gall living and helping behaviour

There are two main reasons why gall living might provide opportunities for altruistic behaviour. First, although galling may originally have evolved to provide protection from natural enemies, especially parasitoids (there is not space here to consider the selective advantages of galling; for discussion see Price *et al.* (1987)), gall aphids face serious fitness costs from attack by specialist predators. For example, Dunn (1960) found that between 90 and 100 per cent of *Pemphigus bursarius* galls at a site had been attacked by the heteropteran bug *Anthocoris*. There is now reasonably good evidence that soldier aphids are able to defend the gall against natural enemies. Many reports of soldier defensive behaviour are anecdotal and remarkably few establish that the soldiers are effective against natural enemies. Aoki (1979*a*) showed that soldiers of *Astegopteryx styracicola* attacked syrphid larvae placed on the gall and also pierced human skin and Foster (1990) showed that soldiers of *Pemphigus spyrothecae* were able to kill the specialist gall predator *Anthocoris minki* (Hemiptera: Cimicidae). In many aphid galls, there are only a very small number of entrances to the gall and these can readily be defended by the soldiers: in more diffuse aphid colonies on leaves, it would be considerably more difficult for the aphids to defend each other. In *P. spyrothecae*, the soldiers wait at the gall entrance and can repel natural enemies, in the same way as worker stingless bees or wasps or soldier termites defend their nest entrances.

The second way in which gall living might provide an opportunity for altruism is by setting the scene for the evolution of housekeeping behaviour. Although gall life may have originally ameliorated the microclimate around the aphid colonies (for example, Kennedy 1951, Forrest 1987), enclosure in a gall brings with it the novel and potentially fatal problem of death by drowning in honeydew (liquid faeces). A partial solution to this problem, adopted by the Pemphigidae, is to wrap the honeydew in wax to form non-wetting droplets: however, these droplets must be removed from the gall. There is now good evidence that soldier aphids from both the Pemphigidae and Hormaphididae remove honeydew from the gall. First instars of *Pemphigus dorocola* (Aoki 1980*a*) and soldiers of *Astegopteryx styraci* (Aoki and Kurosu 1989*b*) and *Ceratovacuna japonica* (Kurosu *et al.* 1990) push honeydew droplets out of the galls with their heads, which are armed with spines. A slightly different type of gall-cleaning behaviour was described by Kurosu and Aoki (1991*d*) in *Hormaphis betulae*: the honeydew is pushed out of the gall by the aphids (mostly first instars) like toothpaste from a tube. The most detailed account of gall cleaning is provided by Benton and Foster's (1992) observations on *Pemphigus spyrothecae*. They established experimentally that the soldier's housekeeping behaviour, which includes removing cast skins and dead aphids as well as honeydew droplets, is costly to the soldiers and essential for gall survival. These observations show that several gall aphid species have a complex repertoire of altruistic behaviours and also suggest that gall cleaning may have been a precursor of defensive behaviour. In pushing droplets out of the gall, the aphids may have come into contact with predators trying to get in. It is interesting that there are some gall aphids that show gall-cleaning behaviour but do not attack predators: this is the case in *Hormaphis betulae* and apparently in two *Hamamelistes* species (Kurosu and Aoki 1991*d*) and may also apply to some species of *Pemphigus* (W.A. Foster and T.G. Benton, unpublished).

2. Gall living and relatedness

Genes for altruistic behaviour will spread in a population only if such behaviour is directed towards other individuals likely to be carrying the same genes. Clonal organisms would therefore seem to provide the ideal setting for the evolution of altruism. And yet altruism, at least in species made up of clones of separated individuals, is very rare. Indeed, as Hamilton (1987) has pointed out, the real question is Why is cooperation in clonal animals, such as aphids, so uncommon? The general answer is that cooperation is undermined by cheating, that is, unrelated individuals can enter the clone and profit from the help received but make no contribution in return. In a simple-minded way, the role of the gall in this context is to provide a ring-fence around the clone. Enclosure in walls of vegetable matter may originally have protected the clone from predation and climatic extremes, but it also, incidentally, helped to maintain the integrity of the clone. An aphid in a gall can be more confident that its neighbours are clone-mates than can an aphid on an open leaf.

Is there any evidence that aphids in a gall are likely to be from the same clone? In some aphids, for example, species of *Pemphigus*, the gall is closed at the beginning of the season and all the colony members will therefore be a clone. However, the gall must eventually open, either to allow winged aphids to emigrate or to allow honeydew and debris to be removed. In species that make looser galls (for example, *Eriosoma* spp.), dispersing aphids could enter at any time. There is mounting evidence that individuals do indeed migrate between galls. Setzer (1980) analysed allozyme variation within galls of *Pemphigus populitransversus* and *P. populicaulis* and found that as many as 25 per cent of the aphids were not of the same phenotype as the fundatrix. Galls that were experimentally isolated with cloth bags showed greatly reduced variation. However, Hebert *et al.* (1990) failed to detect allozyme variation in pairs of aphids from 690 galls of the sumac aphid *Melaphis rhois*, suggesting that migration was not occurring in this species. Aoki (1979*b*) described a specialized migratory morph in *Pachypappa marsupialis*, which was apparently able to enter other galls of the same species. In a study of the effect of premature leaf abscission on *Pemphigus betae*, Williams and Whitham (1986) demonstrated that immature aphids migrate out of galls, both on abscizing and healthy leaves. These observations indicate that a thorough study of intergall migration, involving measurements of intergall related-ness, using modern biochemical techniques and direct observations and experiments on the aphids themselves is needed.

If intergall migration is widespread, what implications does this have for the evolution of altruism? It is possible that one of the functions of the soldiers might be to defend the gall against non-kin migrating aphids. Foster (1990) confined soldiers from different colonies of *Pemphigus spyrothecae* but never observed attacks on conspecifics, even though poten-tial predators were readily attacked under equivalent conditions. Similarly, soldiers of *Pseudoregma bambucicola* did not attack conspecific intruders (Sakata and Itô 1991). The only reported example of a discrimination mechanism is that shown by gall-dwelling soldiers of *Ceratoglyphina bambusae* (Aoki *et al.* 1991): soldiers always attack conspecific non-soldiers, regardless of kinship, when they are on the gall surface. Since most soldiers live within the gall and presumably do not attack their clone-mates there, the authors suggest that this is a context-dependent behaviour which does not rely on kin recognition. This accords with Grafen's (1990) argument that kin recognition in nature is unlikely and that most kin discrimination utilizes *group* recognition cues, such as colony odour. The gall might be of crucial importance here in providing a simple contextual 'decision-rule' for the soldiers: aphids outside the gall or at the gall entrance should be attacked, aphids in the gall ignored.

Is gall living important in the evolution of aphid soldier castes?

It is clear that gall living is neither a necessary nor a sufficient condition for the evolution of social behaviour in aphids: soldier aphids on the secondary hosts do not live in galls and many gall-living species do not have soldiers. It could be argued that the soldiers of the *Colophina-* and *Styrax-*gall-type that occur on the secondary host originally evolved for defence in the gall on the primary host, but Aoki (1987) makes a strong case that the horned soldiers of the Cerataphidinae represent a completely independent origin of soldier behaviour. What factors might have driven the evolution of these horned soldiers? Aoki (1987) has argued con-vincingly that the horns were probably originally devices for prising conspecific aphids away from their feeding sites. However, it is not at all clear, from a kin-selection standpoint, why butting should have evolved in pure clones and if the colonies are not pure clones, the evolution of cooperative defensive behaviour seems unlikely. The critical factor here might be the fact that hormaphidid and pemphigid aphids do not, in general, disperse very effectively on and between the secondary hosts. The winged forms produced on the secondary hosts are usually, but not always, sexuparae that fly back to the primary hosts. As a result, it is difficult for them to escape predation by flight or by producing more alates when predation risks increase: instead, they must remain on the host and defend themselves against predation.

A really high level of predation could provide an immediate fitness cost associated with 'cheating', such that cooperation might persist, even if there were some mixing between colonies. A non-soldier-producing female might initially enjoy greater reproductive success in a mixed-clone colony, but if she and her descendants are ultimately wiped out because of inadequate defences, then altruism would be being selected for by differential survival of colonies rather than of individuals. This would be an example of group selection (*sensu* Wilson 1977). If predation was sufficiently intense, it might benefit clones both to produce migrant aphids and to allow alien migrants into their own galls on a kind of tit-for-tat basis (see also Aoki 1982*a*).

A final problem is to try to explain why many gall-forming aphids do *not* produce soldiers. A possible explanation, which seems to be supported from the very limited available data, is that soldiers are only produced in galls that are relatively long-lived and remain open for a long time (see also Akimoto, in preparation). We are arguing that the only galls worth defending are those that will persist for a long time: short-lived colonies do not merit the costs associated with soldier production. This argument is analogous to that used in the apparency theory of optimal plant defence (for example, Feeny 1976): only plants that are long-lived merit expensive, quantitative defences against herbivores. Observations on two aphid genera support this idea. *Eriosoma moriokense* is the only species of *Eriosoma* in Japan which shows defensive behaviour and it is also the species with the longest-lived galls (Akimoto 1988). In *Pemphigus*, defensive larvae have been observed in *P. spyrothecae* (Aoki and Kurosu 1986; Foster 1990) and *P. dorocola* (Aoki 1978, 1980*a*), both of which have long-lasting (at least 5 months) galls. *Pemphigus monophagus*, like *P. spyrothecae*, is non-host alternating and has long-lived galls with mor-phologically distinct first instars that look as though they might play a defensive role (Aoki and Kurosu 1988). *Pemphigus bursarius* (L), which has shorter-lived galls, does not appear to have defensive first instars (W.A. Foster and T.G. Benton, unpublished), but these other species need to be looked at in greater detail.

Further work

An essential prerequisite for further work on the evolution of social behaviour in aphids is to establish a phylogenetic classification of the Hormaphididae and Pemphigidae and their relationship to the other aphids. Secondly, we need direct evidence, using electrophoretic and DNA fingerprinting techniques, of the clonal structure of aphid societies. Finally, we need more observations and, in particular, more experiments, on the defensive behaviour of aphids, especially comparative studies of closely related species and of hitherto neglected groups, such as the Fordinae and the Nipponaphidinae.

Acknowledgments

We are very grateful to David Stern, Tim Benton, Roger Blackman, and Victor Eastop for their comments on the manuscript and to John Rodford for drawing the figure.

References

Akimoto, S. (1983) A revision of the genus *Eriosoma* and its allied genera in Japan (Homoptera: Aphidoidea). *Insecta Matsumurana*, new series, **27,** 37–106.

Akimoto, S. (1985). Occurrence of abnormal phenotypes in a host-alternating aphid and their implications for genome organisation and evolution. *Evolutionary Theory*, **7,** 179–93.

Akimoto, S. (1988). Competition and niche relationships among *Eriosoma* aphids occurring on Japanese elm. *Oecologia*, **75,** 44–53.

Aoki, S. (1977*a*). *Colophina clematis* (Homoptera, Pemphigidae), an aphid species with 'soldiers'. *Kontyû, Tokyo*, **45,** 276–82.

Aoki, S. (1977*b*). A new species of *Colophina* (Homoptera, Aphidoidea) with soldiers. *Kontyû, Tokyo*, **45,** 333–7.

Aoki, S. (1978). Two pemphigids with first instar larvae attacking predatory intruders (Homoptera, Aphidoidea). *New Entomologist*, **27,** 7–12.

Aoki, S. (1979*a*). Further observations on *Astegopteryx styracicola* (Homoptera: Pemphigidae), an aphid species with soldiers biting man. *Kontyû, Tokyo*, **47,** 99–104.

Aoki, S. (1979*b*). Dimorphic first instar larvae produced by the fundatrix of *Pachypappa marsupialis* (Homoptera: Aphidoidea). *Kontyû Tokyo*, **47,** 390–8.

Aoki, S. (1980*a*). Occurrence of a simple labor in a gall aphid, *Pemphigus dorocola* (Homoptera, Pemphigidae). *Kontyû, Tokyo*, **48,** 71–3.

Aoki, S. (1980*b*). Life cycles of two *Colophina* aphids (Homoptera, Pemphigini) producing soldiers. *Kontyû, Tokyo*, **48,** 464–76.

Aoki, S. (1982*a*). Soldiers and altruistic dispersal in aphids. In *The biology of social insects*, (ed. M.D. Breed, C.D. Michener, and H.E. Evans), pp. 154–8. Westview Press, Boulder, CO.

Aoki, S. (1982*b*). Pseudoscorpion-like second instar larvae of *Pseudoregma shitosanensis* (Homoptera, Aphidoidea) found on its primary host. *Kontyû, Tokyo*, **50,** 445–53.

Aoki, S. (1983). A new Taiwanese species of *Colophina* (Homoptera, Aphidoidea) producing large soldiers. *Kontyû, Tokyo*, **51**, 282–8.

Aoki, S. (1987). Evolution of sterile soldiers in aphids. In *Animal societies: theories and facts*, (ed. Y. Itô, J.L. Brown, and J. Kikkawa), pp. 53–65. Japan Scientific Societies Press, Tokyo.

Aoki, S. and Kurosu, U. (1985). An aphid doing a headstand: butting behaviour of *Astegopteryx bambucifoliae* (Homoptera: Aphidoidea). *Journal of Ethology*, **3**, 83–7.

Aoki, S. and Kurosu, U. (1986). Soldiers of a European gall aphid, *Pemphigus spyrothecae* (Homoptera: Aphidoidea): Why do they molt? *Journal of Ethology*, **4**, 97–104

Aoki, S. and Kurosu, U. (1987). Is aphid attack really effective against predators? A case study of *Ceratovacuna lanigera*. In *Population structure, genetics and taxonomy of aphids and Thysanoptera*, Proceedings of International Symposia, Smolenice, Czechoslovakia, (ed. J. Holman, J. Pelikán, A.F.G. Dixon, and L. Weismann), pp. 224–32. SPB Academic Publishing, The Hague.

Aoki, S. and Kurosu, U. (1988). Secondary monoecy of a North American gall aphid, *Pemphigus monophagus* (Homoptera, Aphidoidca). *Kontyû, Tokyo*, **56**, 394–401.

Aoki, S. and Kurosu, U. (1989a). Two kinds of soldiers in the tribe Cerataphidini (Homoptera: Aphidoidea). *Journal of Aphidology*, **3**, 1–7.

Aoki, S. and Kurosu, U. (1989b). Soldiers of *Astegopteryx styraci* (Homoptera, Aphidoidea) clean their gall. *Japanese Journal of Entomology*, **57**, 407–16.

Aoki, S. and Kurosu, U. (1989c). A bamboo horned aphid attacking other insects with its stylets. *Japanese Journal of Entomology*, **57**, 663–5.

Aoki, S. and Kurosu, U. (1989d). Host alternation of two Taiwanese cerataphidines (Homoptera: Aphidoidea). *Akitu*, **107**, 1–11.

Aoki, S. and Kurosu, U. (1990). Biennial galls of the aphid *Astegopteryx styraci* on a temperate deciduous tree, *Styrax obassia*. *Acta Phytopatholgica et Entomologica Hungarica*, **25**, 1–4.

Aoki, S. and Kurosu, U. (1991a). Host alternation of the aphid *Pseudoregma koshunensis* (Homoptera) in Taiwan. *New Entomologist*, **40**, 31–3.

Aoki, S. and Kurosu, U. (1991b). Discovery of the gall generations of *Ceratovacuna japonica* (Homoptera: Aphidoidea). *Akitu*, **122**, 1–6.

Aoki, S. and Kurosu, U. (1992). Gall generations of the soldier-producing aphid *Pseudoregma bambucicola* (Homoptera). *Japanese Journal of Entomology*, **60**, 359–68.

Aoki, S. and Miyazaki, M. (1978). Notes on the pseudoscorpion-like larvae of *Pseudoregma alexanderi* (Homoptera, Aphidoidea). *Kontyû Tokyo*, **46**, 433–8.

Aoki, S. and Usuba, S. (1989). Rediscovery of '*Astegopteryx*' *takenouchii* (Homoptera, Aphidoidea), with notes on its soldiers and hornless exules. *Japanese Journal of Entomology*, **57**, 497–503.

Aoki, S., Yamane, S., and Kiuchi, M. (1977). On the biters of *Astegopteryx styracicola* (Homoptera, Aphidoidea). *Kontyû, Tokyo*, **4**, 563–70.

Aoki, S., Akimoto, S., and Yamane, S. (1981). Observations on *Pseudoregma alexanderi* (Homoptera, Pemphigidae), an aphid species producing pseudo-scorpion-like soldiers on bamboos. *Kontyû, Tokyo*, **49**, 355–66.

Aoki, S., Kurosu, U., and Usuba, S. (1984).First instar larvae of the sugar-cane woolly aphid, *Ceratovacuna lanigera* (Homoptera, Pemphigidae) attack its predators. *Kontyû, Tokyo,* **52,** 458–60.

Aoki, S., Kurosu, U., and Stern, D.L. (1991). Aphid soldiers discriminate between soldiers and non-soldiers, rather than between kin and non-kin, in *Ceratoglyphina bambusae. Animal Behaviour,* **42,** 865–6.

Arakaki, N. (1989). Alarm pheromone eliciting attack and escape responses in the sugar cane woolly aphid, *Ceratovacuna lanigera* (Homoptera, Pemphigidae). *Journal of Ethology,* **7,** 83–90.

Benton, T.G. and Foster, W.A. (1992). Altruistic housekeeping in a social aphid. *Proceedings of the Royal Society of London. B,* **247,** 199–202.

Dunn, J.A. (1960). The natural enemies of the lettuce root aphid *Pemphigus bursarius* (L.). *Bulletin of Entomological Research,* **51,** 271–8.

Feeny, P. (1976). Plant apparency and chemical defense. *Recent Advances in Phytochemistry,* **10,** 1–40.

Forrest, J.M.S. (1987). Galling aphids. In *Aphids and their biology, natural enemies and control,* (ed. A.K. Minks and P. Harrewijn), pp. 341–52. Elsevier, Amsterdam.

Foster, W.A. (1990). Experimental evidence for effective and altruistic colony defence against natural predators by soldiers of the gall-forming aphid *Pemphigus spyrothecae* (Hemiptera: Pemphigidae). *Behavioral Ecology and Sociobiology,* **27,** 421–30.

Foster, W.A. and Benton, T.G. (1992). Sex ratio, local mate competition and mating behaviour in the aphid *Pemphigus spyrothecae. Behavioral Ecology and Sociobiology,* **30,** 297–307.

Grafen, A. (1990). Do animals really recognise kin? *Animal Behaviour,* **39,** 42–54.

Grafen, A. (1991). Modelling in behavioural ecology. In *Behavioural ecology: an evolutionary approach,* (3rd edn) ed. J.R. Krebs and N.B. Davies), pp. 5–31. Blackwell, Oxford.

Hamilton, W.D. (1964). The genetical evolution of social behaviour I, II. *Journal of Theoretical Biology,* **7,** 1–52.

Hamilton, W.D. (1987). Kinship, recognition and disease: constraints of social evolution. In *Animal societies: theories and facts,* (ed. Y. Itô, J.L. Brown, and J. Kikkawa), pp. 81–102. Japan Scientific Societies Press, Tokyo.

Harvey, P.H. and Pagel, M.D. (1991). *The comparative method in evolutionary biology.* Oxford University Press, Oxford.

Hebert, P.D.N., Finston, T.L., and Foottit, R. (1991). Patterns of genetic diversity in the sumac gall aphid, *Melaphis rhois. Genome,* **34,** 757–62.

Heie, O.E. (1980). The Aphidoidea (Hemiptera) of Fennoscandia and Denmark 1. *Fauna Entomologica Scandanavica,* **9,** 1–239.

Heie, O.E. (1987). Palaeontology and phylogeny. In *Aphids and their biology, natural enemies and control,* (ed. A.K. Minks and P. Harrewijn), pp. 367–91. Elsevier, Amsterdam.

Hölldobler, B. and Wilson, E.O. (1990). *The ants.* Springer-Verlag, Berlin.

Ilharco, F.A. and van Harten, A. (1987). Systematics. In *Aphids and their biology, natural enemies and control,* (ed. A.K. Minks and P. Harrewijn), pp. 51–78. Elsevier, Amsterdam.

Itô, Y. (1989). The evolutionary biology of sterile soldiers in aphids. *Trends in Ecology and Evolution,* **4,** 69–73.

Kennedy, J.S. (1951). Benefits to aphids from feeding on galled and virus-infected leaves. *Nature, London,* **168,** 825.

Kurosu, U. and Aoki, S. (1988*a*). First-instar aphids produced late by the fundatrix of *Ceratovacuna nekoashi* (Homoptera) defend their closed gall outside. *Journal of Ethology,* **6,** 99–104.

Kurosu, U. and Aoki, S. (1988*b*). Monomorphic first instar larvae of *Colophina clematicola* (Homoptera, Aphidoidea) attack predators. *Kontyû, Tokyo,* **56,** 867–71.

Kurosu, U. and Aoki, S. (1990). Formation of a 'cat's-paw' gall by the aphid *Ceratovacuna nekoashi* (Homoptera). *Japanese Journal of Entomology,* **58,** 155–66.

Kurosu, U. and Aoki, S. (1991*a*). The gall formation, defenders and life cycle of the subtropical aphid *Astegopteryx bambucifoliae* (Homoptera). *Japanese Journal of Entomology,* **59,** 375–88.

Kurosu, U. and Aoki, S. (1991*b*). Molting soldiers of the gall aphid *Ceratoglyphina bambusae* (Homoptera). *Japanese Journal of Entomology,* **59,** 576.

Kurosu, U. and Aoki, S. (1991*c*). Incipient galls of the soldier-producing aphid *Ceratoglyphina bambusae* (Homoptera). *Japanese Journal of Entomology,* **59,** 663–9.

Kurosu, U. and Aoki, S. (1991*d*). Gall cleaning by the aphid *Hormaphis betulae*. *Journal of Ethology,* **9,** 51–5.

Kurosu, U., Stern, D.L., and Aoki, S. (1990). Agonistic interactions between ants and gall-living soldier aphids. *Journal of Ethology,* **8,** 139–41.

Lampel, G. (1968–1969). Untersuchungen zur morphenfolge von *Pemphigus spirothecae* Pass. 1860 (Homoptera, Aphidoidea). *Bulletin Naturforsch. Ges. Freiburg,* **58,** 56–72.

Moran, N.A. (1992). The evolution of aphid life cycles. *Annual Review of Entomology,* **37,** 321–48.

Noordam, D. (1991). Hormaphidinae from Java (Homoptera: Aphididae). *Zoologische Verhandeligen,* **270,** 1–525.

Ôhara, K. (1985*a*). Observations on the oviposition behaviour of *Metasyrphus confrater* (Diptera, Syrphidae) and the defensive behaviour of soldiers of *Pseudoregma bambucicola* (Homoptera, Pemphigidae). *Esakia,* **23,** 99–105.

Ôhara, K. (1985*b*). Observations on the prey–predator relationship between *Pseudoregma bambucicola* (Homoptera, Pemphigidae) and *Metasyrphus confrater* (Diptera, Syrphidae), with special reference to the behaviour of the aphid soldiers. *Esakia,* **23,** 107–10.

Peeters, C. and Crozier, R.H. (1988). Caste and reproduction in ants: not all egg-layers are queens. *Psyche,* **95,** 283–8.

Price, P.W., Fernandes, G.W., and Waring, G.L. (1987). Adaptive nature of insect galls. *Environmental Entomology,* **16,** 15–24.

Remaudière, G. and Stroyan, H.L.G. (1984). Un *Tamalia* nouveau de Californie (USA) discussion sur les *Tamaliinae* Subfam. Nov. (Hom. Aphididae). *Annales de la Société Entomologique de France,* New Series, **20,** 93–103.

Ridley, M. (1983). *The explanation of organic diversity: the comparative method and adaptations for mating.* Oxford University Press, Oxford.

Sakata, K. and Itô, Y. (1991). Life history characteristics and behaviour of the bamboo aphid, *Pseudoregma bambucicola* (Hemiptera: Pemphigidae), having sterile soldiers. *Insectes Sociaux,* **38,** 317–26.

Sakata, K., Itô, Y., Yukawa, J., and Yamane, S. (1991). Ratio of sterile soldiers in the bamboo aphid, *Pseudoregma bambucicola* (Homoptera: Aphididae), colonies in relation to social and habitat conditions. *Applied Entomological Zoology*, **26**, 463–8.

Schütze, M. and Maschwitz, U. (1991). Enemy recognition and defence within trophobiotic associations with ants by the soldier caste of *Pseudoregma sundanica* (Homoptera: Aphidoidea). *Entomologie General*, **16**, 1–12.

Setzer, R.W. (1980). Integrall migration in the aphid genus *Pemphigus*. *Annals of the Entomological Society of America*, **73**, 327–31.

Williams, A.G. and Whitham, T.G. (1986). Premature leaf abscission: an induced plant defence against gall aphids. *Ecology*, **67**, 1619–27.

Wilson, D.S. (1977). Structured demes and the evolution of group-advantageous traits. *American Naturalist*, **111**, 157–85.

Wool, D. (1984). Gall-forming aphids. In *Biology of gall insects*, (ed. T.N. Ananthakrishnan.), pp. 11–58. Edward Arnold, Delhi.

11. A breadfruit amongst the dipterocarps: galls and atavism

R.M. JENKINS and D.J. MABBERLEY

Department of Plant Sciences, University of Oxford, South Parks Road, Oxford, UK

Abstract

For nearly three centuries the spinose gall of the Indian tree, *Hopea ponga* (Dennst.) Mabberley (Dipterocarpaceae) was mistaken for a fruit, especially one of a species of *Artocarpus* Forster & Forster f. (Moraceae). The history of the taxonomic confusion is related and the significance of the superficial resemblance of the gall to a fruit is discussed. Fieldwork in Malabar has led to an understanding of the development and ecology of the gall and these are considered in relation to the life cycle of the cecidozoan, an undescribed species in the *Beesonia–Gallacoccus* complex (Homoptera: Coccoidea).

The echinate gall is of further interest in that the spine is a feature otherwise unknown in the Dipterocarpaceae. From a study of literature and herbarium collections, other examples that are of geographical and taxonomic significance from across the family are described. It is suggested that the spine is an atavistic feature whose origins lie in the spinose-fruited dipterocarp ancestors of Gondwanaland. Whilst modern dipterocarps lack spines, some related taxa, thought to have arisen from the same stock, retain this ancestral trait, amongst which are durians *Durio* Adans. (Bombacaceae) and breadfruits (*Artocarpus*).

Introduction

Hopea ponga (Dennst.) Mabberley (Dipterocarpaceae) is most commonly encountered as the dominant component of cleared forest sites in the wet zone of South West India, where it persists in dense coppiced stands, rarely exceeding 50 cm high, which are grazed regularly and cut for firewood. In the field, this species is characterized by its abundant spinose axillary and terminal bud galls which reach 3 cm across and are often in clusters of up to 20 or more, giving the impression of fruits. On dissection,

Plant Galls (ed. Michèle A. J. Williams), Systematics Association Special Volume No. 49, pp. 183–99. Clarendon Press, Oxford, 1994. © The Systematics Association, 1994.

the gall is shown to comprise a vertically reniform central 'core', permeated by sclerenchyma and vascular tissue, from which radiates a mass of lignifed and vascularized spines of generally obtrullate outline. With the core, these surround the large single cecidozoan, a female coccid of the *Beesonia–Gallacoccus* complex (Homoptera: Coccoidea). Whilst the spines are closely packed near their apices, small gaps form between their tapering bases, which expand as the gall develops. These are occupied by offspring of the female. The following account describes the taxonomic misinterpretation that arose when early collectors confused the gall with a fruit. With an understanding of its true identity, the significance of the resemblance is reassessed.

A history of taxonomic confusion

Published in twelve volumes in Amsterdam, van Reede's *Hortus Malabaricus* (1678–1693) is one of the earliest works of significance to be published. Linnaeus (1737, p. 12) noted of it in his *Genera Plantarum*: 'I have not put my whole trust in any other author, excepting the work *Hortus Malabaricus* by the illustrious van Reede' (Vaczy 1980). Nevertheless, the work contains several idiosyncracies including nomenclatural inconsistencies, doubtless a consequence of using at least five different languages (Hamilton 1835), an apparent lack of any herbarium records (Johnston 1970), and accounts of new taxa whose descriptions are based upon galled material (Manilal *et al.* 1980). Of the latter, perhaps of greatest interest is *Hopea ponga*, which appears in *Hortus Malabaricus* as '*Ponga*' (Vol. 4, 1683, t. 35), a Malayalam (the language of Malabar) name and '*Pongu*' (or '*Ilapongu*'), the Tamil counterpart. The accompanying engraving is of a non-flowering branch, in the axils of which are clusters of 'fruits' (Fig. 11.1a). Figure 11.1b shows one of the clusters enlarged. But the specimen is sterile and in the absence of true flowers and fruits, van Reede's judgement was distorted by an extraordinary disguise—the gall as a fruit (Fig. 11.1c). Van Reede was not trained as a botanist and familiarity with fruit details might have prevented his mistake, but fortunately it did not and, consequently, he unwittingly drew attention to a gall of remarkable properties.

The similarity of the spinose gall is with the inflorescences and compound fruits of the genus *Artocarpus* Forster & Forster f. (Moraceae), whose economic importance would have been familiar to van Reede as the jackfruit (*A. heterophyllus* Lam.) (Fig. 11.1d). Indeed, he wrote of *Ponga*: 'the fruits are attached to the branches themselves, and arise not otherwise than '*Tsjaka*'. *Tsjaka* is the Malayalam name for jackfruit. The 'fruits' are described as having 'echinate appendages' and many 'oblong-rounded, pointed, red-tipped female parts'—the spines being confused with flowers

Fig. 11.1. (a) Van Reede's engraving of *Ponga* for *Hortus Malabaricus*, 1683, (b) detail of one of the clusters of 'fruits', (c) the gall of *Hopea ponga* collected at Udupi, Karnataka, India (13.3°N 74.8°E), (d) engraving of jackfruit (*Artocarpus heterophyllus*) in *Hortus Malabaricus*, 1685.

(a)

(c)

HOPEA WIGHTIANA *(Wall.)*

Fig. 11.2. (a) Annotated notes by Fischer (1926) on a sterile galled specimen of *Hopea ponga* held at the Royal Botanic Gardens, Kew (Meebold 1908), (b) Wight's (1840) colour illustration of *Hopea wightiana (H. ponga)*, showing flowers, fruits, and an axillary gall— whole, bisected, and a single spine in detail, (c) the spinose gall drawn by Beddome (1871).

(Manilal *et al.* 1980). The 'seeds' that he describes are undoubtedly, as Jarrett (1960) notes, 'pupae of the insects'. The misinterpretation of the gall was concreted when, based on van Reede's description, Dennstedt (1818) named it *Artocarpus ponga.*

The first description of *Hopea wightiana* (= *H. ponga*) made from flowering material was provided by Wight and Walker-Arnott (1834). No mention is made of the gall and *Artocarpus* is not treated in their work. Hamilton (1837), in a key to *Hortus Malabaricus*, argued that 'Reede's figure of the fruit has little resemblance to *Artocarpus*' and considered the 'dissected capitulum to represent the female flower before the singular receptaculum has elevated the seed'. Again, the spines are mistaken for a mass of flower buds. Hamilton concluded that *Ponga* might be a species of *Broussonetia* L'Hérit. ex Vent. (Moraceae), a genus nevertheless closely allied to *Artocarpus*. Following Hamilton, *Hopea ponga* had names in three genera, but only one, *H. wightiana* Wall. ex Wight & Arn., was based on fertile material.

Herbarium collections made at around this time are also valuable. Of the specimens held at the Royal Botanic Gardens, Kew, seven sheets are galled and all of these are sterile—although collectors may have thought otherwise. Fischer (1928), who succeeded Gamble in writing the *Flora of the Presidency of Madras* (1915–36), has annotated a sheet collected by Meebold in 1908 (Fig. 11.2a). Whilst it is only a seedling, Meebold identifies it as *Artocarpus*. Fischer details the 'perianth' which is 'completely closed over the swollen style' with the 'ovary apparently two-celled', but his description is of a spine! Both Wight (1840), who presented the first colour plate of *H. wightiana* and Beddome (1871) based their descriptions upon flowering and fruiting material and recognized that the gall is not a product of normal development (Fig. 11.2b and c). Whilst Beddome considers it to be 'much like the fruit of a Spanish chestnut', both authors refer to it as being insect-induced and maintain the supposition that the gall is an abortive inflorescence. Even Hooker (1874) notes that 'the inflorescence is often diseased and condensed into a globular mass'. Oddly, he also states that Roxburgh's (1832) description of *Artocarpus lancaefolia* is based upon this species, a view repeated by Cooke (1903). However, Roxburgh gives the distribution of his species as what is now Malaysia and Jarrett (1960) considers that *A. lancifolius* (or *lancaefolia*) is the valid name for another species.

From Wight and Walker-Arnott (1834) and Wight (1840) onwards, the name of *Hopea wightiana* was established and mention sometimes made of the very abundant gall. Dennstedt's *Artocarpus ponga*, which antedates *H. wightiana*, was overlooked until, in 1960, Jarrett became the first to identify that van Reede's *Ponga* was a species of *Hopea* and not *Artocarpus*, thus ending 277 years of misconception. Normally the identification of an earlier specific name would have required a nomenclatural change, but as Dennstedt's name was based upon a gall, that is, a monstrosity, the International Code dictated that such a name was invalid. A final change was to occur in 1975, when, due to a reversion of Article 71, on monstrosities, Dennstedt's specific name became valid once more and Mabberley (1979) published *Hopea ponga* (Dennst.) Mabberley. It is perhaps fitting that Dennstedt's *Ponga* has been retained, for that Malayalam name is now directly creditable to its great source, *Hortus Malabaricus*.

The remarkable echinate gall has continued to confuse. Mani (1973) still viewed it as a 'flower gall', despite the fact, as herbarium collections and field studies show, the gall can be induced on seedlings and coppiced shoots. This raises the question of why the gall should resemble a fruit.

The significance of gall structure to development

There is good reason, perhaps, for confusing the gall with a fruit. Its axillary position, spherical shape, and green coloration all resemble an immature fruit, but the deception is founded upon the spine, an uncommon character that is usually associated with large tropical fruits (Corner 1949, 1954a) and rarely with galls. Of equal importance to the presence of the spine itself, is its shape and how this changes with age is critical to the biology of the gall.

Figure 11.3a shows the sequence development of a gall which is described in detail in Jenkins (1992). The gall is induced when a first instar female nymph inserts her stylet into the cortical tissues of the bud. Normal growth is arrested and the bud grows over and around the female, partially enclosing her. The cortical tissues greatly enlarge, forming the central nutritive 'core', whose cells exhibit hyperplasia, nuclear hypertrophy, and metaplasy (Kingsley 1985). The spines appear to originate from the unicellular trichomes that densely cover the bud and young stem. These undergo a longitudinal division, with the whole of the spine originating from the lowermost cell only. The internal structure is complex: the spine becomes lignified, with the addition of sclerenchyma and vascular tissues and, on the outer surface, a layer of 'tertiary' unicellular trichomes develops. Instars feeding upon the spines frequently induce new galls to develop, producing clusters of up to 20 or more galls of various developmental stages.

In young galls the spines are closely-packed along most of their length, with the exception of a generally acuminate apex. Coinciding with the reproduction of the female, gaps start to form between the bases of neighbouring spines. These are caused by a combination of continued spine-widening near their apex and the enlargement of the core. However, the spines at the basal end of the gall, surrounding the female, remain tightly adpressed, with very few gaps forming (Fig. 11.3b). The first instar

(a)

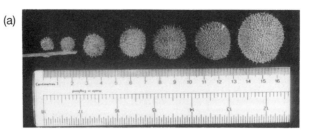

(b)

(c)

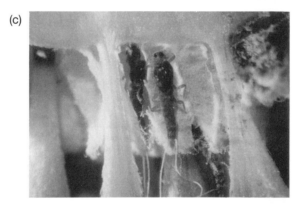

(d)

male nymphs migrate to these spaces where they settle, feed, and pass through three stages. At a specific juncture, when the adult winged males emerge from their pupae (Fig. 11.3c), the radial growth of the upper portion of the spine ceases and an acceleration in the enlargement of the core causes the spines to separate along their whole length. This allows the males to depart, which they do at night, and mate with the first instar females that live exclusively on the surface of the host. Lacking mouthparts, the males are very short-lived (Beardsley 1984; Jenkins 1992). The effectiveness of the spines is such that as soon as they fully part, other insects are able to gain access to the nutrient rich core, which is rapidly consumed (Fig. 3d).

The spines form a dynamic protection, with the gaps between the spine bases forming a three-dimensional matrix and an optimum volume for instar occupation, yet the closely packed upper regions seal them from the outer surface. Indeed, the arrangement may be likened to a cathedral, with its congregation between the columns whose connecting vaults support the roof above.

The cuckoo syndrome

Induced by a parasite to surround and protect it during growth and reproduction, the spine of the gall evokes comparison with the spine of the fruit—the insect causes the plant to develop a tissue that would normally be used for the protection of its own offspring. It is a character 'hijack': the gall is behaving as a fruit and as such it is the ultimate extension of the cuckoo syndrome.

The most significant predators of tropical fruits appear to be large birds such as parrots and herbivorous mammals including marsupials, especially squirrels, bats, and monkeys, which consume fruits before they are mature, with a consequent waste of seeds. Protection of fruits may involve a combination of camouflage, generally by greenness amongst the foliage, chemical defences, which induce unpalatability, and mechanical protection, such as woody fruit walls, persistent sepals, hairs, and spines.

Fig. 11.3. (a) A series of the spinose gall of *H. ponga*, from early development to maturity, (b) reproductive stage in section, with the large basal female partially enclosed by the core and radiating mass of spines. Gaps between the bases of the spines become occupied by male nymphs, but the spines around the female are more numerous and gaps are few (× 7), (c) adult males before departure from the gall (×32), (d) mature gall—the spines having separated and the males departed, other insects are able gain access to the core, which is rapidly consumed by phytophagous larvae (× 7).

Massive armature is closely associated with large fleshy seeds. Such constructions are expensive, but they maximize the survival potential of the individual seed.

In contrast to the fruit, the potential predators of the insect are more likely to be specialized parasites and inquilines, with the former, especially, gaining access using their ovipositors (Washburn 1984; Weis *et al.* 1985, Abrahamson and McCrea 1986; Price and Clancy 1986; Weiss and Abrahamson 1986; Meyer 1987; Price *et al.* 1987; Price and Pschorn-Walcher 1988). Particular attention has been given to the cynipine (Cynipinae—gall wasps) galls on species of *Quercus*. In Britain, there are some 30 cecidogenous species on oaks, with 15 associated inquiline species and a further 45 chalcidoid (Chalcidoidea—gall wasps) species which attack these (Askew 1961). In North America the numbers are more than 10-fold higher (Cornell 1983*a,b*). Preventing the access of these insects is of paramount importance to the cecidozoan and the relatively large size of the female (up to 5 × 7 mm) makes her especially vulnerable to attack. As Cornell (1983*a*) writes 'the primary ecological role of complex gall morphology is to provide protection from parasitoid attack'. Furthermore, 'diameter alone precludes attack by many para-sitoids' and hard, compact tissues have been demonstrated to increase parasitoid drilling time by up to 20-fold (Cornell 1983*a,b*). This may explain why the spines at the female end of the gall are more densely packed, completely enclosing her body and inhibiting insects, etc., from crawling down between the spines. The upward-pointing trichomes on the spines appear to enhance this defence. In addition, the position of the female—near the stem on the underside of the gall—and the arrangement of spine apices, make it difficult without dissection, to find the female beneath the surface. It appears as if all the stages of gall development are dependent upon the active feeding of the female (not of the instars) and as she does not reproduce until the gall has reached a relatively advanced state of development, her protection is all the more important. The distribution of male nymphs may reflect the necessary trade-off between reproductive potential and female defence. The length of the spine is clearly important to ensure that a sufficient distance is established between the cecidozoan and the surface of the gall.

Whether it is insects or seeds that the spine protects, the resemblance of the gall to an immature fruit, with its spherical shape, green coloration, and axillary position, probably deters foraging vertebrates. Its final and ultimate hoax is to deceive the botanical collector (van Reede 1683; Dennstedt 1818; Hamilton 1835). When a cuckoo lays its egg in the nest of a surrogate parent, the victim treats it as its own, rearing and feeding the offspring and protecting it from predators.

Perhaps the most remarkable feature of the spine is that it is unknown as a part of normal ontogeny in the family Dipterocarpaceae. This appears to be one of the few examples in which a novel structure is expressed as part of a gall (Mani 1964; Cornell 1983*a,b*). The question arises, therefore, from where does the cecidozoan obtain the information to induce its expression? What is the origin of the spine? The complexity of structure and ontogeny of the gall suggest that the source for the genetic information to induce the gall and the spine in particular, are likely to be inherent in the host: 'Plant cells probably have the latent genetic capability of developing in a wide variety of abnormal directions that are suppressed during normal development' (Cornell 1983*a*). If the spine is an atavistic expression, do the dipterocarps have spinose ancestors?

The gall at the family level

In extending our attention from a single Indian example, we attempted to discover other examples of spinose galls across the family, noting any patterns of taxonomic and geographical distribution. The work was based upon a survey of literature and herbarium collections made on the genera *Dipterocarpus* (69 species; Mabberley 1987), *Shorea* (357 spp.), *Hopea* (102 spp.), and *Vatica* (65 spp.), which together account for some 87 per cent of dipterocarp species. Whilst the survey of published material yielded eight galled species, extensive work at the herbaria of Kew, Leiden, Singapore, and Bogor enabled a further 60 galls to be described (Jenkins 1992). Whilst showing considerable variation in spine shape and size, all appeared to be induced by a similar coccid. Figure 11.4a–d shows some examples. Galls were recorded from each of the four main genera in numbers which reflect the size of each genus. Their geographical range was very considerable, extending across virtually the entire range of each genus, but it is concentrated in the region of maximum host species diversity—the aseasonal forests of S.E. Asia.

There is strong evidence for a Gondwanic origin for the dipterocarps (Ashton 1982,1988), with the largest modern subfamily, the Asian Dipterocarpoideae, formerly inhabiting the eastern tropical part of Gondwanaland. The more 'primitive' offshoot Monotoideae, has remained in a central position in Africa. It is most significant that from this latter branch, three galled species have been described, all apparently induced by a coccid similar to those from Asia. The spines of *Monotes hypoleucas* (Welw.) Gilg and *M. discolor* R.E. Fr. are broad and very hirsute, whereas those of *M. angolensis* De Wild. are narrow and almost 'wiry' (Fig. 11.5a and b). The occurrence of the spinose gall within the widely separated subfamilies Monotoideae and Dipterocarpoideae suggests that the parasitism existed before the division of Gondwanaland. Alternatively, the cecidozoan must have been able to bridge not only any taxonomic gap

Fig. 11.4. Spinose galls on other species of Dipterocarpaceae. (a), *Dipterocarpus acutangulus* Vesque from Sabah, Borneo (Leiden, Mail 3907), (b) *D. acutangulus*, mature gall in section (Kew, Toipin SAN 40714), (c) *D. obtusifolius* Teysm. var. *subnudus* Ryan & Kerr from Thailand, crushed gall showing the long tapering spine shafts covered in whitish wax produced by the male instars (Kew, Marcan *s.n.*), (d) *Shorea fallax* Meijer from Sabah, Borneo, remarkable galls described by the collector as 'green fruits', developing upon terminal and axillary buds, leaf midrib and lateral veins (Leiden, Nooteboom 1164).

Fig. 11.5. (a) *Monotes hypoleucas* (Welw.) Gilg from Angola, with unusually hirsute spines (Kew, J. Gossweiller 1804), (b) *M. angolensis* De Wild from Angola, with a mass of narrow 'wiry' spines (Kew, *Anon.* 9663), (c) *Sarcolaena eriophora* (Sarcolaenaceae), with its terminal hirsute bud gall and large basal cavity which appears to have been occupied by a cecidozoan similar to those of dipterocarp galls (Kew, May 2704) (×1.1).

that evolved subsequently, but also to have travelled the distance between Africa and India. This latter argument is very unlikely and it would appear, therefore, that the association between the *Beesonia–Gallacoccus* complex and modern dipterocarps has endured since the early Cretaceous period, approximately 140 million years ago. Support for this view may come from a coccid gall on *Sarcolaena eriophora* Thouars, whose family Sarcolaenaceae, endemic to Madagascar, is considered to be derived from the same dipterocarpaceous stock. However, the appendages of the gall are hirsute and not echinate (Fig. 11.5c).

If the ability of the cecidozoan to induce spines has persisted for millions of years, can the spine be traced to the ancestors of the modern family or to related modern taxa?

Taxonomic affinities and evolution

In 1949 Corner published his revolutionary Durian Theory, which he later expanded upon (Corner 1953, 1954*a,b*). Concerned with the tropical fruit, he argues that the ancestral form bore a combination of characters: large arillate seeds enclosed within a massive spinose dehiscent capsule, the 'type' example being the Durian (*Durio zibethinus* Murray, Bombacaceae). Such trees had a relatively massive, sparsely branching architecture to support megaphyllous compound leaves and the heavy fruits were held close to the branches or even the main trunk (cauliflory, for example, jackfruit). Whilst only a few distantly related taxa are similarly constructed, many families bear a few 'Durian' characters alongside more derived ones. The fruits of the dipterocarps are typical: whilst lacking spines and bearing unilocular, indehiscent capsules, several ancient taxa have fruits with residual arils (for example, *Upuna*) and two- or three-seeded fruits (*Vatica*). Ashton (1982) states that the dipterocarp fruit is 'indehiscent', that is, 'no known dipterocarp has fruit dehiscent on the tree', however, several genera have fruits which germinate 'along three loculicidal sutures' (*Marquesia, Stemonoporus,* and *Upuna*). The ancestral dipterocarpaceous fruit appears to have been a multilocular capsule, with several large arillate seeds within a thick pericarp. But what of the spine? Corner (1963) described a dipterocarp gall collected from Sabah, Borneo. As with the galls that have been described elsewhere (Jenkins 1992), it is a coccid-induced spinose bud gall. Corner (1963) concluded: '*Dipterocarpus* has lost all its tendency to form a spinose fruit, but that in its stellately hairy epidermis it retains one primitive property that can be evoked biochemically by the action of the insect'.

There is strong evidence for the association of Dipterocarpaceae with the Malvales, specifically Malvaceae, Tiliaceae, and Bombacaceae (Maguire *et al.* 1977, Maguire and Ashton 1980; Ashton 1982, 1988; Swarupanandan 1988). In the search for the 'origin' of the spine, it is perhaps most interesting that the Bombacaceae should be mentioned, for it is in this family that the 'type' for the Durian Theory, *Durio*, is placed. Other notable spiny-fruited genera in Bombacaceae include *Cullenia* Wight, *Coelostegia* Benth., and, in the allied Elaeocarpaceae, *Sloanea* L.

It may be hypothesized that most, if not all of the Gondwanic malvalean/dipterocarpaceous ancestors bore large, capsular fruits with an impressive armour of spines, from which have evolved those with indehiscent, non-spiny pericarps, containing exarillate seeds. Only a few modern taxa in this group display these ancestral characters, but it appears that, in the Dipterocarpaceae at least, the genetic potential to express the spine remains—realized by a cecidogenous parasite. How many other families have the same latent potential to express 'lost' characters, 'locked' within the generations of evolving ancestors? What is the potential of other gall formers if these can produce a breadfruit amongst the dipterocarps?

Conclusions

The history of taxonomic confusion that surrounded the gall of *Hopea ponga* has led us to investigate an association between the *Beesonia–Gallacoccus* complex and Dipterocarpaceae that is remarkable in its taxonomic and geographical magnitude. The potential for further research is very considerable: we are undertaking formally to identify the Indian coccid, which should provide the framework upon which to address the taxonomy of those on other dipterocarps, especially in Africa. With the exception of *Hopea ponga*, the biology of none of these galls appears to have been described in any detail. As Corner (1967) wrote: 'botany needs help from the tropics; its big plants will engender big thinking'.

Note added in proof

The coccid has recently been named *Mangalorea hopeae* Takagi, gen. et sp. nov. (Raman and Takagi 1992).

References

Abrahamson, W.G. and McCrea, K.D. (1986). The impacts of galls and gallmakers on plants. *Proceedings of the Entomological Society of Washington,* **88,** 364–7.

Ashton, P.S. (1982). Dipterocarpaceae. *Flora Malesiana*, **9**(2), 237–553.

Ashton, P.S. (1988). Dipterocarp biology as a window to the understanding of tropical forest structure. *Annual Review of Ecology and Systematics*, **19**, 347–70.

Askew, R.R. (1961). On the biology of the inhabitants of oak galls of Cynipidae (Hymenoptera) in Britain. *Transactions of the Society for British Entomology*, **14**, 237–47.

Beardsley, J.W. (1984). Gall-forming Coccoidea. In *Biology of gall insects*, (ed. T.N. Anathanakrishnan). Oxford & IBH Publishing Co., New Delhi.

Beddome, R.H. (1871). *Flora sylvatica for Southern India*, Vol. 1. Gantz Bros., Madras.

Cooke, T. (1903). *The flora of the Presidency of Bombay*, Vol. 1. Taylor & Francis, London.

Cornell, H.V. (1983*a*). Why and how gall wasps form galls; cynipids as genetic engineers. *Antenna*, **7**, 53–8.

Cornell, H.V. (1983*b*). The secondary chemistry and complex morphology of galls formed by the Cynipinae (Hymenoptera): why and how? *The American Midland Naturalist*, **110**, 225–34.

Corner, E.J.H. (1949). The Durian Theory or the origin of the modern tree. *Annals of Botany*, **27**, 367–414.

Corner, E.J.H. (1953). The Durian Theory extended (I). *Phytomorphology*, **3**, 465–76.

Corner, E.J.H. (1954*a*). The Durian Theory extended (II). The arillate fruit and the compound leaf. *Phytomorphology*, **4**, 152–65.

Corner, E.J.H. (1954*b*). The Durian Theory extended (III). Pachycauly and megaspermy—conclusion. *Phytomorphology*, **4**, 263–74.

Corner, E.J.H. (1963). A dipterocarp clue to the biochemistry of durianology. *Annals of Botany*, **27**, 339–41.

Corner, E.J.H. (1967). On thinking big. *Phytomorphology*, **17**, 24–8.

Dennstedt, A.W. (1818). *Schlüssel zum Hortus Indicus Malabaricus*. Verlage des Landes, Weimar.

Fischer, C.E.C. (1928). *Flora of the Presidency of Madras*, Vol. 3, pp. 1368–9. Adlard & Son, London.

Hamilton, F. (1837). A commentary on the fourth part of the *Hortus Malabaricus*. *Transactions of the Linnean Society of London*, **17**, 147–252.

Hooker, J.D. (1874). *The flora of British India*, Vol. 2. L. Reeve & Co., London.

Jarrett, F.M. (1960). Studies in *Artocarpus* and allied genera. IV. A revision of *Artocarpus* subgenus *Pseudojaca*. *Journal of the Arnold Arboretum*, **41**, 111–40.

Jenkins, R.M. (1992). Taxonomic character change and the significance of galls. D. Phil Thesis deposited at the University of Oxford.

Johnston, M.C. (1970). Still no herbarium records for *Hortus Malabaricus*. *Taxon*, **19**, 655.

Kingsley, S.J. (1985). Studies on the developmental morphology and histochemistry of some galls induced by coccids (Coccoidea: Insecta) from Southern India. D. Phil thesis deposited at the University of Madras.

Linnaeus, C. (1737). *Genera Plantarum*. Holmiae (Stockholm), [Impensis Laurentii Salvii].

Mabberley, D.J. (1979). The Latin name of the '*Ilapongu*' tree (Dipterocarpaceae). *Taxon*, **28**, 587.

Mabberley, D.J. (1987). *The plant-book.* Cambridge University Press, Cambridge.

Maguire B. and Ashton, P.S. (1980). *Pakaraimaea dipterocarpacea* II. *Taxon,* **29,** 225–31.

Maguire, B., Ashton, P.S., De Zeeuw, C., Giannasi, D.E., and Niklas, K.J. (1977). Pakaraimoideae, Dipterocarpaceae of the Western Hemisphere. *Taxon,* **26,** 341–85.

Mani, M.S. (1964). *Ecology of plant galls.* W. Junk, The Hague.

Mani, M.S. (1973). *Plant galls of India.* Macmillan, Madras.

Manilal, K.S., Suresh, C.R. and Sivarajan, V.V. (1980). Galled plants described under distinct names in *Hortus Malabaricus.* In *Botany and history of Hortus Malabaricus,* (ed. K.S. Manilal). Oxford & IBH Publishing Co., New Delhi.

Meyer, J. (1987). Plant galls and gall inducers. Gebrüder Borntraeger, Berlin and Stuttgart.

Price, P.W. and Clancy, K.M. (1986). Interactions among three trophic levels: gall size and parasitoid attack. *Ecology,* **67,** 1593–600.

Price, P.W. and Pschorn-Walcher, H. (1988). Are galling insects better protected against parasitoids than exposed feeders? A test using tenthredinid sawflies. *Ecological Entomology,* **13,** 195–205.

Price, P.W., Fernandes, G.W., and Waring, G.L. (1987). Adaptive nature of gall insects. *Environmental Ecology,* **16,** 15–24.

Raman, A. and Takagi, S. (1992). Galls induced in *Hopea ponga* (Dipterocarpaceae) in southern India and the gall-maker belonging to the Beesoniidae (Homoptera: Coccidea). *Insecta Matsumurana, new series,* **47,** 1–32.

Roxburgh, W. (1832). *Flora Indica.* Thacker & Co, Calcutta and Parbury, Allen & Co., London.

Swarupanandan, K. (1988). Seedling morphology and some contemporary thoughts on the phylogeny and cirumscription of the family Dipterocarpaceae. Dissertation deposited at the Kerala Forest Research Institute, Peechi, Trichur.

Vaczy, C. (1980). *Hortus Indicus Malbaricus* and its importance for botanical nomenclature. In *Botany and history of Hortus Malabaricus,* (ed. K.S. Manilal). Oxford & IBH Publishing Co., New Delhi.

Van Reede, H.A. (1678–1693). *Hortus Indicus Malabaricus.* 12 Vols. Amstelodami, (sumptibus Joannis van Someren et Joannis van Dyck).

Washburn, J.O. (1984). Mutualism between a cynipid gall wasp and ants. *Ecology,* **65,** 654–6.

Weis, A.E. and Abrahamson, W.G. (1986). Evolution of host-plant manipulation by gall makers: ecological and genetic factors in the *Solidago–Eurosta* system. *The American Naturalist,* **127,** 681–95.

Weis, A.E., Abrahamson, W.G., and McCrea, K.D. (1985). Host gall size and oviposition success by the parsitoid *Eurytoma gigantea. Ecological Entomology,* **10,** 341–8.

Wight, R. (1840). *Illustrations of Indian botany.* J.B. Pharoah, Madras.

Wight, R. and Walker-Arnott, G.A. (1834). *Prodromus (Florae Peninsulae Indiae Orientalis),* Vol. 1. Neill & Co., Edinburgh.

12. Gall midges (Cecidomyiidae): classification and biology

KEITH M. HARRIS

International Institute of Entomology, 56 Queen's Gate, London, UK.

Abstract

The Cecidomyiidae probably originated in the Mesozoic age from a group of Diptera with mycophagous larvae. The family is almost certainly monophyletic and of the three extant subfamilies, the Lestremiinae and Porricondylinae are still essentially mycophagous, with larvae feeding on macrofungi and on fungal mycelium in soil, in decaying wood, and in other dead organic matter. The third subfamily, the Cecidomyiinae, contains the greatest number of species and exhibits the greatest biological diversity. It must have evolved rapidly with flowering plants in the Cenozoic. Most of the included species are phytophagous and many induce galls, especially on the aerial parts of flowering plants. Gall induction seems to have evolved on a number of separate occasions within the main phytophagous tribes (Asphondylini, Cecidomyiini, Clinodiplosini, Lasiopterini, and Oligotrophini). Other tribes contain secondary mycophagous and zoophagous species, including predators on aphids, mites, and other invertebrates and parasitoids of aphids and psyllids.

The biological and ecological strategies of the main groups of gall midges are summarized, possible mechanisms of gall induction are reviewed, and probable routes for the evolution of gall induction are indicated.

Introduction

Many biologists know that gall midges, belonging to the dipterous family Cecidomyiidae, induce galls on plants but few have any detailed knowledge of their classification and biology. There are currently probably less than a dozen specialists with any substantial interest in the family on a world-wide basis and interest at national and local levels is similarly limited. Yet the Cecidomyiidae comprise one of the largest families of flies (Diptera), with approximately 5000 described species and many

Plant Galls (ed. Michèle A. J. Williams), Systematics Association Special Volume No. 49, pp. 201–11. Clarendon Press, Oxford, 1994. © The Systematics Association, 1994.

times that number of species yet to be recognized and described, let alone studied in any detail. These flies are also most interesting biologically as they show great diversity of larval feeding strategies, ranging from mycophagy, through highly specialized phytophagy, (including gall induction), to zoophagy, with many species predaceous on insects and other invertebrates and some species developing as endoparasitoids of aphids and psyllids. Some phytophagous species are major pests of crops, especially cereal crops and some of the zoophagous species are of importance as natural enemies of pest species.

The published literature is extensive, mainly as a result of research late last century and early this century by Kieffer and Rübsaamen, based in Europe and by Felt, based in the USA. Barnes, based in the UK, specialized on species of economic importance. Current and recent workers include Sylvén in Sweden, Nijveldt and Roskam in the Netherlands, Meyer and Rohfritsch in France, Möhn and Bühr in Germany, Solinas in Italy, Skuhravá and Skuhravý in Czechoslovakia, Mamaev, Krivosheina and Marikovski in the former USSR, Mani and Grover in India, Yukawa in Japan, and Gagné in the USA. Of these, Gagné is currently the foremost world specialist on the taxonomy of the family and has recently published an excellent account of *The plant-feeding gall midges of North America* (Gagné 1989), which includes summaries of biology, anatomy, classification, and distribution, as well as illustrated keys to galls and an extensive bibliography.

Although many species of Cecidomyiidae induce galls, many do not. The common name 'gall midge' is therefore not entirely appropriate for members of the family as a whole, for which the terms 'cecidomyiid' or 'cecid' are often used.

All cecidomyiids are relatively small flies, usually with wing lengths of 1–2 mm, but with some larger species having wing lengths of up to 15 mm. Wing venation is reduced and adults are generally weak fliers. Compound eyes are generally well developed and the long, many-segmented antennae bear thread-like sensoria, the circumfila, which are often complex. They probably assist mate location by males and host location by females. The female abdomen is often modified to form a tapered, telescoping ovipositor which is used to place eggs in crevices in buds, flowers, and developing fruits. In some genera, such as *Asphondylia*, a needle-like chitinized ovipositor is used to insert eggs into plant tissues.

Larvae are usually less than 5 mm long when fully fed and are, therefore, relatively inconspicuous, but many are orange, yellow, or red in colour, which makes them easier to see on plants or in the soil. Some, but not all, larvae have a median, ventral, thoracic sternal spatula which is unique to the family and therefore diagnostic.

Cecidomyiids mainly feed as larvae, although some adults may occasionally imbibe fluids. In all cases, larvae also feed on fluids, such as plant sap or invertebrate body fluids. Digestion is external, through secretion of digestive enzymes and larvae are unable to ingest solids. Few direct observations have been made but most larvae probably feed by making small incisions in host tissues with their minute, but functional, mandibles. They then secrete enzymes and imbibe the resultant fluid, possibly applying some suction to damaged tissues to enhance the flow of fluids and this method probably applies whether larvae are feeding on fungal mycelium, on tissues of higher plants, or on invertebrate hosts.

Classification

Despite some claims to the contrary, the family is monophyletic, as indicated by the unique larval sternal spatula. No parallel or precursor of this structure is known within the Diptera or elsewhere within the Insecta and, although it has been secondarily lost on many occasions, it is present in all three subfamilies.

Current classification includes the family Cecidomyiidae in the infraorder Bibionomorpha, along with the fungus gnats (Mycetophilidae and Sciaridae), to which they are closely related. The family is divided into the following three subfamilies.

1. Lestremiinae

This is the most primitive group. All species are detritus feeders in decaying organic matter, where they feed mainly on fungal mycelium and none of them are associated with living plants. Genera are mostly species-poor and wide-ranging and approximately 400 species are known.

2. Porricondylinae

Most species of this subfamily also feed on fungi in decaying organic matter. Only two genera, *Asynapta* and *Camptomyia* are associated with plant tissues but neither induces galls.

3. Cecidomyiinae

This is the largest and most highly evolved subfamily. It contains most of the known species and all of the gall-inducing species. A stable higher classification is yet to be established within this subfamily but the latest position is indicated by Gagné (1989) for North America. Table 12.1 is based on Gagné's publication, amended to indicate the most important

Table 12.1. Summary classification of the Cecidomyiinae, including some of the major gall-inducing genera in the Palaearctic region[a]

Cecidomyiinae	Number of species
Cecidomyiidi	
Asphondyliini	
Asphondylia	51
Asteralobia	10
Cecidomyiini	
Ametrodiplosis	12
Contarinia	153
Halodiplosis	37
Harmandia	5
Planetella	26
Resseliella	23
Clinodiplosini	
Ametrodiplosis	12
Lasiopteridi	
Lasiopterini	
Baldratia	23
Lasioptera	44
Stefaniella	5
Stefaniola	67
Oligotrophini	
Arnoldiola	9
Cystiphora	6
Dasineura (including *Rabdophaga*)	288
Jaapiella	31
Janetiella	15
Macrolabis	31
Mayetiola	26
Oligotrophus	10
Psectrosema	10
Rhopalomyia	48
Wachtliella	9

[a] Genera with less than five included species are not listed.
Based on Gagné (1989) and Skuhravá (1986).

gall-inducing genera in the Palaearctic region, as indicated in the catalogue prepared by Skuhravá (1986).

Biology and ecological strategies of gall midges

Although some species of Lestremiinae and Porricondylinae reproduce parthenogenetically, all known members of the subfamily Cecidomyiinae reproduce sexually. Adult life is relatively brief, often no longer than a few hours or days and the priorities are mate location by males, (assisted by release of pheromones by females) and then host location by ovipositing females. Eggs are generally deposited singly on host plants but some species have cutting or piercing ovipositors that facilitate insertion of eggs into plant tissues. Eggs generally hatch within a few days or weeks, followed by larval development, which is the main feeding stage. This may also be completed rapidly, in a few days or weeks, but in many univoltine species it takes the best part of a year or even longer. Adverse conditions may induce a lengthy larval diapause and the record for longevity, observed in the artificial circumstances of an insectary at Rothamsted Experimental Station, relates to one of the wheat blossom midges, *Sitodiplosis mosellana* (Géhin), adults of which emerged 13 years after larvae had entered soil to overwinter (Barnes 1956). Most gall midges are highly host-specific, often being restricted to development during a brief period in the growth of a single plant species or a few closely related species. Synchronization of adult emergence to coincide with the appropriate stage of host plant development is, therefore, critical and the ability of larvae to persist in diapause over a number of years provides a useful insurance against adverse seasons, when unfavourable conditions could threaten the survival of midge populations.

Gall midge larvae may remain on their plant hosts throughout their development, either in galls or within buds, flower-heads, and other plant structures or they may complete their development in soil or leaf litter. Many species leave galls and work their way into the soil or similar substrates by jumping, but some species remain in dehiscent galls which fall to the ground and are subsequently covered by falling leaves.

Pupal development on plants or in soil is relatively short and is usually completed in 1–2 weeks. Pupae are often armed with spines that facilitate limited movements prior to adult emergence and in some cases enable pupae to cut exit holes through plant tissues.

Emergence of adults is generally very closely geared to the seasonal development of the host plant so that emergence coincides with the availability of host tissues in the right state for gall induction. Most adults emerge either at dawn or dusk, when adult activity is favoured by higher humidity and daily emergences may continue for a week or so. Males usually emerge first and remain near emergence sites. They are rapidly attracted to emerging females, probably by pheromones and after brief mating the females depart in search of host plants. They are weak fliers and disperse passively in wind currents until a suitable host plant is located. Egg laying then starts and usually continues over a period of some hours.

In ecological terms, gall midges generally exhibit extreme K strategies, including restrictive food specializations, symbiosis, lack of mobility, close association with particular host plants and ecosystems, and considerable evolution of species complexes *in situ*. These strategies have operated even within extreme terrestrial environments, notably through association with halophytes in deserts and with sclerophyll plants in other dry ecosystems.

Roskam (1992) analysed host plant data for 556 species of cecidomyiid and recorded 29.9 per cent on Asteridae, 25.5 per cent on Rosidae and 16.7 per cent on Dilleniidae, compared with 1.6 per cent on ferns and gymnosperms and none on Magnolidae. This indicates a strong association with woody and herbaceous angiosperms, which reflects recent coevolution.

The strategies of univoltine Japanese gall midges have been summarized recently by Yukawa (1987). This analysis covers 58 species for which sufficient information is available and four main strategies are indicated, outlined below.

Strategy IA species develop rapidly and larvae leave galls and drop to the ground to pass the summer, autumn, and winter in cocoons before pupating in the following spring. They are mostly associated with deciduous trees or herbaceous plants. Galls develop rapidly, mainly on leaves, as simple pouch, fold, or roll galls from which the relatively active larvae can easily escape.

Strategy IB species also develop rapidly but remain in galls after completing their development. Galls fall to the ground in autumn and the larvae overwinter in them and pupate in the following spring. They mainly cause leaf or fruit galls on deciduous trees and herbaceous plants. Galls are more complex, take longer to develop, and lack openings.

Strategy IIA species develop slowly and complete their development by autumn. They remain in galls on plants and pupate in them in the following spring. They induce leaf or bud galls on evergreen trees or stem galls on deciduous trees and on herbs.

Strategy IIB larvae develop very slowly and overwinter as first instars in galls on plants, completing their development and pupating in the galls in the following spring. They induce galls on fruits and the fruits remain attached to the host plants until the following flowering season.

These strategies are interpreted by Yukawa as adaptations to host plant attributes (deciduous/evergreen) and phenology (leaf/fruit development) and to avoidance of parasitoids and predators. The tendency for some taxonomically related groups of gall midges to exhibit a particular strategy is also noted. Some of the most species-rich genera can be categorized in this way, for example, *Contarinia* (IA), *Lasioptera* (IIA), and *Asphondylia* (IIB), although there are exceptions, such as *Dasineura* (IA and IIA).

Galls

Gagné (1989) adopted a broad definition of a gall as 'any predictable and consistent plant deformation that occurs in response to feeding or other stimulus by foreign organisms' and noted that this definition encompasses almost all damage done to plants by gall midges. He also used a simplified terminology in which galls are either simple (leaf spots, leaf blisters, leaf rolls, vein folds, etc.) or complex. Complex galls involve complete reorganization of tissues into shapes that differ fundamentally from any structure normally found on the host plant. Galls induced by gall midges mostly develop on aerial parts of flowering plants, especially on leaves, stems, buds, flowers, and fruits, but some develop on aerial roots of epiphytic orchids and a few develop on fungi. In most cases, each species induces a particular gall and groups of closely related species may produce a range of distinct galls on the same host plant. In such cases the species are often more readily recognized on the basis of the galls that they induce than by any morphological characters of the adult or immature stages.

During the past 20 years illustrated keys to gall midge galls have been published for North America (Gagné 1989), Central and Northern Europe (Bühr 1965), and The Netherlands (Wiebes-Rijks and Houtman 1982) and should be consulted for further information. Gagné is currently working on the Neotropical plant-feeding gall midges. Earlier publications are available for North America, Europe, India, and Indonesia but are now substantially out of date.

Most galls are relatively simple but it is the more complex galls that are of greatest general interest since it is more difficult to provide adequate explanations of the control mechanisms that operate to produce them. One of the most interesting of these is the 'cylinder pistol gall' of a *Contarinia* sp. (= *Lobopteromyia*) on *Acacia ferruginea* DC, described by Mani (1964) as 'the most remarkable of leaf galls from the world'. Rohfritsch (1971, 1974) has studied it and a similar gall induced by *Contarinia ramachandrani* Mani in detail. The gall is formed from two adjacent leaflets of the *Acacia*, the inferior leaflet forming a cylinder and the superior leaflet forming a piston that fits into it, with the larva occupying the terminal cavity. In one instance, illustrated by Rohfritsch (1971), one leaflet contributes the cylinder to one gall and the piston to another, which emphasizes the very different effects that the gall midge larva has on the upper and lower surfaces of the leaflet. Unfortunately, this Indian gall is rare and opportunities of studying the mechanisms that induce it are therefore limited.

Gall induction

Rohfritsch (1992) has published an excellent detailed review of available information on gall induction by gall midges and other insects in which she recognizes three basic models: cecidomyiid, sawfly, and cynipid. The cecidomyiid model is considered to be the most basic or general type of prosoplasmic gall development and the simplest cases in the cecidomyiid model are covering galls. In these, the first instar larvae locate feeding sites on epidermal cells and, within a few hours, initiate gall development through physiological and cytological changes in the cells around them which are induced directly by the feeding activities of the larvae. These activities involve secretion of saliva and direct damage of individual epidermal cells by puncture and suction. This first critical stage of gall initiation, which is referred to as metaplasy, isolates the adjacent cells from control by the host plant and these physiologically modified cells serve as the source for nearly all of the cells that form the gall. Metaplasy is quickly followed by a second phase within the second or third days of larval feeding when cellular hyperplasy produces a wall of cells around the feeding larva. This establishes a new growth polarity perpendicular to the surface of the attacked organ. During the third phase of development the growth of the gall tissues is amplified and redirected and in most cases the gall tissues grow over the larvae. During a fourth stage of gall development the tissues may become further differentiated over a period of several weeks.

In the initial stages of development, nutrititive tissue is usually formed, under the influence of the young larva. This tissue has prominent nuclei, with large nucleoli, a large number of mitochondria, and abundant endosplasmic reticulum and serves as a short-distance transport system between the vascular tissue of the gall and the larva. In some galls a second type of nutritive tissue develops and acts as an attractor and converter of nutrients for the later stages of larval development.

All events in the development of cecidomyiid covering galls require the presence of larvae and development proceeds without disorganization or proliferation of cells. Once the cells have been removed from the normal control of the host plant, they are engaged in a new system which is controlled by the larvae, mainly by light mechanical wounding of cells and by salivary enzymes.

The development of other types of cecidomyiid galls, notably pouch and mark galls, are also the direct result of localized larval activity and Rohfritsch notes that 'The cecidomyiid model is the most common but also the most plastic. It is the model that best demonstrates the variability in gall induction. Better than any other insect group, the cecidomyiids illustrate the development potentials not normally expressed in plants and that can be solicited by externally applied factors.'

Asphondylia species and some species of Lasiopterini induce galls in which a symbiotic fungus ('ambrosia fungus') develops instead of nutritive cells. Fungal spores are probably introduced during oviposition but the mechanisms are not fully understood. In these, and some other cases, galls may be initiated by ovipositing females rather than by first instar larvae.

Evolutionary trends

Rodendorf (1964), in his account of the historical development of the Diptera, states that the Cecidomyiidae are undoubtedly derived from some Mesozoic representative of the Fungivoridea, a group which, as the name suggests, were mainly fungivorous. Gagné (1989) noted that, although the family has ancient roots, it obviously diversified with the flowering plants during the Cenozoic, so that most of its members are very young. He also noted that this diversification is analagous to that of the acalypterate Diptera.

The oldest known fossil cecidomyiids are from the Cretaceous but none of them belongs to plant-feeding groups (Gagné, 1977). Others are known from Upper Oligocene–Lower Miocene strata, which contain representatives of extant genera and indicate that the cecidomyiid fauna of 30 million years ago was similar to that found today (Gagné 1973).

Mamaev (1965) postulated evolution of the family by larval adaptations to utilize a sequence of resources ranging from fungal mycelium in soil, leaf litter, and decomposing wood to feeding in unmodified plant organs, such as buds and flower-heads and, finally, to feeding in plant galls, to which may be added additional specializations such as inquilines, predators, and parasitoids. Most of the adaptations have been histological and biochemical, rather than morphological and the external structure of the larva has generally remained relatively constant throughout a long period of evolution.

Roskam (1992) has recently reviewed the evolution of the main gall-inducing insect groups and supports Mamaev's conclusions that gall induction evolved on a number of separate occasions within the subfamily Cecidomyiinae. Extra intestinal digestion was obviously an appropriate pre-adaptation facilitating transitions from mycetophagy by invasion of wounds, leading to cambial feeding (as in the extant genera, *Ledomyia* and *Profeltiella*) and through utilization of flowers and flower buds, possibly assisted by fungal infections. Gall midge evolution certainly seems to have radiated with the rise of the angiosperms, probably as a result of adaptations of the digestive system for feeding on cambial sap flows. Fewer species are associated with monocotyledons, conifers, and ferns and such associations are probably of more recent origin.

Price *et al.* (1987) reviewed the major hypotheses on the adaptive significance of insect gall formation and indicated that the two main routes to gall formation from free-feeding active herbivores were through mining plant tissues and through sedentary surface feeding. They included the cecidomyiids in this second category and argued that 'In the sedentary surface feeders, persistent feeding in one location is likely to result in differential growth of tissues, depressions and rolling in leaves, and an immediate effect on the boundary layer of the leaf. The herbivore becomes better protected from physical factors, especially moisture stress. The evolution of more potent stimulating mechanisms would cause deepening of depressions and ultimate gall formation. Such a scenario may be easily envisaged for aphids and cecidomyiid larvae. Here the major change of adaptive value seems to be primarily in relation to physical factors and secondarily to nutrition and enemies.' They also note that, although these herbivores would have been well adapted to the nutrition of their host plants, 'Nutrition may well have improved as more cells retained the parenchymatous type, and as meristematic activity and the associated nutrient sink were prolonged.'

The relatively high incidence of cecidomyiid galls in dry habitats suggest that water conservation may be an important factor in the evolution of galls although the primary function must be nutritional.

Conclusions

Much is already known about the classification, biology, and evolution of the gall midges but most of the world's gall midge fauna is still poorly known. European and North American gall-inducing species are relatively well studied but the gall midges of Africa, Asia, Central and South America, Australia, and the Pacific have not been adequately studied. The total numbers of species described from three of the major zoogeographical regions are Afrotropical 154, Oriental 332, and Australasian–Oceanian 208, compared with a total list of at least 630 species in the British Isles and 2200 for the Palaearctic.

With such areas of ignorance, host associations cannot be fully understood and much of our present information is based on the Palaearctic and Nearctic faunas. Further refinements of classification will be needed, especially at generic and suprageneric levels and molecular methods should be used to try to clarify phylogenies, especially where adaptations to phytophagy and zoophagy have followed ecological rather than phylogenetic pathways.

There is also much basic descriptive work to be done and too few cecidologists to do it.

References

Barnes, H.F. (1956). *Gall midges of economic importance*. Vol. 7: *Gall midges of cereal crops*. Crosby Lockwood & Son, London.

Bühr, H. (1965). *Bestimmungstabellen der Gallen (Zoo- und Phytocecidien) an Pflanzen Mittel- und Nordeuropas*, Vols I and II. G. Fischer, Jena.

Gagné, R.J. (1973). Cecidomyiidae·from Mexican Tertiary amber (Diptera). *Proceedings of the Entomological Society of Washington*, **75**, 169–71.

Gagné, R.J. (1977). Cecidomyiidae (Diptera) from Canadian amber. *Proceedings of the Entomological Society of Washington*, **79**, 57–62.

Gagné, R.J. (1989). *The plant-feeding gall midges of North America*. Cornell University Press, Ithaca.

Mamaev, B.M. (1965). *Evolution of gall forming insects: gall midges*. Akademia Nauk, Leningrad (trans. 1975). British Library, Boston Spa, UK.

Mani, M.S. (1964). *Ecology of plant galls*. Monographiae Biologicae XII. Dr W. Junk, The Hague.

Price, P.W., Fernandes, G.W., and Waring, G.L. (1987). Adaptive nature of insect galls. *Environmental Entomology*, **16**, 15–24.

Rodendorf, B.B. (1964). Historical development of dipteran insects. *Trudy paleontologischeskogo Instituta, Akademiya Nauk USSR Moscow*, **100**, 1–312 (in Russian).

Rohfritsch, O. (1971). Etude d'une galle de *Lobopteromyia* sp. sur *Acacia (Acacia feruginea* D.C.). *Marcellia*, **34**, 171–82.

Rohfritsch, O. (1974). Etude de la galle du *Lobopteromyia ramachandrani* Mani et de son développement sur l'*Acacia* ferruginea D.C. *Marcellia*, **38**, 67–75.

Rohfritsch, O. (1992). Patterns in gall development. In *Biology of insect-induced galls*, (ed. J.D. Shorthouse and O. Rohfritsch), pp. 34–49. Oxford University Press, New York, Oxford.

Rohfritsch, O. and Shorthouse, J.D. (1982). Insect galls. In *Molecular biology of plant tumors*, (ed. G. Kahl and J. Schell), pp. 131–52. Academic Press, New York.

Roskam, J.C. (1992). Evolution of the gall-inducing guild. In *Biology of insect-induced galls*, (ed. J.D. Shorthouse and O. Rohfritsch), pp. 34–49. Oxford University Press, New York, Oxford.

Skuhravá, M. (1986). Family Cecidomyiidae. In *Catalogue of Palaearctic Diptera*, Vol. 4 *Sciaridae-Anisopodidae*, (ed. A. Soós and L. Papp), pp. 72–297. Akadémiai Kiadó, Budapest.

Wiebes-Rijks, A.A. and Houtman, G. (ed.) (1982). *W.M. Docters van Leeuwen Gallenboek*. W.J. Thieme & Cie, Zutphen.

Yukawa, J. (1987). Life history strategies of univoltine gall making Cecidomyiidae (Diptera) in Japan. *Phytophaga*, **1**, 121–39.

13 Risk of parasitism on *Taxomyia taxi* (Diptera: Cecidomyiidae) in relation to the size of its galls on yew, *Taxus baccata*

MARGARET REDFERN and ROBERT CAMERON

School of Continuing Studies, University of Birmingham, Birmingham, UK

Abstract

Galls of *Taxomyia taxi* (Inchbald) on yew are attacked by two chalcid parasitoids, *Mesopolobus diffinis* (Walker) (Pteromalidae) and *Torymus nigritarsus* (Walker) (Torymidae). Earlier work suggested that risk of attack by either species varied with the size of the galls.

We consider here an array of more than 18 000 galls obtained both from many trees sampled over 2 years and from three trees sampled for over 20 years. While the position of galled buds (terminal or axillary) appears not to affect the risk of parasitism, the size of the galls has significant and different effects on the risk of attack by each parasitoid.

Both overall and in individual samples the risk of attack by *M. diffinis* increases the more gall size deviates from the mean. Where galls in a particular sample have a high mean size the risk is greatest in even larger galls; in samples with a lower than normal mean, the risk is greatest in very small galls.

The risk of attack by *T. nigritarsus* shows an opposite trend, being greatest at the mean and least at the extremes. This trend remains even amongst those galls which have not been attacked earlier by *M. diffinis*.

These results are discussed in relation to the complex life cycles shown by *T. taxi*, to the population dynamics of *T. taxi* and both parasitoids, to the size and shape of the ovipositors in both parasitoids, and to the structure of the galls. We suggest that the results support the theory that the size of galls is influenced by the protection from parasitism they afford.

Introduction

Gall-forming herbivorous insects are important and convenient subjects for studies on plant–herbivore and herbivore–predator–parasitoid inter-

Plant Galls (ed. Michèle A. J. Williams), Systematics Association Special Volume No. 49, pp. 213–30. Clarendon Press, Oxford, 1994. © The Systematics Association, 1994.

actions, because they are easy to sample and preserve evidence of most of the interactions involved for some time after they have happened (Redfern and Askew 1992).

In addition to studies on population dynamics and trophic interactions, they have also proved useful in increasing our understanding of evolutionary processes as, for example, in debates about the adaptive significance of galls (Price *et al.* 1987), of life cycle patterns (Redfern and Cameron 1978; Weis *et al.* 1988), and of variation in gall size and structure (Price and Clancy 1986; Weis and Abrahamson 1986; Weis *et al.* 1989).

The gall midge *Taxomyia taxi* induces leafy galls on yew *Taxus baccata* L. (Redfern 1975*a*). Compared to many other systems, especially those involving cynipids on oaks *Quercus* spp. (Askew 1961*b*), the community of interacting species associated with this gall midge is simple: two parasitoid insects and a few non-specific predators (Redfern 1973; Redfern and Cameron 1978). Yew trees are relatively uncommon, taxonomically isolated, and protected by toxins, all of which may constrain the number of herbivores capable of exploiting them (Strong *et al.* 1984).

As is usual with insect herbivores (Strong *et al.* 1984), it is unlikely that the numbers of *T. taxi* are often constrained by lack of food. Parasitoids and predators are a more likely source of regulation, although conclusive evidence that this happens is not yet available (Redfern and Cameron 1978). The parasitoids, *Mesopolobus diffinis* (Walker) and *Torymus nigritarsus* (Walker) show differing patterns of attack and have different life cycles, suggesting that competition between them may be occurring or has occurred in the past (Hassell and Anderson 1989).

The *T. taxi* system has particular advantages for combining ecological and evolutionary studies. Its galls are unilocular, so that gall size and shape can be attributed to the interaction between the plant and single individuals, unlike systems in which multilocular galls are formed and interactions between individuals are important (Redfern and Cameron 1985, 1989; Zwölfer 1985; Redfern 1988). The parasitoids both attack after gall formation is complete, removing a confounding effect when interaction between gall size and occupant survival is considered.

Although simple in species composition, the *T. taxi* system is unusual in the complexity of life cycle patterns shown by its components (Cameron and Redfern 1978; Redfern and Cameron 1978). This study was stimulated by the observation that the risks of attack by each parasitoid varied between life cycles of *T. taxi* and by indications that even within a life cycle, risk of attack varied with size of gall. Variation in risk of parasitism or predation with gall size is now known for a number of systems (Weis and Abrahamson 1985; Weis *et al.* 1985; Price and Clancy 1986; Sitch *et al.* 1988) and adds to the growing body of evidence that non-random risk of attack by predators and parasites is a widespread feature of herbivore populations and one which has important consequences for their population dynamics (Hassell and Anderson 1989).

In this paper, we consider the incidence of attacks on *T. taxi* by the two parasitoids in relation to the life cycles of their host, to the size and position of host galls, and to the ovipositor lengths of the parasitoids. We relate our findings both to the dynamics of the system and to its evolution.

Life cycles

Taxomyia taxi has two basic life cycles. While most individuals take 2 years to develop from egg to adult, a small number develop in 1 year (Redfern 1975*a*).

In both cycles, adults emerge in late May or early June. They mate immediately and lay eggs on leaves of new shoots. Eggs hatch in 1–3 weeks and the hatchlings crawl down leaves to the stem and then towards the tip of the shoots to enter buds. Terminal buds are preferentially infested, but the proportion of axillary buds attacked may be high when several eggs are laid on one shoot.

In the 2-year life cycle, leaf proliferation in the galled bud starts in the following March and the gall reaches its full size (between 0.4 and 3.3 cm in length) in July, 13 months after infestation. The larva remains in the first instar until its second August and in the second instar until October when it molts into its final instar. Growth stops over the winter and resumes in spring. The larva is fully fed in late April and pupates approximately 4 weeks before adult emergence.

In the 1-year life cycle, the larvae remain in the first instar for only 2–3 months. The molt to the second instar correlates with enlargement of the meristem on which the larva lies, in the first August of life (as opposed to the second August in the 2-year cycle). Thereafter, the 1-year cycle repeats the processes seen in the 2-year cycle and produces pupae and adults of similar size.

One-year galls never develop beyond swollen buds (approximately 5 mm high; Redfern 1975*a*); once the meristem enlarges new leaves are not made and the structure of the whole 1-year gall resembles that of the inner cluster of leaves which surrounds the larva in a 2-year gall.

Buds are infested and develop into galls every year, but the 2-year generations infesting 'even-year' buds are nearly independent of those infesting 'odd-year' buds: the generations 'leap-frog'. The only connection is via the smaller number of individuals emerging in 1 year. These usually contribute less than 20 per cent of the adults emerging in any one year, but occasionally more when the density of 2-year galls is very low (Redfern and Cameron 1978).

Mesopolobus diffinis has several generations a year: on *T. taxi* in winter and spring and on other cecidomyiid hosts in summer (Askew 1970; Redfern 1975*b*; Redfern and Cameron 1978). On *T. taxi*, eggs are laid in October and November on third instar host larvae which are not fully fed but are enclosed in fully developed galls. Adults from this over-wintering generation emerge in March and lay eggs on host larvae which are considerably larger. These develop quickly, producing a second spring generation which attacks fully grown larvae and pupae in April and early May. Adults emerge in early summer and attack other hosts.

Mesopolobus diffinis attacks *T. taxi* in both 1 and 2-year life cycles. Both functional and numerical responses to host availability fall on the same generation of hosts and mortality caused by *M. diffinis* tends to be density dependent, especially in *T. taxi* with a 1-year life cycle (Redfern and Cameron 1978).

Torymus nigritarsus has one generation a year. Except for the 1-month period when it is adult, it spends all its life inside its host's gall (Redfern 1973; Redfern and Cameron 1978). Eggs are laid on host pupae in May, usually later than the last attacks by *M. diffinis*. The larvae grow rapidly and are fully fed by July. They pupate in August, the pupae overwinter, and adults emerge the following spring.

Torymus nigritarsus attacks *T. taxi* in 1-year life cycles only very rarely, but can achieve very high levels of parasitism in hosts with 2-year cycles. It is the last major cause of mortality for *T. taxi* prior to emergence. Its numerical response to its host's density falls on a subsequent host generation which is independent of that in which the parasitoids bred and there is no evidence of direct density-dependent mortality (Redfern and Cameron 1978).

In both parasitoids only one larva can develop on a single host. Superparasitism occurs occasionally, as does hyperparasitism by the same and by the other species. Although *T. nigritarsus* usually attacks last there are a few instances in which it is successfully attacked by *M. diffinis*. Overall, there is evidence for incomplete but significant avoidance of galls already parasitized (Redfern 1973).

The starting points for this study are the observations that *M. diffinis* regularly attacks 1-year galls which *T. nigritarsus* does not and that the variances in size of 2-year galls attacked by each parasitoid appear to differ (Redfern 1973).

Methods

The data are based on dissections of *T. taxi* galls collected in late June after adult *T. taxi* have emerged. The length of each gall was measured to the nearest 1 mm and dissection revealed the fate of the contents. Mortality due to the two parasitoids was recognized by the presence of eggs, larvae, pupae, or pupal skins. Identification is straightforward (Redfern and Askew 1992). Although nearly all 2-year galls in each sample were measured, measurement and cause of death were only recorded together for parasitoid attack.

Two sets of samples are considered. The first comes from each of three trees (numbered 5, 63, and 78 in Redfern and Cameron (1978)) sampled every year from 1970 to 1991. The second comes from 16 additional trees sampled in 1970 and 1971. Each tree appears to support a separate population; adult *T. taxi* are short-lived and weak fliers (Redfern 1973). Although each population (and year) is considered separately in some analyses, combinations are also analysed to increase sample sizes, with checks on the consistency of overall trends.

Sampling methods are described in Redfern (1975*a*). In any one year, samples contain 2-year galls originating 2 years before the sampling date and 1-year galls originating 1 year previously (Fig. 13.1), thus including all galls available for attack by parasitoids in the year of sampling.

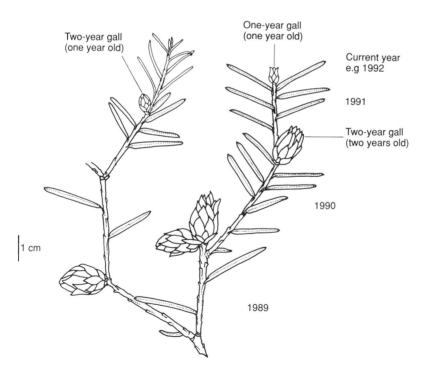

Fig. 13.1. Yew twig with 4 years growth, illustrating 1-year and 2-year galls in buds of different ages, including an example of the dating system used. Adult *T. taxi* from the mature 1-year and 2-year galls emerge at the same time, that is, late May 1992.

The data analysed refer to 2-year galls. One-year galls are almost always smaller than 2-year galls, are few in number, and show consistently high levels of attack by *M. diffinis* and no (or, occasionally, very low levels of) attack by *T. nigritarsus*.

From these and other samples made in March and April in several years, adult parasitoids were reared in the laboratory and the lengths of the ovipositor sheaths were measured to the nearest 0.1 mm. Measurements were made on dry-mounted specimens and may underestimate lengths in living females.

Results

Table 13.1 summarizes the size distribution of 2-year galls and the incidence of parasitism for all samples combined. The incidence of parasitism by *M. diffinis* is greater at the extremes of the size distribution than near the mean. This difference is most marked and consistent amongst smaller than average galls. The mean size of galls parasitized by *M. diffinis* is very slightly less than the mean for all galls and the variance is greater.

Galls parasitized by *T. nigritarsus* show an opposite trend. The incidence of parasitism is greatest near the mean and declines towards the extremes. The mean length of parasitized galls is again very slightly smaller than the overall mean, but the variance is less, being only half of that recorded for galls parasitized by *M. diffinis*.

Since *T. nigritarsus* usually attacks *T. taxi* after *M. diffinis*, avoidance of galls already attacked could result in a concentration around the mean. The last column of Table 13.1 shows the incidence of parasitism by *T. nigritarsus* on galls not attacked by *M. diffinis*. While the effect is reduced slightly it remains clear, with the incidence of parasitism close to the mean being two to three times that at the extremes.

Table 13.1 amalgamates samples from many trees and years, between which mean gall length, sample size, and incidence of parasitism all vary. Table 13.2 shows a partial disaggregation of the results: for all trees sampled in 1970 and 1971 and for each of the trees sampled continuously from 1970–1991. For these three trees (5, 63, and 78) additional columns show the incidence of parasitism by *T. nigritarsus* only for those years in which it was present (*T. nigritarsus* became very rare in the period 1973–1984). Although there are minor irregularities, the trends seen in Table 13.2 are the same as those seen in Table 13.1.

Results for individual samples are necessarily more variable, but there is evidence that they also conform (Table 13.3). For all samples with at least five galls attacked by the appropriate parasitoid, variance in gall size is greater for galls parasitized by *M. diffinis* than for all galls in a

Table 13.1. Size distribution of galls and the incidence of parasitism for all trees in all years. Note the combination of size classes at extremes

Gall length (mm)	Total number	Galls with *Md* number	%	Galls with *Tn* number	%	% of galls without *Md*
4–7	113	37	32.7	3	2.7	3.9
8–9	390	130	33.3	16	4.1	6.1
10–11	1595	463	29.0	139	8.7	12.3
12–13	4384	887	20.2	464	10.6	13.3
14–15	5389	875	16.2	603	11.2	13.4
16–17	3766	608	16.1	321	8.5	10.2
18–19	1858	335	18.0	101	5.4	6.6
20–21	752	156	20.7	33	4.3	5.5
22–23	254	44	17.3	5	2.0	2.4
24 +	143	38	26.6	4	2.8	3.8
Total number	18644	3573	19.2	1689	9.1	11.2
Mean	14.76	14.40		14.35		
Variance	8.83	10.69		5.40		

Md, Mesopolobus diffinis; Tn, Torymus nigriarsus.

Table 13.2. Size distribution of galls and the incidence of parasitism: (a) for all trees sampled in 1970 and 1971, (b)–(d) on trees 5, 63, and 78, respectively over the period 1970–1991. In (b)–(d), additional columns show the incidence of parasitism by *Tn* only in those years in which it was present

	Gall length (mm)	Total number	Galls with *Md* number	%	Galls with *Tn* number	%	Years when *Tn* present Total number	Galls with *Tn* number	%
(a)	4–7	59	21	35.6	2	3.3			
	8–11	735	222	30.2	73	9.9			
	12–15	2764	432	15.6	519	18.8	*Tn* was present in 1970 and 1971		
	16–19	1543	245	15.9	185	12.0			
	20–23	353	74	21.0	17	4.8			
	24 +	83	21	25.3	4	4.8			
	Total number	5537	1015	–	800	–			
	Mean	14.69	14.28	–	14.30	–			
	Variance	11.101	14.270	–	5.505	–			
(b)	4–7	9	2	22.2	0	0.0	1	0	0.0
	8–11	795	225	28.3	38	4.8	314	38	12.1
	12–15	4397	861	19.6	321	7.3	1522	321	21.1
	16–19	900	183	20.3	39	4.3	234	39	16.7
	20–23	38	19	50.0	3	7.8	17	3	17.6
	24 +	2	0	0.0	0	0.0	1	0	0.0
	Total number	6141	1290	–	401	–	2089	401	–
	Mean	13.61	13.49	–	13.55	–	13.41	13.55	–
	Variance	4.079	4.885	–	3.003	–	3.859	3.003	–

<table>
<thead>
<tr><th></th><th></th><th></th><th></th><th></th><th></th><th></th><th></th><th></th></tr>
</thead>
<tbody>
<tr><td>(c)</td><td></td><td></td><td></td><td></td><td></td><td></td><td></td><td></td></tr>
<tr><td>4–7</td><td>48</td><td>13</td><td>27.1</td><td>1</td><td>2.1</td><td>31</td><td>1</td><td>3.2</td></tr>
<tr><td>8–11</td><td>433</td><td>137</td><td>31.6</td><td>47</td><td>10.9</td><td>273</td><td>47</td><td>17.2</td></tr>
<tr><td>12–15</td><td>2155</td><td>417</td><td>19.3</td><td>289</td><td>13.4</td><td>1455</td><td>289</td><td>19.9</td></tr>
<tr><td>16–19</td><td>1875</td><td>363</td><td>19.4</td><td>192</td><td>10.2</td><td>1166</td><td>192</td><td>16.5</td></tr>
<tr><td>20–23</td><td>249</td><td>62</td><td>24.9</td><td>17</td><td>6.8</td><td>162</td><td>17</td><td>10.5</td></tr>
<tr><td>24+</td><td>22</td><td>8</td><td>36.4</td><td>1</td><td>4.5</td><td>16</td><td>1</td><td>6.2</td></tr>
<tr><td>Total number</td><td>4782</td><td>1003</td><td>—</td><td>547</td><td>—</td><td>3103</td><td>547</td><td>—</td></tr>
<tr><td>Mean</td><td>15.11</td><td>14.84</td><td>—</td><td>14.88</td><td>—</td><td>15.06</td><td>14.88</td><td>—</td></tr>
<tr><td>Variance</td><td>8.385</td><td>10.812</td><td>—</td><td>6.062</td><td>—</td><td>8.421</td><td>6.062</td><td>—</td></tr>
<tr><td>(d)</td><td></td><td></td><td></td><td></td><td></td><td></td><td></td><td></td></tr>
<tr><td>4–7</td><td>2</td><td>1</td><td>50.0</td><td>0</td><td>0.0</td><td>0</td><td>0</td><td>0.0</td></tr>
<tr><td>8–11</td><td>99</td><td>24</td><td>24.2</td><td>7</td><td>7.1</td><td>45</td><td>7</td><td>15.5</td></tr>
<tr><td>12–15</td><td>1008</td><td>123</td><td>12.2</td><td>73</td><td>7.2</td><td>461</td><td>73</td><td>15.8</td></tr>
<tr><td>16–19</td><td>1720</td><td>227</td><td>13.2</td><td>49</td><td>2.8</td><td>688</td><td>49</td><td>7.1</td></tr>
<tr><td>20–23</td><td>497</td><td>68</td><td>13.7</td><td>5</td><td>1.0</td><td>175</td><td>5</td><td>2.9</td></tr>
<tr><td>24+</td><td>58</td><td>15</td><td>25.9</td><td>0</td><td>0.0</td><td>24</td><td>0</td><td>0.0</td></tr>
<tr><td>Total number</td><td>3385</td><td>458</td><td>—</td><td>134</td><td>—</td><td>1393</td><td>134</td><td>—</td></tr>
<tr><td>Mean</td><td>16.82</td><td>16.89</td><td>—</td><td>15.25</td><td>—</td><td>16.56</td><td>15.25</td><td>—</td></tr>
<tr><td>Variance</td><td>8.456</td><td>11.288</td><td>—</td><td>4.819</td><td>—</td><td>8.346</td><td>4.819</td><td>—</td></tr>
</tbody>
</table>

Md, Mesopolobus diffinis; Tn, Torymus nigriarsus.

Table 13.3. Number of samples in which variance of parasitized galls exceeds or is lower than that for the sample overall (all samples with 5 + parasitized galls for each) and in which per cent parasitism in the three 1 mm size classes nearest the overall mean exceeds or is less than the incidence for the whole sample (samples from 1970 and 1971 with at least 100 galls)

Variance of gall size	Greater than overall	Less than overall	χ_1^2
With *M. diffinis*	61	28	12.2***
With *T. nigritarsus*	10	48	24.9***
Incidence of parasitism in galls 14–16 mm			
With *M. diffinis*	3	17	9.8**
With *T. nigritarsus*	13	4	4.8*

*** $p. <0.001$ ** $p. <0.01$ * $p. <0.05$

significant excess of cases, while the reverse is true for *T. nigritarsus*. Furthermore, the incidence of parasitism by *M. diffinis* is lower near the mean than overall in a significant excess of the trees sampled in 1970–1971 and again the reverse is true for *T. nigritarsus*.

These results raise a number of questions of interpretation. Galls may be formed in terminal or in axillary buds. Axillary buds are, on average, slightly smaller than terminal buds (0.63 mm difference in means on tree 5) and size and position effects may be confounded. However, the difference in mean size is small relative to the ranges of size over which differential parasitism occurs; galls occur much more frequently in terminal than axillary buds [usually less than 10 per cent of galls are in axillaries; Redfern (1973)] and there is little evidence of differential parasitism between terminal and axillary buds regardless of size (Table 13.4).

Since mean gall length varies between trees and between years on the

Table 13.4. Tests of significance (χ_1^2) on the incidence of parasitism in galls in terminal and axillary buds for all samples from trees 5, 63, and 78 in which no expected value is less than 3

	Number of samples	Not significant	Significant excess In terminal buds	In axillary buds
M. diffinis	31	26	2	3
T. nigritarsus	12	12	0	0
Total	43	38	2	3

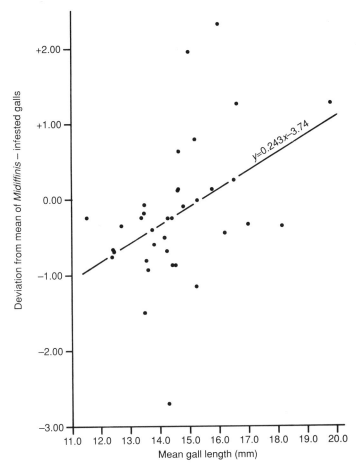

Fig. 13.2. Deviation of *M. diffinis*—infested galls from the mean length of all galls in a sample compared to the sample mean, in all trees sampled in 1970 and 1971. Despite the wide scatter, the regression is significant; $r = 0.4363$ (SE$b = \pm 0.086$, $t_{34} = 2.826$, $p < 0.01$).

same tree, one can ask whether variation in the incidence of parasitism relates to absolute size or to deviations from the local sample means. Individual samples are generally too small to make comparisons of incidence around the overall and sample means useful. Figure 13.2 shows the regression of deviations of mean gall length in galls attacked by *M. diffinis* relative to sample means against the sample means themselves. There is a significant positive slope, indicating that when the sample mean is greater than average, it is the upper tail of the size distribution which bears the brunt of parasitism and vice versa. A similar trend can be seen in the three trees studied over many years (Table 13.2). Tree 78

Table 13.5. The distribution of deviations of the mean size of galls parasitized by *M. diffinis* from their sample means for trees 5, 63, and 78 (samples with 5 + galls attacked by *M. diffinis*) in relation to (a) whether the sample mean lies above or below the mean for the tree and (b) whether the sample mean lies above or below the overall mean, using data for all 19 trees sampled in 1970 and 1971

(a)	Deviation of mean for *M. diffinis* from sample mean		(b)	Deviation of mean for *M. diffinis* from sample mean	
	+	−		+	−
Sample mean > tree mean	15	12	Sample mean > overall mean	18	15
Sample mean < tree mean	9	20	Sample mean < overall mean	6	17
	$\chi_1^2 = 3.43$ ns			$\chi_1^2 = 4.48*$	

*p <0.05, ns = not significant

Table 13.6. The distribution of deviations of the mean size of galls parasitized by *T. nigritarsus* from their sample means in relation to whether the sample mean lies above or below the overall mean for 1970 and 1971. Data for all trees and years with 4 + galls attacked by *T. nigritarsus* (many samples from 1973–1984 are therefore excluded)

	Deviation of mean for *T. nigritarsus* from sample mean	
	+	−
Sample mean > overall mean	8	15
Sample mean < overall mean	21	9
	$\chi_1^2 = 6.52*$	

*p <0.05

has a large mean and galls attacked by *M. diffinis* are larger still, while the opposite is true for the other two. Furthermore, although the directions of deviations of the means of *M. diffinis*-attacked galls from their sample means do not relate to variation in sample means about the mean (for all samples) for each tree considered separately, they do relate to variation in sample means about the overall mean for all trees sampled in 1970 and 1971 (Table 13.5).

Similarly, there is a significant trend for the mean size of galls parasitized by *T. nigritarsus* to be smaller than the sample mean when that mean is above the overall mean and vice versa (Table 13.6). Again,

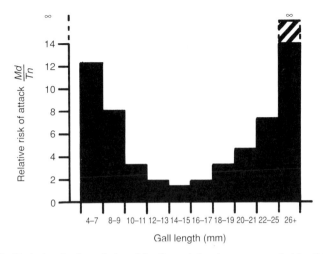

Fig. 13.3. Variation in the relative risk of attack by the two parasitoids. (*M. diffinis: T. nigritarsus* for the whole data set, see Table 13.1).

in tree 78, in which mean gall size is exceptionally large, the mean size of galls attacked by *T. nigritarsus* is significantly smaller.

The effects of these patterns of parasite attack on relative risk of attack by each parasite are shown in Figure 13.3. The particular values clearly depend on the overall levels of infestation by each parasite, but the shape of the relationship holds across most samples.

There are no significant relationships between the mean gall lengths, variances or deviations of parasitized galls, and either host density or the incidence of parasitism. *Mesopolobus diffinis* has a much shorter ovipositor sheath than has *T. nigritarsus* (Table 13.7). There is no overlap in length between the species.

Table 13.7. Length of ovipositor sheath in dry specimens of *M. diffinis* and *T. nigritarsus*

	Md	*Tn*
Number measured	164	45
Mean length (mm)	0.57	2.32
SE (standard error) of mean	0.017	0.034

Discussion

Evidence of selective attack by predators or parasites is of interest in two contexts:

(1) evolutionary, in which the nature of the forces which shape and maintain a system might be determined;

(2) ecological, in which the consequences of differential availability of prey or hosts on the system's dynamics are explored (Hassell and Anderson 1989; Strong *et al.* 1984).

In evolutionary terms, the effects of selection are dependent on the heritability of the character selected. We have no evidence of heritability in the *Taxus–Taxomyia* system, but other studies on gall-based systems reveal what one would expect: that gall size and morphology are the product of a complex interaction between hereditary variation in both plant and gall former which is also influenced by the environment (Fritz *et al.* 1986; Price and Clancy 1986; Price *et al.* 1987; Weis *et al.* 1988; Weis *et al.* 1989).

In terms of immediate effect, attacks by *M. diffinis* exert stabilizing selection on the size of 2-year galls of *T. taxi*. Risk of parasitism is least at or near the general mean. In particular populations in which the population mean deviates substantially from an overall norm, *M. diffinis* may exercise directional selection in favour of a return to the overall value.

An opposite effect arises with *T. nigritarsus*. In populations with mean gall sizes close to the norm, it causes disruptive selection. In particular in populations with deviant population means it will exercise directional selection to increase the deviation from the overall value.

The combined effects of these two parasitoids will clearly depend on the absolute and relative intensities of their attacks, which no doubt vary between populations. The two tend to cancel each other out (stabilizing both mean and variance), but until we have data on the interaction between gall size and the incidence of lepidopteran and bird predation on *T. taxi*, we cannot estimate the overall selective effects of mortality as have Abrahamson *et al.* (1989) on gall size in the *Eurosta–Solidago* system (see below).

Many other studies on mortality in gall-forming insects have shown potential selective effects on gall size exercised by predators and parasites. Where galls are multilocular or where parasitoids oviposit before gall formation is complete, the interpretation of such effects can be complex (Weis 1982; Michaelis 1985; Redfern and Cameron 1985). There are, however, a number of examined cases in which single occupant galls are attacked after gall formation is complete. Cases of parasitoid selection against smaller than average galls are reported by Price and Clancy (1986), Weis and Abrahamson (1985), and by Weis *et al.* (1985). Selection against larger than average galls is also reported Clancy and Price 1987; Sitch *et al.* 1988).

Abrahamson *et al.* (1989) have conducted a particularly thorough study, based on 20 natural populations of the gall fly *Eurosta solidaginis*, which makes single-celled stem galls on goldenrod *Solidago altissima*. Of the parasitoids involved, one (*Eurytoma gigantea*) attacked smaller than average galls, while another (*E. obtusiventris*) attacked larger than average galls, as did birds acting as predators. The inquiline beetle *Mordellistena unicolor* attacked both gall tissue and the gall fly larvae. Overall its effect was selective against small galls but it had a partly stabilizing effect in that risk of attack was least at the mean, thus paralleling the effect of *M. diffinis* in this study. Overall, the effect of *E. gigantea* outweighed all other causes of mortality, so that the populations studied by Abrahamson *et al.* (1989) appeared to be subject to continuous directional selection for increase in gall size. Longer-term studies would be needed to determine whether size does in fact increase and what countervailing forces might confer stability. In none of these examples is there a parallel for the disruptive effects displayed by *T. nigritarsus*.

In many of the cases reported above and in others selective attack by parasitoids is influenced by the length of their ovipositors. Those with very short ovipositors frequently attack before gall formation is complete (Claridge 1961) while others [for example, *E. gigantea*: Askew (1961*a*,*b*) and Weis and Abrahamson (1985)] are limited to smaller galls by a modest ovipositor. It would seem that parasitoids with long ovipositors avoid very small galls and this appears to be so for *T. nigritarsus* in this study; it avoids both 1-year galls and the smallest 2-year galls, although also appearing constrained by the largest of the latter.

Mesopolobus diffinis has a short ovipositor; appropriately it attacks the smallest galls. However, the intensity of its attack also increases on the largest galls and this bimodality is so far unique for a chalcid parasitoid. We have yet to determine the causes, but we suspect that attacks on very large galls are made possible by the increased separation in the largest galls of the leaves of which the gall is composed, allowing the parasitoid to walk some way into the gall before ovipositing. Any explanation, for either parasitoid, must take into account the fact that there is no size class of gall in which parasitism appears to be impossible. There is no perfect defence.

In ecological terms, the *Taxus–Taxomyia* system is simple in terms of numbers of species involved, but complex and unusual in the life history patterns of its components Cameron and Redfern 1978; Redfern and Cameron 1978, 1993).

Mesopolobus diffinis is not host-specific and it is multivoltine. It appears to have the potential to devastate *Taxomyia* because both functional and numerical responses can fall on the same generation of hosts. This sometimes happens in 1-year galls considered separately, where the mortality inflicted by *M. diffinis* can be both heavy and strongly density-dependent, but when 2-year galls are considered such mortality is neither of these. Although there is variation from year to year (and tree to tree) the mean mortality inflicted by *M. diffinis* is approximately 20 per cent of available hosts and the highest recorded is 58 per cent. Clearly, not all galls are available and size of gall alone can be but part of the explanation.

By contrast, *T. nigritarsus* is probably host-specific and is univoltine. For a specialist, it is in a remarkably precarious position for, without the 1-year cycle in its host, the interaction with 2-year galls would lead first to the extinction of one of the two 'leap-frogging' generations of *T. taxi*, followed immediately by its own demise (Bulmer 1977). As it is, its numbers fluctuate greatly and, in the trees studied here, it appeared locally extinct for a period of 8–10 years; this may be part of a multigenerational cycle of abundance and scarcity (Redfern and Cameron 1978, 1993). At the peak it can remove 90 per cent or more of the available *T. taxi*. The concentration of its attacks on galls close to the mean suggest a very close adaptive dependence.

Finally, we come to the host itself. The existence of 2-year life cycles in a species capable of developing in 1 year is in itself an oddity, since it reduces reproductive capacity. In existing populations, the balance of advantage appears to depend on the proportions of the two parasitoids attacking (Redfern and Cameron 1978). Given the differential risk of attack by each associated with gall size, it is tempting to view the life cycle pattern as an adaptive response. An original 1-year cycle *T. taxi*, heavily attacked by *M. diffinis*, produced occasional sports in which induction of the gall was delayed for a year, with the accidental consequence of causing the leafy artichoke gall typical of the 2-year cycle. This conferred some protection from *M. diffinis*, but suited the longer ovipositor of an ancestral *Torymus*.

Acknowledgement

Thanks are due to English Nature (formerly the Nature Conservancy Council) for annual permission to collect galls of *Taxomyia taxi* at Kingley Vale National Nature Reserve, West Sussex.

References

Abrahamson, W.G., Sattler, J.F., McCrea, K.D., and Weis, A.E. (1989). Variation in selection pressures on the goldenrod gall fly and the competitive inter-actions of its natural enemies. *Oecologia*, **79**, 15–22.

Askew, R.R.A. (1961*a*). A study of the biology of species of the genus *Mesopolobus* Westwood (Hymenoptera: Pteromalidae) associated with cynipid galls on oak. *Transactions of the Royal Entomological Society of London*, **113**, 155–73.

Askew, R.R.A. (1961*b*). On the biology of the inhabitants of oak galls of

Cynipidae (Hymenoptera) in Britain. *Transactions of the Society for British Entomology*, **14**, 237–68.

Askew, R.R.A. (1970). Observations on the hosts and host food plants of some Pteromalidae (Hym. Chalcidoidea). *Entomophaga*, **15**, 379–85.

Bulmer, M.G. (1977). Perioidical insects. *American Naturalist*, **111**, 1099–117.

Cameron, R.A.D. and Redfern, M. (1978). Population dynamics of two hymenopteran parasites of the yew gall midge *Taxomyia taxi* (Inchbald). *Ecological Entomology*, **3**, 265–72.

Clancy, K.M. and Price, P.W. (1987). Rapid herbivore growth enhances enemy attack: sublethal plant defences remain a paradox. *Ecology*, **68**, 733–7.

Claridge, M.F. (1961). Biological observations on some eurytomid (Hymenoptera: Chalcidoidea) parasites associated with Compositae, and some taxonomic implications. *Proceedings of the Royal Entomological Society of London, Series A*, **36**, 153–8.

Fritz, R.S., Saachi, C.F., and Price, P.W. (1986). Competition versus host plant phenotype in species composition: willow sawflies. *Ecology*, **67**, 1608–18.

Hassell, M.P. and Anderson, R.M. (1989). Predator–prey and host–pathogen interactions. In *Ecological concepts. The contribution of ecology to an understanding of the natural world*, 29th Symposium of the British Ecological Society, (ed. J.M. Cherrett), pp. 147–196. Blackwell Scientific Publications.

Michaelis, H. (1985). Nahrungsnetzuntersuchungen an *Urophora stylata* F., einer gallbildenden Tephritide (Diptera) in den Blütenköpfen von *Cirsium vulgare*. *Verhandlungen der Gesellschaft für Ökologie*, **13**, 587–91.

Price, P.W. and Clancy, K.M. (1986). Interactions among three trophic levels: gall size and parasitoid attack. *Ecology*, **67**, 1593–600.

Price, P.W., Wilson Fernandes, G., and Waring, G.L. (1987). Adaptive nature of insect galls. *Environmental Entomology*, **16**, 15–24.

Redfern, M. (1973). Studies on the life history and population dynamics of the yew gall midge, *Taxomyia taxi* (Inchbald), and its chalcid parasites. Unpublished PhD thesis, University of Reading.

Redfern, M. (1975a). The life history and morphology of the early stages of the yew gall midge *Taxomyia taxi* (Inchbald) (Diptera: Cecidomyiidae). *Journal of Natural History*, **9**, 513–33.

Redfern, M. (1975b). The identity of *Mesopolobus* species (Hym., Pteromalidae) parasitizing *Taxomyia taxi* (Inchbald) (Dipt., Cecidomyiidae) on yew *Taxus baccata* L. *Entomologist's Monthly Magazine*, **111**, 201–4.

Redfern, M. (1988). Interaction between the gall fly *Urophora stylata* (Diptera: Tephritidae) and spear thistle *Cirsium vulgare* (Compositae). *Journal of Biological Education*, **22**, 88–90.

Redfern, M. and Askew, R.R.A. (1992). *Plant galls*. Naturalists' Handbook No. 17, 99 pp. Richmond, Slough.

Redfern, M. and Cameron, R.A.D. (1978). Population dynamics of the yew gall midge *Taxomyia taxi* (Inchbald) (Diptera: Cecidomyiidae). *Ecological Entomology*, **3**, 251–63.

Redfern, M. and Cameron, R.A.D. (1985). Density and survival of *Urophora stylata* (Diptera: Tephritidae) on *Cirsium vulgare* (Compositae) in relation to flower head and gall size. *Proceedings of the VI International Symposium on*

Biological Control of Weeds, 1984, Vancouver, Canada, (ed. E.S. Delfosse), pp. 453–77. Agriculture Canada, Ottawa.

Redfern, M. and Cameron, R.A.D. (1989). Density and survival of introduced populations of *Urophora stylata* (Diptera: Tephritidae) in *Cirsium vulgare* (Compositae) in Canada, compared with native populations. *Proceedings of the VII International Symposium on Biological Control of Weeds, 1988, Rome, Italy,* (ed. E.S. Delfosse), pp. 203–10. Istituto Sperimentale per la Patalogia Vegetale (MAF), Rome.

Redfern, M. and Cameron, R.A.D. (1993). Population dynamics of the yew gall midge *Taxomyia taxi* and its chalcid parasitoids: a 24-year study. *Ecological Entomology,* **18**.

Sitch, T.A., Grewcock, D.A., and Gilbert, F.S. (1988). Factors affecting components of fitness in a gall-making wasp (*Cynips divisa* Hartig). *Oecologia,* **76,** 371–5.

Strong, D.R., Lawton, J.H., and Southwood, Sir R. (1984). *Insects on plants, community patterns and mechanisms.* Blackwell Scientific Publications, Oxford.

Weis, A.E. (1982). Resource utilization patterns in a community of gall-attacking parasitoids. *Environmental Entomology,* **11,** 804–15.

Weis, A.E. and Abrahamson, W.G. (1985). Potential selective pressures by parasitoids on a plant-herbivore interaction. *Ecology,* **66,** 1261–9.

Weis, A.E. and Abrahamson, W.G. (1986). Evolution of host-plant manipulation by gall makers: ecological and genetic factors in the *Solidago–Eurosta* system. *American Naturalist,* **127,** 681–95.

Weis, A.E., Abrahamson, W.G., and McCrea, K.D. (1985). Host gall size and oviposition success by the parasitoid *Eurytoma gigantea*. *Ecological Entomology,* **10,** 341–8.

Weis, A.E., Walton, R., and Crego, C.L. (1988). Reactive plant tissue sites and the population biology of gall makers. *Annual Review of Entomology,* **33,** 467–86.

Weis, A.E., Wolfe, C.L., and Gorman, W.L. (1989). Genotypic variation and integration in histological features of the goldenrod ball gall. *American Journal of Botany,* **76,** 1541–50.

Zwölfer, H. (1985). Energieflußsteuerüng durch informationelle Prozesse—ein vernachlässigtes Gebiet der Ökosystemforschung. *Verhandlungen der Gesellschaft für Ökologie,* **13,** 285–94.

14. Studies on host-plant resistance in rice to gall midge *Orseolia oryzae*

I. DAVID R. PERIES

Department of Biological Sciences, University of Durham, South Road, Durham, UK

Abstract

Necrotic hypersensitive reaction in rice as a mechanism of host plant resistance to the rice gall midge, *Orseolia oryzae*, is a new phenomenon. This reaction in the highly resistant seedlings starts as a slight yellowing of the tissues of the shoot apex following the arrival of the midge larva. This leads to the browning and final death of the apical meristem and of the midge larva feeding on it.

A biochemical explanation for this extreme reaction to infestation was sought in analysing four resistant, four moderately resistant, and two susceptible rice varieties. Phenolic acids were estimated using HPLC on ethanolic extracts of healthy rice seedlings. Reversed-phase analysis was done using a C18 column and a gradient elution with methanol–water mixture.

Seven phenolic acids and three unknown compounds were detected in the chromatograms. *para*-Coumaric acid and unknown compound 3 were the major constituents, although the amounts of these acids could not be used to identify the resistant varieties. However, discriminant function analysis and principal component analysis of the data indicated that protocatechuic acid, unknown compound 2, and *para*-hydroxybenzoic acid were the main variables permitting grouping of the varieties as resistant, moderately resistant, or susceptible. The group of resistant varieties showed higher levels of vanillic acid and unknown compound 1 than the group of susceptible varieties.

This study indicated the complex processes involved in host plant resistance mechanisms and possible areas of further research.

Introduction

Rice is one of the most important food crops of the world with approximately 2000 million people relying on it for 25–50 per cent of

Plant Galls (ed. Michèle A. J. Williams), Systematics Association Special Volume No. 49, pp. 231–43. Clarendon Press, Oxford, 1994. © The Systematics Association, 1994.

their diet. Unfortunately approximately 30 per cent of the potential rice yield is lost to insects (Cramer 1967), among which the Asian rice gall midge, *Orseolia oryzae* (Wood-Mason) ranks as a major pest causing 50–100 per cent maximum crop losses in some countries (Reddy 1967).

Until the mid-1960s the rice gall midge was considered a minor pest only occasionally inflicting serious damage in certain endemic areas in south and southeast Asia. The shift to major pest status coincided with the onset of the 'Green Revolution' in rice in Asia. The new 'miracle rices' (high yielding varieties) are highly susceptible to rice gall midge, have a different plant type, and require higher levels of management (fertilizers, irrigation, pesticides, etc.). This led to a drastic change in the plant microhabitat in which the pest thrived very well. Plant damage is caused by the rice gall midge larvae which hatch from eggs laid on rice leaves. They migrate to the terminal shoot apex and begin to feed on the meristematic tissue to trigger gall formation. Larval development and pupation are completed within the gall, with the adult emerging from the gall 18–22 days later.

Rice gall midge control in the 1960s and 1970s had relied heavily on the use of a range of insecticides. Attempts were also made then to use host plant resistance, following large-scale varietal screening programmes. Thousands of varieties had been subjected to these programmes and approximately 180 of them identified as resistant by 1980 (Heinrichs and Pathak 1981).

Although many resistant hybrids derived from these donors are being widely cultivated, the nature of the resistance mechanism itself has not been adequately examined. The author was involved in such a screening programme in Sri Lanka for several years. On examination of these highly resistant varieties a very characteristic reaction to infestation was observed: the terminal shoot apex attacked by rice gall midge larvae showed brown discolouration and finally became necrotic. Larval development was adversely inhibited; the first-instar lasted beyond the normal complete larval period. The larva died in due course. In contrast, there was no hint of browning or necrosis of plant tissues in uninfested highly resistant varieties. Gall formation and insect development proceeded without any hindrance in the susceptible varieties. This hypersensitive plant reaction resulting in the development of necrotic lesions, prompted an investigation of the mechanism of host plant resistance to rice gall midge. The field of study was narrowed to histology and biochemistry of resistance since previous workers had disproved the involvement of ovipositional, physical, or mechanical resistance (Modder and Alagoda 1972).

Materials and methods

1. Histology

Seedlings of three varieties (Ob677 (highly resistant, R), W1263 (moderately resistant, M), and IR8 (susceptible, S)) were grown in a glasshouse, in Sri Lanka, for 7 days and infested with laboratory-reared rice gall midge adults (Perera and Fernando 1967). The infested seedlings were uprooted 9–10 days later for histological preparation. The outer two or three leaf sheaths were removed and a 5 mm piece of the leaf whorl containing the terminal shoot apex was picked for fixation. A set of 10 seedling segments of any one variety was fixed together in Carnoy–LeBrun's fixative. The fixed material was embedded in molten paraffin wax for serial sectioning and the sections stained with Ehrlich's hematoxylin.

A parallel set of infested seedlings of each variety was dissected the same day to study the nature of gall formation and any sign of hypersensitive reaction. These observations were later compared with the microscopic preparations.

2. Biochemical analysis

An HPLC analysis for phenolic acids was performed on plant samples at the University of Wales, Cardiff, adapting a method described by Cole (1984). Rice seedlings required for the analysis were raised in a growth chamber (26°C, RH 80 ± 10 per cent, 12 h light). Seeds of 10 rice varieties (4 R(Ob677, Ptb18, Bg400–1, Muey Nawng 62M), 4 M (Eswarakora, W1263, Ptb21, RD4), and 2 S (Siam 29, IR8)) were each sown in two 10 cm diameter plastic pots filled with compost. Six-week-old seedlings were uprooted from the pots and 10 cm long bases were cut. Fresh material (5 g) was extracted in boiling ethanol (95 per cent, 50 ml) for 15 min. All 20 extractions were done on the same day.

The extract was filtered and the filtrate evaporated to near dryness in a rotary evaporator at 55°C. The residue was reconstituted in 10 ml water, adjusted to pH8, and centrifuged at 3600 rpm for 15 min. Sample clean-up was effected by passing the supernatant through a disposable ion exchange column (Tippins 1987).

The above samples were analysed using a chromatography system based on an octadecyl silane reversed-phase (ODS C18) column that permits a good separation of mixtures of moderately polar constituents (Cole 1984). The eluting solvent was methanol:water, and the pump was programmed for a gradient elution of 10–50 per cent methanol over a period of 20 min. at a flow rate of 1.5 ml min^{-1}. The two solvents were initially filtered through membrane filters and constantly degassed with helium. An excess of sample was injected into the valve ensuring the $20\,\mu l$ sample loop was thoroughly flushed free of any trace of previous sample. The column effluent was monitored using a variable wavelength UV detector linked to a chart recorder and an integrator to provide both a hard copy of the chromatogram and an integrated chart with a table of retention times, heights, and areas of peaks.

A mixture of known phenolic acids was used to serve as standards with each day's runs. The following acids were used: caffeic (CAF), cinnamic (CIN), chlorogenic (CHL), *para*-coumaric (CUM), ferulic (FER), gallic (GAL), gentisic (GEN), *para*-hydroxybenzoic (BEN), protocatechuic (PRT), salicylic (SAL), syringic (SYR), and vanillic (VAN).

Each sample was analysed at two UV detector settings, 265 and 320 nm wavelengths, as some of the acids had their peak absorbance around 265 nm and others around 320 nm. The chromatogram traced on the recorder contained several peaks, each corresponding to an individual component in the original sample. These were identified by comparing their retention times with those of standards processed under identical experimental conditions. There were three peaks representing the unknown compounds UC1, UC2, and UC3. The area of each peak of a chromatogram is related to the quantity of the component. The results of HPLC analysis were subjected to multivariate analysis using an SPSSX® statistical package.

Results

1. Histology

A total of 516 microscope slides, each bearing 20–30 sections, were examined. The midge larvae were found at both terminal and axillary shoot apices. The development of the gall primordium was seen to have progressed well in susceptible varieties, with no sign of necrosis (Fig. 14.1a). The tissues of the leaf primordium and the growth cone appeared to be quite normal, with no sign of cellular damage. In the moderately resistant variety, gall formation was distorted and incomplete. In contrast, in highly resistant varieties (Fig. 14.1b) there was no sign of gall formation. Wherever larvae were present there was browning of the tissue in a restricted area around them. This led to the collapse of cells, especially the meristematic tissue of the leaf primordium. Terminal shoot apices were severely affected by necrosis to the extent that all of them were killed. This reaction was localized only in the immediate vicinity of the affected area. Uninfested axillary shoot apices found nearby were not affected at all. These axillary shoot apices are capable of taking over the seedling growth.

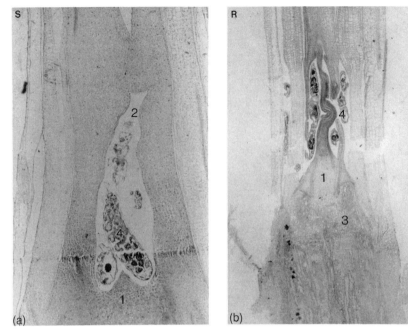

Fig. 14.1. Photomicrograph of L.S. of infested seedlings of (a) susceptible (S) and (b) resistant (R) rice varieties. Key: 1, growth cone; 2, gall cavity; 3, necrosis; 4, larvae.

In the hypersensitive area itself, extensive cellular damage was observed. A large number of cells appeared empty, surrounded by heavily pigmented cells. These pigmented cells are found towards the proximal end of leaf sheaths.

2. HPLC

Four chromatograms were obtained for each variety—two detected at 265 nm wavelength (for GAL, PRT, CAF, VAN, SYR, and CIN acids) and two at 320 nm (for GEN, CAF, CUM, FER, UC1, UC2, and UC3). Some of the acids included in the standard mixture were found to be either very low or absent in the samples. The acids CAF, CHL, CIN, GEN, and SAL were detected in very low amounts and were excluded from subsequent calculations.

Examination of the acid values for the 10 varieties did not reveal any apparent correlation of acid values with host plant resistance. Drastic differences in any one set of acid values could be seen even within a particular (R, M, or S) group of varieties. For example, SYR acid values varied from 200 to 740 arbitrary units within highly resistant varieties, with values for moderately resistant and susceptible varieties falling within

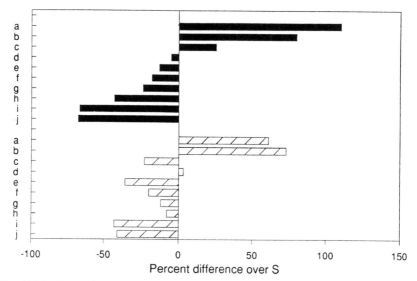

Fig. 14.2. Proportional difference of phenolic acid values in the resistant (R) and moderately resistant (M) variety groups over that of the susceptible (S) group. Key: a, unknown compound 1; b, vanillic; c, unknown compound 3; d, *para*-coumaric; e, unknown compound 2; f, *para*-hydroxybenzoic; g, ferulic; h, syringic; i, gallic; j, protocatechuic acids.

this range. It was thus not possible to identify a variety as highly or moderately resistant or susceptible on the basis of any one acid value. Total acid values for each variety were also not useful.

The individual acid values for each group were then compared by calculating a mean value for an acid among the R, M, and S groups. A clear picture emerged when the acid value of, for example, S was kept stable and values for R and M calculated proportional to it (Fig. 14.2). For example, the proportional differences of VAN for R and M were 73 and 80 per cent more than S, respectively. An examination of Fig. 14.2 indicates a trend within these groups. The acids UC1 and VAN occurred in proportionally greater quantities in the R and M groups than in the S group. The reverse is true for PRT, GAL, SYR, FER, BEN, and UC2. While proportional differences in CUM values were low in R and M, it was inconclusive for UC3.

3. Discriminant function analysis (DFA)

All 10 HPLC acid values for each variety were entered as variables, at the end of which five of them were picked as the best set of discriminating

Table 14.1. Phenolic acid variables selected at the end of discriminant function analysis (DFA) on 10 rice varieties

Step	Variable entered	Variable removed	Variable in	Wilks' lambda	Significance
1	PRT	x	1	0.18552	0.0000
2	UN3	x	2	0.11135	0.0000
3	BEN	x	3	0.06936	0.0000
4	VAN	x	4	0.05600	0.0000
5	UN2	x	5	0.04810	0.0000

PRT, protocatechuic acid; BEN, *para*-hydroxybenzoic acid; VAN, vanillic acid; UC2 and UC3, unknown compounds 2 and 3; x, no variable removed in the step-wise analysis.

variables (Table 14.1). Two functions were derived which together accounted for 100 per cent of the between-group variation (Table 14.2). The most powerful discriminating variables were PRT, UC2, BEN, and VAN in function 1 which explain 97 per cent of the total variance.

This discrimination was clearly seen in the scatter diagram of individuals (Fig. 14.3). The three groups of rice varieties could be distinctly separated in function 1, but not in function 2. The classification results also showed that the varieties were distinct in the three groups, with 100 per cent of varieties correctly assigned to the three groups.

Table 14.2. Per cent total variance explained by two discriminant functions derived from the DFA of phenolic acids measured in 10 rice varieties

Function	Eigen value	Per cent variance	Cumulative per cent	Correlation	Wilks' lambda	Significance
1	14.0328	97.34	97.34	0.9661	0.0481	0.0000
2	0.3829	2.66	100.00	0.5262	0.7231	0.3016

4. Principal components analysis (PCA)

For principal components analysis all the HPLC data on the 10 rice varieties were entered without grouping the varieties in any manner. A correlation matrix of variables derived in this analysis indicated a high degree of correlation between CUM and FER, CUM and VAN, and UC2 and UC3. The original set of variables, transformed into independent components, indicated that the first component accounted for 42.2 per cent of the total variance of the correlation matrix. The next four components raised the cumulative variability to 90 per cent (Table 14.3).

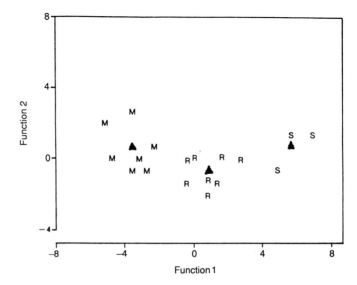

Fig. 14.3. Plot of first two canonical functions of the discriminant function analysis of 10 rice varieties. Key: M, moderately resistant; R, resistant; S, susceptible varieties. Triangle indicates a group centroid.

When the factor scores (index of the relationship of the original data to the principal component) of each variable were plotted as a two-dimensional scattergram using components 1 and 2 as axes (Fig. 14.4) they were found to be clustered in three groups representing highly

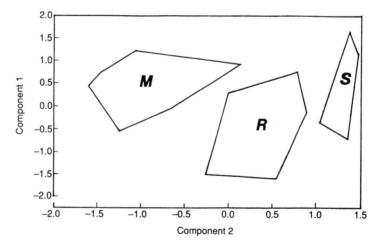

Fig. 14.4. Plot of first two principal components for phenolic acid values measured in 10 rice varieties. Key as in Fig. 14.3.

Table 14.3. Principal components analysis: eigenvalues of the correlation matrix of the 10 phenolic acids

Component	Eigenvalue	Per cent variability	
		Component	Cumulative
1	4.224	42.2	42.2
2	1.840	18.4	60.6
3	1.174	11.7	72.4
4	1.091	10.9	83.3
5	0.743	7.3	90.6

resistant, moderately resistant, and susceptible varieties. Among the five components only component 2 was able to separate the groups adequately; components 1, 3, 4, or 5 did not result in any clustering of the factor scores. An analysis of the relative contribution of each variable to the variance expressed by a component is given by its factor loading. The highest factor loadings for component 2 are by BEN, PRT, GAL, and UC2.

Discussion

Rice gall midge larvae were found in all the varieties, confirming that ovipositional preference or physical barriers could not be responsible for the host plant resistance mechanism. Though Modder and Alagoda (1972) reported an observation similar to the one reported here on moderately resistant varieties, necrotic host plant resistance to insect infestation in rice is reported here for the first time. Such cellular changes in other plants have been well-documented by several workers, with one of the earliest reports on the occurrence of red pigments in an insect gall by Nierenstein (1930). He suggested that the sawfly, *Pontania proxima* (*Lepeletier*), counters the resistance of willow, *Salix caprea* L., by producing an enzyme that converts the plant tannin into insoluble red pigment. More recently, Beardmore *et al.* (1983) reported on the hypersensitive necrotic reaction induced by the rust fungus *Puccina graminis* Pers. in highly resistant wheat cultivars. Their work indicated cellular accumulation of phenolic compounds following disorganization of the infected cells. The release of peroxidases and other enzymes that are normally discrete, catalyses an irreversible polymerization of phenols to tannin and lignin. Similar severe necrosis has been reported in tobacco (Edreva *et al.* 1972) and tomato (Brueske and Dropkin 1973).

Miles (1968) discussed at length the involvement of insect secretions with plant resistance. Injuries to plants, whatever their cause, are liable to induce an increase in the content of toxic phenolic compounds in the injured cells, the effective toxins usually being quinones. This transformation is under the influence of polyphenoloxidase. This enzyme could be produced either by the plant or by the invading organism. However, Suzuki (1965) reported its absence in rice leaves. Therefore, it is probable that this enzyme is produced either by the midge larva or by the rice plant in response to infestation. These two possible routes of production of the enzyme have not been determined. Miles (1968) reported that polyphenoloxidase is widespread in insect secretions and a polyphenoloxidase has been detected in the salivary glands of larval *Drosophila*.

The presence of phenolic acids in rice was well-established by several Japanese researchers (see Suzuki 1965) and we can conclude that these phenolic substances are acted upon by the polyphenoloxidase secreted by the midge larvae to produce the toxic quinones. The injured area is sealed off as a consequence of the polymerization and tanning action of the phenol–polyphenoloxidase system. The fact that such a reaction does not take place in a susceptible variety could mean either low levels of phenolic compounds or the presence of other enzymes that inhibit this enzyme system.

The chromatographic analysis of rice varieties indicated that the phenolic acids CUM and UC3 were to be found in all varieties in much greater quantities than other phenolic acids. On average these two acids were found at four times or more greater concentrations than any other compound. CUM is known as one of the key phenolic acids involved in host plant resistance mechanisms (Farkas and Kiraly 1962; Kosuge 1969). It is reported to be found (along with FER) in larger quantities in varieties of maize resistant to the maize weevil, *Sitophilus oryzae* (L.) (Niemeyer 1988). Nevertheless, there were no significant differences when CUM and UC3 values of the varieties were compared. Discriminant function analysis also did not identify these characters as variables contributing to the discrimination of the groups (Table 14.1). Similarly the factor loadings in the principal components analysis did not pick these two compounds as major contributing variables in component 2 along which the three groups of varieties were separated. In contrast, DFA picked three minor constituents (PRT, UC2 and BEN) as the three main variables in function 1. The principal components analysis also picked the same three acids (along with GAL) with highest factor loadings in principal component 2. This might indicate that these minor constituents play a major role in the separation of varietal groups.

Does this mean that the two components (CUM and UC3) occurring in greatest amounts in the rice varieties tested are not important in the host plant resistance mechanism under investigation? Or, conversely, does this mean that only the minor constituents are important? It is easy to draw such conclusions, but they may not be correct. What has been determined here is the levels of the various phenolic acids found in healthy rice seedlings. But hypersensitive reaction follows infestation of the rice seedlings by the gall midge and it is certain that a series of biochemical reactions are then set in motion by the midge larva feeding on the meristematic tissue. Therefore, the range of phenolic acids (both qualitatively and quantitatively) of the varieties after infestation may be quite different from that of the healthy seedlings tested.

Although CUM levels did not vary among the test varieties, the final products of this phenolic acid could be different in different varieties, depending on their enzyme levels and composition. Hence, it is probable that, while in the resistant varieties this acid is acted upon by the oxidative enzymes leading to the production of quinones and, finally, the polymers, in the susceptible variety this series of enzymatic processes is prevented from proceeding normally for one of the following reasons (Overeem 1976):

1. The presence of less phenolic substrates.

2. The presence of a polymerization inhibitor (enzymes known as dehydrogenases were reported by Farkas and Kiraly (1962) to act antagonistically to the oxidative enzymes keeping the phenolics in the reduced state and, thus, preventing necrosis of plant tissues).

3. The presence of other plant biochemical constituents. An example of this was provided by Corcuera *et al.* (1982) who reported that toxic effects of DIMBOA were reduced by the addition of the amino acid cysteine to a synthetic diet. Likewise, Bloem *et al.* (1989) found the addition of cholesterol to synthetic media removed the toxicity of tomatine to *Heliothis zea*. They were of the opinion that host plant resistance was not a very discrete and definable property of plants; instead it had to be considered in terms of the toxic material with other plant constituents. Therefore, there is a possibility that the susceptible rice varieties in this project contain some other plant biochemical constituent that nullifies the effect of the high levels of toxic CUM detected in them.

The results of the discriminant function analysis and the principal components analysis did separate the varietal groups based on the variables measured in the HPLC analysis. These two statistical analytical methods could be used to identify varieties, possessing resistance to the midge. As seen from the discriminant function analysis, function 1 (which explained 93 per cent of the variability among the groups) is composed of variables PRT, UC2, and BEN. The principal components analysis also picked these constituents as key variables in separating the groups. Considering these results with that of the HPLC analysis, it is readily seen that five acids could be used to identify host plant resistance in rice varieties. The phenolic acids, PRT, VAN, GAL, BEN, and UC2, could be important in separating resistant varieties from susceptible varieties. An immediate benefit of this work is that such biochemical analyses for phenolic acids could be made routinely, processing hundreds of rice varieties quickly, without going through the cumbersome process of mass culturing the midge for laboratory screening. Such evaluations could even be conducted in countries where the pest does not occur. For instance, the International Rice Research Institute (IRRI) could screen the large world rice germplasm for gall midge resistance in The Philippines, where the pest has not been recorded. Additionally, more research leading to a better understanding of the biochemistry of host plant resistant to rice gall midge could be beneficial to situations involving other pests and diseases.

References

Beardmore, J., Ride, J.P., and Granger, J.W. (1983). Cellular lignification as a factor in hypersensitive resistance of wheat to stem rust. *Physiological Plant Pathology*, **22**, 209–20.

Bloem, K.A., Kelly, K.C., and Duffey, S.S. (1989). Differential effect of tomatine and its alleviation by cholesterol on larval growth and efficiency of food utilisation in *Heliothis zea* and *Spodoptera exigua*. *Journal of Chemical Ecology*, **15**, 387–98.

Brueske, C.H. and Dropkin, V.H. (1973). Free phenols and root necrosis in Nematex tomato infected with the root knot nematode. *Phytopathology*, **63**, 329–34.

Cole, R.A. (1984). Phenolic acids associated with resistance of lettuce cultivars to the lettuce root aphid. *Annals of Applied Biology*, **105**, 129–45.

Corcuera, L.J., Argandoña, V.H., Peña, G.F., Pérez, F.J., and Niemeyer, H.M. (1982). Effect of benzoxazinone from wheat on aphids. In *International Symposium on Insect–Plant Relationships*, No. 5, (ed. J.H. Visser and A.K. Minks), pp. 33–9. Pudoc, Wageningen.

Cramer, H.H. (1967). *Plant protection and world crop protection*. Bayer, Leverkusen.

Edreva, A., Bailov, D., and Nikolov, S. (1972). Investigation about the necrosis formation 'sharilka' (aladja) on the leaves of field-grown tobacco plants. *Beitrage zur Tabakforschung*, **6**, 236–48.

Farkas, G.L. and Kiraly, L. (1962). Phenolic compounds in the physiology of plant diseases and disease resistance. *Phytopathologische Zeitschrift*, **44**, 105–50.

Heinrichs, E.A. and Pathak, P.K. (1981). Resistance to the rice gall midge, *Orseolia oryzae*, in rice. *Insect Science and its Application*, **1**, 123–32.

Kosuge, T. (1969). The role of phenolics in host response to infection. *Annual Review of Phytopathology*, **7**, 195–222.

Miles, P.W. (1968). Insect secretions in plants. *Annual Review of Phytopathology*, **6**, 137–64.

Modder, W.W.D. and Alagoda, A. (1972). A comparison of the susceptibility of rice varieties, IR8 and Warangal 1263, to attack by the gall midge, *Pachydiplosis oryzae* (Wood-Mason) (Dip., Cecidomyiidae). *Bulletin of Entomological Research*, **61**, 745–53.

Niemeyer, H.M. (1988). The role of secondary plant compounds in aphid–host interactions. Contribution to the *International Congress of Entomology*, No. 18, 3–9 July 1988, Vancouver.

Nierenstein, M. (1930). Galls. *Nature* (London), **125**, 348–49.

Overeem, J.C. (1976). Pre-existing antimicrobial substances in plants and their role in disease resistance. In *Biochemical aspects of plant–parasitic relationships*, (ed. J. Friend and D.R. Threlfall), pp. 195–206. Academic press, London.

Perera, N. and Fernando, H.E. (1967). Laboratory culture of the rice gall midge *Pachydiplosis oryzae* (Wood-Mason). *Bulletin of Entomological Research*, **58**, 439–54.

Reddy, D.B. (1967). The rice gall midge, *Pachydiplosis oryzae* (Wood-Mason). In *Major insect pests of the rice plant*, IRRI Symposium, pp. 457–91. Johns Hopkins University Press, Baltimore.

Suzuki, N. (1965). Nature of resistance to blast. In *The rice blast disease*, IRRI Symposium, pp. 277–301. Johns Hopkins University Press, Baltimore.

Tippins, B. (1987). Selective sample preparation of endogenous biological compounds using solid-phase extraction. *International Laboratory*, **17** (3), 28, 32–34, 36.

15. Parasitoids as a driving force in the evolution of the gall size of *Urophora* on Cardueae hosts

H. ZWÖLFER and J. ARNOLD-RINEHART

Department of Animal Ecology, University of Bayreuth, D-95440 Bayreuth, Germany

Abstract

In the tephritid genus *Urophora*, gall forming evolved from non-galling ancestors whose larvae mined within the achenes and receptacles of Asteraceae flower heads. In the Palearctic *Urophora* species an evolutionary trend to increase the proportion of lignified gall tissue can be recognized. This includes the transition from unilocular to multilocular galls. The structure and size of *Urophora* galls influences larval survival and parasitization rates by a complex of parasitoids with a distinct guild structure and an ancient and stable association with *Urophora* hosts. The available evidence leads to the conclusion that parasitoid pressure was one of the driving forces in the evolution of *Urophora* galls.

Introduction

Parasitoids are notorious inhabitants of the galls of many insect species (Weis *et al.* 1988). It is therefore an obvious question as to whether parasitoids play a role in the evolution of insect galls. In their critical review of gall literature Price and co-workers (Price *et al.* 1986, 1987) discuss this 'enemy hypothesis' as one of the attempts to explain the adaptive nature of galls. Other major hypotheses are 'plant protection', 'mutual benefit', 'nutrition', and 'microenvironment' and there is also the counter hypothesis 'no adaptive value'. Price *et al.* (1986, 1987) conclude that the enemy hypothesis involves several uncertain issues which need to be resolved. Based on an analysis of data from the literature Hawkins and Lawton (1987) and Hawkins (1988*a*) state that gall formers generally support more parasitoids than external feeders and

Plant Galls (ed. Michèle A. J. Williams), Systematics Association Special Volume No. 49, pp. 245–57. Clarendon Press, Oxford, 1994. © The Systematics Association, 1994.

Hawkins (1988*b*) in a comparison of galling and non-galling diptera finds no support for the hypothesis that galls provide protection from parasitoids. Price and Pschorn-Walcher (1988), on the other hand, provide evidence that parasitoids have been important as a selective factor in the evolution of galling nematine sawflies.

As a contribution to the discussion on the adaptive nature of galls we present here results obtained during an extensive study of the insect inhabitants of Cardueae flower heads (Zwölfer 1988) with the genus *Urophora* (Diptera: Tephritidae) and its parasitoids. For this gall former taxon the host plant and parasitoid relationships are well-known. *Urophora* species exhibit a broad range of gall structures which can be arranged in a series of increasing complexity (Arnold-Reinehart 1989) and there is much information available on the impact of gall sizes on larval survival rates and parasitization effects (Zwölfer 1979; Schlumprecht 1990; Zwölfer and Arnold-Rinehart 1992).

Galls of the genus *Urophora*

With regard to the taxonomy and biology of the western Palearctic species of *Urophora* the reader is referred to White and Korneyev (1989). It is not as yet clear, whether 'Urophora' species recorded from Asteraceae flower heads in North America (Goeden 1987), South America (Lewinsohn 1991), and South Africa (Clark 1988) really belong to the genus *Urophora* or whether they should be placed into a new sister genus. However, like the Mediterranean *U. syriaca* Hendel they can be taken as models of the plesiomorphic, non-galling *Urophora* ancestors whose larvae fed within or on Asteraceae achenes. From this origin evolution of gall formation in *Urophora* can be reconstructed by a number of hypothetical steps leading to a sequence of synapomorphies:

(1) non-lignified achene galls (= ovary galls of *U. quadrifasciata* Meigen (Fig. 15.1a);

(2) lignified, unilocular 'cup-shaped' ovary–receptable galls (Fig. 15.1b);

(3) lignified, multilocular cup-shaped ovary–receptacle galls;

(4) lignified, 'block-shaped' ovary–receptacle galls (Fig. 15.1c and d) and shoot galls.

These gall types are discussed in detail by Arnold-Rinehart (1989) and Zwölfer and Arnold-Rinehart (1992), who also cite the literature on the subject. Cup-shaped galls (for example, those of *U. affinis* Frfld (Fig. 15.1b), *U. siruna-seva* Hering, *U. jaculata* Rond., *U. jaceana* Hering, or *U. cuspidata* (Meigen)) are found in *Urophora* species associated with

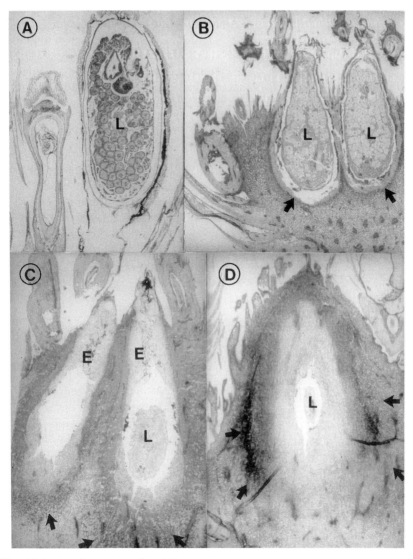

Fig. 15.1. (a) Mature larva (L) of *U. quadrifasciata* within the remainders of an ovariole gall of *Centaurea jacea*. Note that there are no lignified tissues. At the left side is an atrophied ovariole (length of the larva, 4 mm). (b) Two unilocular, cup-shaped, ovariole–receptacle galls of *U. affinis* in a flower head of *Centaurea maculosa*. The larvae (L) are mature (length of the larvae, 3.5 mm). The arrows point towards the thin lignified walls of the galls. (C) Early stage of a multilocular ovariole–receptacle gall of *U. stylata* in a *Cirsium vulgare* flower head. Two larval chambers, one with an immature third-instar larva(L) (length, 3 mm). The arrows point to areas where the formation of the lignified, protective zone starts. E, loose callus tissue in the pre-formed exit channel. (D) Early stage of a block-shaped ovariole–receptacle gall of *U. mauretanica* Macqu. in the flower head of *Carthamus lanatus*. Larva (L) in the late second instar (length, 2.5 mm). Arrows indicate zones where the formation of sclerenchyma cells starts.

Centaureinae host plants, whereas the great majority of block-shaped galls (*sensu* Arnold-Rinehart 1989) are found in *Urophora* species attacking host plants of the subtribe Carduinae (for example, *U. stylata* F. (Fig. 15.1c), *U. solstitialis* (L.), *U. congrua* Loew, and *U. terebrans* (Loew)). *Urophora mauretanica* Macqu. forms block-shaped galls on *Carthamus* spp. (Fig. 15.1d), a host plant genus which belongs to the subtribe Centaureinae.

Pönisch and Brandl (1992) investigated the structure of the salivary gland chromosomes of *Urophora* species. They found a correspondence between the number of chromosomal inversions and the allozyme distances between species pairs. They conclude from the observed cytogenetic and allozyme patterns, that the Centaureinae are the ancestral host plant taxon of *Urophora* and that the Carduinae were colonized by multiple host transfers. This hypothesis means that the type 'multilocular block-shaped ovary–receptable gall' must have evolved independently in two or more ancestral lineages. The evolution of block-shaped galls may have been an adaptation to flower head size which in the genera *Carduus*, *Cirsium*, and *Carthamus* is on average distinctly larger than in *Centaurea*.

The parasitoid guild of the genus *Urophora*

The parasitoid complexes of *Urophora* spp. have a characteristic composition: they consist of one highly specialized endoparasitoid and two to three species of less specialized ectoparasitoids. The endoparasitoid always belongs to the genus *Eurytoma* (Hymenoptera: Eurytomidae). In the guild of *U.cardui* (L.) the endoparasitoid is *E.serratulae* F. and in other *Urophora* species it is *E. tibialis* Bohem., which probably constitutes a cluster of sibling species (Schlumprecht 1990). These endoparasitic *Eurytoma* species parasitize the early larval instars of *Urophora* and allow the host to complete its larval feeding period. As they benefit from the continued life of their host, they can be classified as 'koinobionts' (*sensu* Askew and Shaw 1986). The ectoparasitoids are less specialized, they mainly attack mature *Urophora* larvae, which are killed immediately, that is, these parasitoids are 'idiobionts' (*sensu* Askew and Shaw 1986). Among the ectoparasitoids of *Urophora* there always occurs—and often as a dominant species—a representative of another species group of the genus *Eurytoma, E. robusta* Mayr, a morphospecies which also probably represents a cluster of sibling species (Schlumprecht 1990, personal communication). The other ectoparasitoids of *Urophora* belong to genera of the Chalcidoidea family Pteromalidae *Pteromalus* [for example, *Pteromalus elevatus* (Walker)] and *Torymus* (for example, *Torymus chloromerus* (Walker)).

The competitive capacities of the endoparasitoid and the ecto-parasitoids of *Urophora* are balanced: if the ectoparasitoids attack an endoparasitized host larva early (that is, before the endoparasitoid

Eurytoma larva starts its larval feeding period), the endoparasitoid larva is consumed together with its host, that is, the ectoparasitoids are in fact secondary parasitoids. However, after the endoparasitoid has started feeding, it induces a sclerotization of the skin of the host larva, which protects it against competitors (Varley and Butler 1933). If the *Urophora* gall is in a developmental phase in which it contains trophic tissues, *E. robusta* larvae without sufficient hosts have another option as they can switch to phytophagy and complete their development as inquilines (Zwölfer 1979).

The available data (Varley 1947; Claridge 1961; Zwölfer 1979; Redfern 1983) show that the basic structure of the parasitoid guilds of *Urophora* galls on Cardueae host plants has a stereotyped pattern. A personal 30-year study of the insect fauna of Cardueae in Europe and Asia provides evidence that the combination of two *Eurytoma* species, one an endoparasitic koinobiont and the other an ectoparasitic idiobiont, occurs in *Urophora* food webs from the Atlantic coast to Japan. Zwölfer and Arnold-Rinehart (1992) list 25 cases of such *Eurytoma* associations for 14 *Urophora* host species and 21 Cardueae host plants originating from Europe, Jordan, Pakistan, and Japan. Moreover, an ecologically homologous *Eurytoma* combination occurs in North America in the food web of the gall-forming tephritid *Eurosta solidaginis* Fitch (the endoparasitic *Eurytoma obtusiventris* Gahan and the ectoparasitic *Eurytoma gigantea* Walsh.) (Uhler 1951). This widespread geographic occurrence suggests that the parasitoid guilds of *Urophora* evolved as species systems with a simple pattern of opposed strategies of resource exploitation and that the parasitoid species are evolutionarily adjusted modular units.

Gall size, parasitization, and survival rates of *Urophora*

Schlumprecht (1990) investigated the question of whether the gall size or flower head size provides a partial refuge for tephritid larvae for 10 *Urophora–Eurytoma* associations and three other Tephritid–Eurytoma systems. He measured the length of the parasitoid's ovipositor and the distances which ovipositing females have to overcome to reach the host larva. He showed that there is no correlation between the length of the ovipositor of *Eurytoma tibialis* and the diameter of the flower heads of the host plants which harbour hosts of this parasitoid. Schlumprecht (1990) calculated the theoretical 'refuge effect' in different *Urophora–Eurytoma* systems and found empirical evidence for his hypothesis. This is shown in Fig. 15.2 which has been redrawn from Schlumprecht's data.

In Fig. 15.3 we summarize results of an analysis of survival and mortality rates of *U. cardui* in gall material ($n = 1370$ galls) collected in

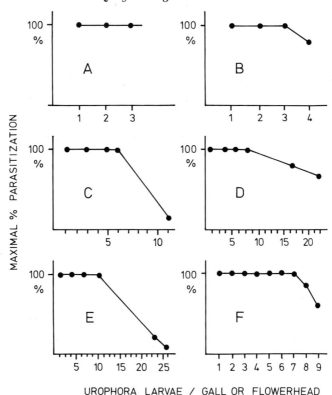

Fig. 15.2. Maximal parasitization rates as a function of the number of *Urophora* larvae per gall (redrawn from Schlumprecht 1990). (A) *Urophora affinis* galls in *Centaurea diffusa* heads. (B) *Urophora affinis* galls in *Centaurea maculosa* heads. (C) *Urophora jaceana* galls in *Centaurea jacea* heads. (D) *Urophora congrua* galls in *Cirsium erisithales* heads. (E) *Urophora stylata* galls in *Cirsium vulgare* heads. (F) *Urophora cardui* galls on *Cirsium arvense* stems.

1989 and 1990 in the Belfort-Sundgau region (eastern France). An increase in gall diameter as well as an increase in the number of cells per gall enhances the survival rates of *U. cardui* larvae from below 20 per cent to above 60 per cent. The survival rate increases steadily for gall diameter (Fig. 15.3a), but a rise in the number of cells per gall leads to a plateau in galls with more than five cells (Fig. 15.3b). The improvement in the survival of *U. cardui* is due to a decline in the rate of empty cells (caused by deficiencies in the process of gall formation and partially also by ectoparasitoids) and, in medium-sized and big galls, by a decline in the attack rate of the ectoparasitoids. The parasitization rate by the specialized endoparasitoid increases from small to medium-sized galls and remains constant from medium-sized to large galls. The situation in *U. stylata* galls differs from that of *U. cardui*, as only the gall diameter

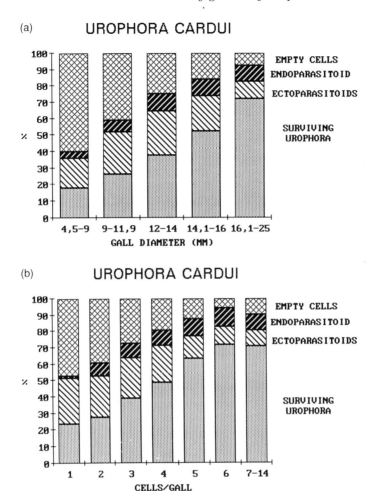

Fig. 15.3. Survival and mortality rates of *Urophora cardui* in stem galls of *Cirsium arvense*. Dissected material = 1370 galls (origin: Belfort-Sundgau region, 1989/1990). Endoparasitoid, *Eurytoma serratulae*; Ectoparasitoids, *E. robusta, Pteromalus elevatus* and *Torymus chloromerus*. (a) Survival and mortality as a function of the gall diameter: differences between survival rates significant for all diameter classes at $p > 0.0000$ (analysis of variance). (b) Influence of cell number/gall. Differences among the survival rates for cells/gall classes 1–5 significant at $p > 0.0000$.

(Fig. 15.4a) not the number of cells per gall (Fig. 15.4b) has a positive influence on the survival of *Urophora* and a negative influence on parasitization by ectoparasitoids. This is consistent with the results of a study of Redfern and Cameron (1985), who found that survival and overall mortality of *U. stylata* did not relate to the number of cells per head but to the size of the flower head (which is highly correlated with gall size) (Fig. 15.4c). They report that for any given number of cells in a gall, mortality of *Urophora* was lower in galls with larger diameter. As in our material, they found that ectoparasitoids caused the greatest mortality in small heads while the endoparasitoid, *E. tibialis*, was rather indiscriminate. Our results also correspond with the data of Michaelis (1984), who analysed the impact of the 'packing density' (that is, the number of *Urophora* larvae occupying a volume unit of a gall) of *U. stylata* on larval survival in galls in *C. vulgare* heads. In contrast to the number of cells per gall packing density had a highly significant positive influence on the mortality caused by the endoparasitoid *E. tibialis* and the ecto-parasitoid *E. robusta* and a significant negative influence of the survival of *U. stylata*.

For different biotypes of *U. solstitialis* (origin: Germany, Austria, Swiss Alps) Knoll (Bayreuth, unpublished data) examined the relationship between survival rates and the number of cells per gall. These were not significant in the biotypes associated with *Carduus nutans*, *C. acanthoides*, and *C. personata*, but survival of *U. solstitialis* larvae increased significantly ($p < 0.001$) with increasing numbers of cells per gall in the heads of *C. deflorata* (Fig. 15.5) and *C. crispus*. In addition total parasitization decreased ($p < 0.05$).

Discussion and conclusions

We present below three lines of evidence that enemy pressure has played a part in the evolution of the size and structure of *Urophora* galls.

1. The evolution of *Urophora* galls can be reconstructed as a sequence of steps which lead from non-galling miners in Asteraceae achenes and receptacles to the formation of simple ovary galls without sclerenchyma tissues and then to lignified, unilocular cup-shaped and multilocular

---

Fig. 15.4. Survival and mortality rates of *Urophora stylata* in flower heads of *Cirsium vulgare*, ($n = 633$ galls; Bayreuth 1989/90). *Eucosma*, mortality caused by larvae of the Pyralid *Eucosma cana* Haw.; endoparasitoid, *Eurytoma tibialis*; ectoparasitoids and empty cells as in Fig. 15.3. (a) Influence of gall diameter. The differences in the survival rates, the ectoparasitoid mortality, and the effect of empty cells are significant at $p > 0.000$ in the first three diameter classes. Parasitization rates by the endoparasitoid and mortality by *Eucosma cana* remains constant. (b) Influence of cell number/gall. Increase of the survival rate of *U. stylata* weak ($p < 0.067$), decrease of the ectoparasitoids significant ($p < 0.0000$). Rates of mortality by endoparasitoid, empty cells, and *Eucosma* constant. (c) Data of Redfern and Cameron (1985), *U. stylata* ($n = 689$ galls from the Bayreuth area and Waldringfield, England, 1982, 1983). Influence of the diameter of *C. vulgare* flower heads on survival and mortality of *U. stylata*.

(a)

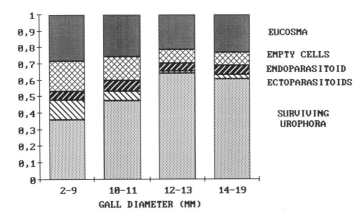

(b)

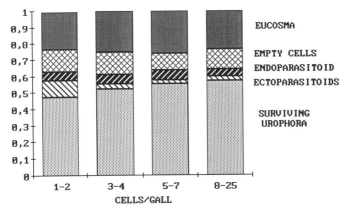

(c)

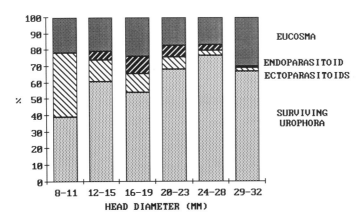

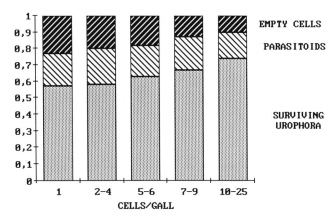

Fig. 15.5. Influence of cells/gall on survival and mortality *Urophora solstitialis* in flower heads of *Carduus defloratus* ($n = 451$ galls from the Swiss Alps). Increasing cell numbers/gall favour the survival rate of *Urophora* ($p < 0.001$) and lead to a decline of the parasitization rate ($p < 0.05$).

block-shaped ovary–receptacle galls and stem galls (Arnold-Rinehart 1989). There is a distinct evolutionary trend to increase the lignified proportion of the galls, which seems to have been favoured by the transfer from Centaureinae to Carduinae host plants (Pönisch and Brandl 1992).

2. Over the whole Palearctic distribution area of *Urophora*, its parasitoid guild has a characteristic composition, which combines a highly specialized, endoparasitic *Eurytoma* sp. (a koinobiont *sensu* Askew and Shaw 1986) with a few less specialized, ectoparasitic species belonging to another species group of *Eurytoma* and to the chalcid genera *Pteromalus* and *Torymus* (= idiobionts *sensu* Askew and Shaw 1986). The wide geographic distribution of the association of *Urophora* with an endoparasitic and an ectoparasitic *Eurytoma* species (Zwölfer and Arnold-Rinehart, 1992) and the fact that in this species constellation competitive superiorities and inferiorities are well-balanced (Zwölfer 1979) suggests that the parasitoid guild of *Urophora* is an ancient and stable one.

3. For three *Urophora* species we could show that larval survival rates increase significantly with increasing number of cells per gall and/or increasing thickness of the lignified part of the gall. One reason for this correlation of gall size and change of larval survival is that larger galls provide more protection against ectoparasitiods (Michaelis 1984; Redfern and Cameron 1985; Zwölfer and Arnold-Rinehart 1992). The specialized

endoparasitoid attacks *Urophora* at an early stage of gall formation (Zwölfer 1979; Schlumprecht 1990) and is therefore better adapted to overcome the mechanical protection provided by the lignified gall tissues.

It seems questionable as to whether the impact of enemies was the initial cause for the evolution of gall formation in *Urophora*, as the primitive, unlignified *Urophora* gall provides hardly any additional protection. It is more likely that the increased nutrient flow (Weis *et al.* 1988), the augmentation of the size of overioles and achenes, and, perhaps also, a protection against abortion or abscission of attacked flower heads were selection factors which started galling in *Urophora*. However, once lignified *Urophora* galls had evolved, enemies were affected and this would have influenced the further evolutionary history of gall forming in *Urophora*.

Cardueae flower heads can provide host refuges even in non-galling tephritids. This has been shown by Schlumprecht (1990) and Romstöck-Völkl (1990). The presence of multilocular *Urophora* galls enlarges Cardueae flower heads (Zwölfer, unpublished) and creates an additional mechanical barrier. In this way lignified *Urophora* galls are a protection against unspecialized enemies and competitors such as weevil larvae in Cardueae heads (Zwölfer 1979). On the other hand, multilocular *Urophora* galls become an apparent target for specialized enemies, that is, the guild of 'niche-specific' (Askew and Shaw 1986) ectoparasitoids. Therefore, the transition to multilocular galls and an increase of sclerenchyma tissues brought a reproductive advantage to *Urophora* as it provided the larvae and pupae with a partial protection against this mortality factor.

There are obviously constraints set by the physiological condition of the host plant (Weis *et al.* 1988) which limit the size and cell number of multilocular *Urophora* galls. Gall forming is a process which involves a high degree of mortality of the gall former and much of this seems to be host plant-mediated. This is suggested by the considerable rates of 'empty cells' (Figs 15.3–15.5) and by our data on egg mortality in *U. cardui*. Under field conditions the mean number of cells (including empty cells) per gall in this species is 3.644 ± 0.051 (mean and standard error of 307 *U. cardui* populations with a total of more than 12 000 galls). In cage experiments we obtained a similar figure. In oviposition experiments, G. Freese (Bayreuth, unpublished) found that an average *U. cardui* females deposit clusters of 11.33 eggs ($n = 42$ egg clusters). Comparable figures (13.96 eggs per cluster) were observed in a laboratory study in Canada (Peschken and Harris 1975). Thus, on average only 26–32 per cent of the deposited eggs develop to a stage where a gall cell is initiated. We do not know if this high egg mortality is also representative of other *Urophora* species. But our data show that constraints mediated by the host plant and not the number of deposited eggs set upper limits to the number of cells per gall and gall sizes of *U. cardui*.

Acknowledgements

We thank Steffi Knoll and Gunther Freese (Department of Animal Ecology, University of Bayreuth) for unpublished data and Marion Preiß and Gabi Lutschinger for technical assistance. Dr T.N. Petney (Heidelberg) kindly corrected our manuscript. We gratefully acknowledge the financial support of the Deutsche Forschungsgemeinschaft (SFB 137).

References

Arnold-Rinehart, J. (1989). Histologie und Morphogenese von *Urophora*- und *Myopites*-Blütenkopfgallen (Diptera:Tephritidae) in Asteraceae. PhD dissertation, University of Bayreuth.

Askew, R.R. and Shaw, M.R. (1986). Parasitoid communities: their size, structure and development. In *Insect parasitoids*, (ed. J. Waage and D. Greathead), pp. 225–264. Academic Press, London.

Claridge, M.F. (1961). Biological observations on some Eurytomid (Hym. Chalcidoidea) parasites associated with Compositae, and some taxonomic implications. *Proceedings of the Royal Entomological Society of London (A)*, **36**, 153–8.

Clark, M.M. (1988). Insect herbivore communities colonising the flower-heads of *Berkheya* in South Africa and Carduoideae in Europe and California. PhD thesis, University of Cape Town, South Africa.

Goeden, R.D. (1987). Host plant relations of native *Urophora* spp. (Diptera:Tephritidae) in southern California. *Proceedings of the Entomological Society of Washington*, **89**, 269–74.

Hawkins, B.A. (1988a). Species diversity in the third and fourth trophic levels: patterns and mechanisms. *Journal of Animal Ecology*, **57**, 137–62.

Hawkins, B.A. (1988b). Do galls protect endophytic herbivores from parasitoids? A comparison of galling and non-galling diptera. *Ecological Entomology*, **13**, 473–7.

Hawkins, B.A. and Lawton, J.H. (1987). Species richness for parasitoids of British phytophagous insects. *Nature*, **326**, 788–90.

Lewinsohn, T.M. (1991). Insects in flower heads of Asteraceae in southeast Brazil: a case study on tropical species richness. In *Plant–animal interactions. Evolutionary ecology in tropical and temperate regions*, (ed. P.W. Price, T.M. Lewinsohn, G. Wilson Fernandes, and W.W. Benson), pp. 525–59. J. Wiley & Sons, New York.

Michaelis, H. (1984). Struktur- und Funktionsuntersuchungen zum Nahrungsnetz in den Blütenköpfen von *Cirsium vulgare*. PhD dissertation, University of Bayreuth.

Peschken, D.P. and Harris, P. (1975). Host specificity and biology of *Urophora cardui* (Diptera: Tephritidae). A biocontrol agent for Canada thistle (*Cirsium arvense*). *Canadian Entomologist*, **107**, 1101–10.

Pönisch, S. and Brandl, R. (1992). Cytogenetics and diversification of the phytophagous fly genus *Urophora* (Tephritidae). *Zoologischer Anzeiger*, **228**, 12–25.

Price, P.W. and Pschorn-Walcher, H. (1988). Are galling insects better protected against parasitoids than exposed feeders?: a test using tenthredinid sawflies. *Ecological Entomology*, **13**, 195–205.

Price, P.W., Waring, G.L. and Wilson Fernandes, G. (1986). Hypotheses on the adaptive nature of galls. *Proceedings of the Entomological Society of Washington*, **88**, 361–3.

Price, P.W., Wilson Fernandes, G. and Waring, G.L. (1987). Adaptive nature of galls. *Environmental Entomology*, **16**, 15–24.

Redfern, M. (1983). *Insects and thistles*, Naturalist's Handbooks 4. Cambridge University Press, Cambridge.

Redfern, M. and Cameron, R.A.D. (1985). Density and survival of *Urophora stylata* (Diptera:Tephritidae) on *Cirsium vulgare* (Compositae) in relation to flower head and gall size. In *Proceedings of the VI International Symposium on Biological Control of Weeds*, Vancouver, BC 19–25 August 1984 (ed. E.S. Delfosse), pp. 453–77. Agriculture Canada, Ottawa.

Romstöck-Völkl, M. (1990). Host refuges and spatial patterns of parasitism in an endophytic host–parasitoid system. *Ecological Entomology*, **15**, 321–31.

Schlumprecht, H. (1990). Untersuchungen zur Populationsökologie des Phytophagen–Parasitoid-Systems von *Urophora cardui* L. (Diptera: Tephritidae). Unpublished PhD thesis, University of Bayreuth, Germany.

Uhler, L.D. (1951). *Biology and ecology of the goldenrod gall fly, Eurosta solidaginis (Fitch)*. Publications of the New York State College of Agriculture, Cornell University memoir, 300. Ithaca, NY.

Varley, G.C. (1947). The natural control of the population balance of the knapweed gall-fly (*Urophora jaceana*). *Journal of Animal Ecology*, **16**, 139–87.

Varley, G.C. and Butler, C.G. (1933). The acceleration of development of insects by parasitism. *Parasitology*, **25**, 263–8.

Weis, A.E., Walton, R., and Crego, C.L. (1988). Reactive plant tissue sites and the population biology of gall makers. *Annual Review of Entomology*, **33**, 467–8.

White, I.M. and Korneyev, A.V. (1989). A revision of the western Palearctic species of *Urophora* Robineau-Desvoidy (Diptera: Tephritidae). *Systematic Entomology*, **14**, 327–74.

Zwölfer, H. (1979). Strategies and counterstrategies in insect population systems competing for space and food in flower heads and plant galls. *Fortschritte der Zoologie*, **25**, (2/3), 331–53.

Zwölfer, H. (1988). Evolutionary and ecological relationships of the insect fauna of thistles. *Annual Review of Entomology*, **33**, 103–22.

Zwölfer, H. and Arnold-Rinehart, J. (1992). The evolution of interactions and diversity in plant–insect systems: the *Urophora–Eurytoma* food web in galls on Palearctic Cardueae. *Ecological Studies*, **99**, 211–33.

16. Life cycle strategies in a guild of dipteran gall formers on the common reed

LUC DE BRUYN

Department of Biology, Evolutionary Biology Group, University of Antwerp (RUCA), Groenenborgerlaan 171, 2020 Antwerpen, Belgium

Abstract

The flies of the genus *Lipara* are strict monophagous parasites of the common reed, *Phragmites australis* (Poaceae). Due to feeding activities and/or metabolic products, the newly formed internodes of the shoot are significantly shortened and a typical cigar-or spike-like gall is formed at the top of the shoot. Reed is a perennial rhizomatous grass that produces fresh shoots every year during spring. Because the shoots dry up and die at the end of the summer, they can only serve as a source of food for the herbivore during a short time of the year. Previous studies have revealed a high between-plant and between-year variation in plant quality caused by several interacting factors, such as the genetic differences between reed clones, the water and nutrient content of the soil, and interspecific competition with other plants. In addition, the annual germination and growth of the reed is strongly influenced by climatic factors. In the course of evolution, two main solutions have evolved to overcome the above mentioned problems within the genus *Lipara*.

1. The *L. pullitarsis* strategy. The female flies emerge early in the season, they deposit a high number of eggs, randomly distributed among the reed shoots, there is a high mortality rate of larvae before gall-formation, and the larvae cause only a slight deformation of the reed shoot.

2. The *L. lucens* strategy. The females emerge later when the young reed shoots are well developed, they carefully select a suitable place to deposit a limited number of eggs, there is a rather low mortality of larvae before gall formation, and a distinct gall-chamber is formed with strongly hardened walls and filled with a dense mass of parenchymatous tissue.

Plant Galls (ed. Michèle A. J. Williams), Systematics Association Special Volume No. 49, pp. 259–81. Clarendon Press, Oxford, 1994. © The Systematics Association, 1994.

Introduction

Compared to more complex plants such as trees, grasses only possess simple plant architecture, relatively low protein concentration, and a low diversity of secondary compounds, which influences the insect community feeding on them (Bernays and Barbehenn 1987; Tscharntke 1988). As a consequence, grass habitats are in general represented by a low species richness (Strong and Levin 1979; Lawton 1983).

Many endophagous herbivore insects are selective feeders, consuming only specific high-quality tissues while rejecting others (Kimmerer and Potter 1987; Reaey and Gaston 1991). Gall makers not only consume these specific tissues, but may also force their host plant to produce improved conditions (Abrahamson and Weis 1987; Roininen *et al.* 1988; Tscharntke 1989; Hartley and Lawton 1992). In addition, the galls may provide improved protection against the natural enemies of the herbivore (Weis *et al.* 1985; Price and Clancy 1986; Price *et al.* 1987*a*; Waring and Price 1989).

The common reed, *Phragmites australis* (Cav.) Trin. ex Steud., is a member of the family Poaceae. Reed beds occur all over the world as large monocultures. Several organisms use these reed beds as a source of food or to raise their offspring. Due to the developmental constraints of the host, reed beds harbour many specialized herbivore parasites (for example, Skuhravý 1981; Vogel 1984; Tscharntke 1988, 1989).

The flies of the genus *Lipara* Meigen (Diptera: Chloropidae) are strict monophagous parasites of the common reed, *Phragmites australis*, whereupon they induce typical cigar- or spike-like galls (Chvála *et al.* 1974). In the present paper we report on the life cycle strategies utilized by the different gal-inducing *Lipara* species.

Organisms studied

1. The hostplant: Phragmites australis

The common reed, *P. australis*, is a member of the family Poaceae, of typical appearance. The erect culms (1.5–3 m high) bear a lanceolate leaf at each node. Above the growing point, only the newly formed enwrapped leaves are present. These leaves are directed upward in line with the shoot. Below the soil surface, the different shoots of a reed clone are connected with a rhizome.

Reed is a perennial rhizomatous grass that produces fresh shoots every year during spring (April–May). The vegetational phase lasts until the end of June, after which an ear is formed (start of the reproductive phase). During autumn, the shoots dry up and the parts of the reed clone above ground die off. Because these shoots dry up and die at the end of the summer, they can only serve as a source of food for the herbivore during a short time of the year. Previous studies have revealed a high between-plant and between-year variation in plant quality caused by several interacting factors such as the genetic differences between reed clones, the water and nutrient contents of the soil, and interspecific competition with other plants (Björk 1967; Haslam 1970*a*, *b*; van der Toorn 1972; Dykyjová and Hradecká 1973; Daniels 1991). In addition, the annual germination and growth of the reed is strongly influenced by climatic factors (Vogel 1984).

2. *The herbivores: the genus Lipara*

The distribution of the genus *Lipara* is restricted to the Palaearctic region (Beschovski 1984). In Belgium, three *Lipara* species are encountered frequently, viz. *Lipara lucens* Meigen, *Lipara pullitarsis* Doskočil & Chvála, and *Lipara rufitarsis* (Loew) (De Bruyn 1985).

Adult *Lipara* flies emerge from the end of May until early July. Oviposition takes place 1–3 days later. The female flies deposit their eggs on the surface of the reed shoot. After approximately 9 days (Ruppolt 1957), the first instar larvae emerge and crawl to the top of the culm. Here they enter the shoot under the edge of the leaf sheath of the uppermost leaf. Next they gnaw their way down through the enwrapped leaves until they reach the growing point where they feed upon the newly emerging leaves. Due to the larval presence, the newly formed internodes do not elongate any more and the species-specific gall chamber forms (Chvála *et al.* 1974). Interesting to note is that only one *Lipara* larva per shoot can develop. After a few weeks, when gall formation is completed, the larvae of *L. lucens* and *L. rufitarsis* gnaw through the growing point and enter the gall chamber where they continue their life cycle. The larvae of *L. pullitarsis* never pass the growing point and live between the enwrapped leaves during their entire larval phase. At the end of August, the last instar larvae stop feeding and go into diapause. Pupation takes place in the next spring.

Gall structure

Lipara larvae are sedentary herbivores and are therefore strongly dependent on the quantity and quality of the available resources in the shoot wherein they develop. To assess the importance of gall formation on the development of the inhabiting larva we carried out a structural and chemical analysis of the *Lipara* galls.

Galls were harvested in early August. At this time, gall formation is complete, but the larvae are still feeding on the young enwrapped leaves above the growing point. The tissues in the gall chamber are still intact.

Gall morphology

The most striking feature of the galls of *L. lucens* is that the gall chamber is completely filled with a dense mass of parenchymatous pith tissue (Fig. 16.1b). The nodia, which bear no septa, are strongly widened (mean = 2.41 ± 0.52 times the normal shoot diameter). In the wall of the gall chamber a number of sclerenchymatous cells have formed. When the larva has penetrated the growing point and starts feeding on the parenchymatous tissue, this sclerotenization process continues until all wall tissues have been transformed into a hard mass, at the end of the summer.

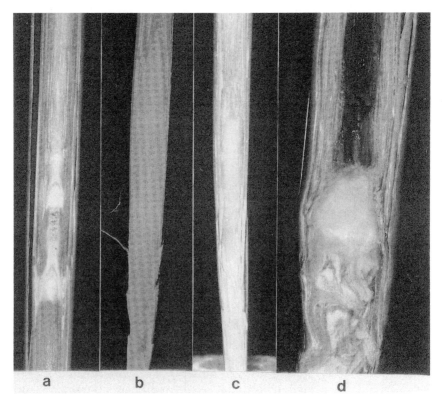

Fig. 16.1. Transverse sections through (a) an uninfested *P. australis* shoot, (b) a *L. lucens* gall, (c) a *L. rufitarsis* gall, (d) a *L. pullitarsis* gall.

Gall formation occurs in a roughly analogous manner in *L. rufitarsis*. A gall chamber is filled with parenchymatous pith and no septa are formed (Fig. 16.1c). However, there are fewer internodia involved, and they are less wide than in *L. lucens* (mean = 1.81 ± 0.59 times the normal shoot diameter). As a consequence, the gall chamber is distinctly smaller. In addition, the wall of the gall chamber does not sclerotenize.

No gall chamber is formed in the galls of *L. pullitarsis*. Only the top internodia, immediately below the growing point, are strongly deformed (Fig. 16.1d). The meristematic tissue is hardened and no parenchyma is formed. There is no apparent widening of the shoot. The place where the larva feeds on the enwrapped leaves is indicated by a dark brown channel with a rotting appearance.

Chemistry of the gall tissues

For the qualitative analysis we carried out a chemical examination of the enwrapped leaves above the growing point and the parenchymatous tissue in the gall chamber of *L. lucens*. The galls were collected in the field and transported on ice to the laboratory. Here they were dissected and lyophilized immediately. Protein and carbohydrate content were assessed by standard laboratory methods. Water content of the tissues was obtained by measuring fresh and oven-dry (temperature 70°C) weight on a microbalance. For a detailed description of the methods used we refer to De Bruyn (1989*a*).

Our results indicate that the parenchymatous tissues were higher in all three measured parameters than the enwrapped leaves (Table 16.1). This was particularly the case for protein (paired-$t = 4.92$, df $= 23$, $p < 0.001$) and water (paired-$t = 5.54$, df $= 78$, $p < 0.001$). The difference in carbohydrate content was less pronounced, but still slightly significant (paired-$t = 2.07$, df $= 24$, $p = 0.05$). Moreover, there is also a histological difference. The parenchymatic pith consists of relatively large, soft cells. The enwrapped leaves, on the contrary, contain numerous parallel fibrous veins which increase toughness and as such hamper or even prevent food uptake and digestion.

Our evidence indicates that *L. lucens* and *L. rufitarsis* have evolved the possibility of improving the quality of their host plant by aggregating highly nutritious tissues in their galls. *L. pullitarsis*, on the contrary, does not change the growth and quality of its host in any conspicuous manner. Analogous observations have been made for other chloropid species which are shoot borers of Poaceae (Wetzel 1967). Therefore, we can consider *L. pullitarsis* more as an ordinary shoot borer than as a real gall inducer on the common reed. Our results are therefore consistent with

Table 16.1. Water, protein, and carbohydrate content of the different gall tissues (data are mean ± SD) in *L. lucens*.

Tissue	water (mgg^{-1} wet wt)	protein (mgg^{-1} dry wt)	carbohydrate (mgg^{-1} dry wt)
Enwrapped leaves	849.70 ± 35.77	242.41 ± 98.74	149.36 ± 97.46
Parenchyma	886.02 ± 57.85	407.36 ± 166.22	190.06 ± 79.61
Paired t-test			
Number of cases	78	23	24
t-value	5.54	4.92	2.07
p-value	< 0.001	< 0.001	= 0.05

the nutrition hypothesis as an explanation of the adaptive nature of insect galls (Price *et al.* 1987*a*).

Habitat distribution

During winter (December–January) we collected *Lipara* galls at 45 different localities in Belgium and The Netherlands (see De Bruyn (1989*a*) for detailed information on the localities, collection dates, and *Lipara* species). Each collection included at least four 0.25 m^2 samples. The galls were dissected and the inhabiting species were identified. Based on the available literature information on the growth of *P. australis* (for example, Allen and Pearsall 1963; Björk 1967; Dykyjová and Hradecká 1973; Haslam 1973) and *Lipara* habitat observations (for example, Wagner 1907; Ruppolt 1957; Mook 1967, 1971; Pokorny 1971), environmental variables which possibly could influence the *Lipara* distribution, were assessed. These were pH, moisture and nitrogen contents of the soil, light exposure, mean shoot density, mean shoot diameter, total size of the reed bed and distance to the nearest other reed bed, and the possible management practice. Next, we calculated a multiple linear regression for each *Lipara* species where the *Lipara* density was the dependent variable and the independent variables were the environmental data. Of these variables, the mean shoot diameter of the reed bed turned out to be the single statistically significant factor influencing the *Lipara* population densities (De Bruyn 1989*a*).

The relationship between the shoot diameter and the *Lipara* density was calculated (De Bruyn 1987*a*). Most galls of *L. lucens* were found on shoots with a diameter between 3 and 3.5 mm. The proportion of galled shoots declined on thicker shoots. As a consequence there was a clear negative relation between the mean shoot diameter of a reed bed and the *L. lucens* densities (Fig. 16.2a). A comparable result was obtained for

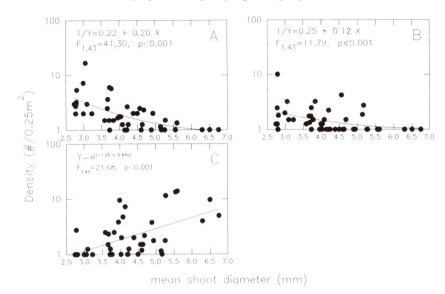

Fig. 16.2. Relationship between the mean shoot diameter in a reedbed and the *Lipara* densities. (A) *L. lucens*, (B) *L. rufitarsis*, (C) *L. pullitarsis*.

L. rufitarsis. Here the highest proportion of galled shoots was found on shoots with a diameter between 2 and 2.5 mm and declined rapidly on thicker shoots, resulting in a negative relationship between shoot density and *L. rufitarsis* density (Fig. 16.2b). The opposite was found for *L. pullitarsis* (Fig. 16.2c). Density increased with increasing mean shoot diameter. Most infested galls were found on the thickest shoots (maximum between 5 and 8 mm; very few galls were found on shoots thinner than 3 mm).

Our results showed that the diameter of a particular reed shoot is of utmost importance for the success of the three *Lipara* herbivores. Former studies have revealed that the shoot diameter is strongly correlated with growth, final shoot length, the possibility to produce an ear, and is as such a measure of the strength of the shoot (Mook 1967; Haslam 1971). Below we assess how the shoot diameter influences the different parts of the *Lipara* life cycle and consequently defines the life cycle strategies of the three species.

Mortality factors

The study of plant resistance to insect herbivory has always been one of the major topics of researchers of plant–insect interactions. Host plants can influence the success of their herbivore parasites in two different ways. First the plant can act directly through its defence systems (Coley *et al.* 1985; Karban and Myers 1989; Smith 1989; Stiling and Simberloff 1989; Gould 1991). Secondly, a plant can evolve certain traits which affect herbivore vulnerability to the herbivores enemies (Price *et al.* 1980; Faeth 1985; Craig *et al.* 1988; Gross and Price 1988; Waring and Price 1989; Andow and Prokrym 1990). Insect-induced plant galls usually form a base for a complex community of parasitoids and predators (Price 1971, 1972, 1973; Force 1974; Weis 1982). According to theory, parasitoids and predators play an important role in regulating insect herbivore densities (Hairston *et al.* 1960; Slobodkin *et al.* 1967). However, field studies revealed that the parasitization rate can be very variable, ranging from negligible to over 99 per cent (Stinner and Abrahamson 1979; Washburn and Cornell 1979, 1981; Abrahamson *et al.* 1983; Price 1985; Weis and Abrahamson 1985).

Finally, several papers have been addressed to the abundant inter- and intraindividual variation of natural plant populations for both direct plant resistance and herbivore mortality due to parasitization. This variation may be due to environmental phenotypic variation, as well as genotypic components (Price *et al* 1987*b*; Collinge and Louda 1988; Fritz and Price 1988; Price 1989; Fritz and Nobel 1990; Alexander 1991; Smith and Rutz 1991).

Here we analysed the mortality factors acting on the three *Lipara* species. First we studied the direct influence of the reed shoots on the possibility of a larva being able to induce a gall. Secondly, we assessed the impact of enemies on the survival of the *Lipara* species. Special attention was paid to the impact of the shoot diameter.

Survival up to gall formation

When the young *Lipara* larvae have emerged, they have to crawl up the shoot, enter it, and initiate a gall. To assess *Lipara* survival during this period a field experiment was set up (De Bruyn 1989*b*). During the oviposition period in June, several reed shoots were selected in our experimental reed bed. Each shoot was provided with one single *Lipara* egg. To achieve this a muslin bag, containing a mated female, was put over the top of each shoot. In general, the female had deposited one or more eggs after 1 day. The bags were removed and the shoots were checked for superfluous eggs, which were removed. Shoots from different diameter classes were chosen to include the complete range of each *Lipara* species. Each shoot was individually labelled so they could be found again and identified later. At the beginning of August, when the

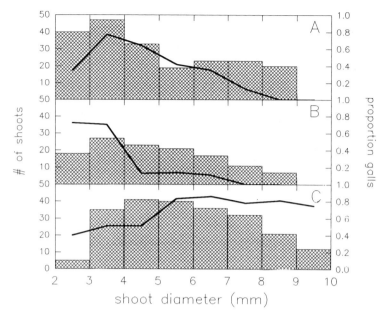

Fig. 16.3. Shoot diameter dependent larval survival up to gall formation: bars, number of shoots tested; line, proportion of shoots with galls. (a) *L. lucens*, (b) *L. rufitarsis*, (c) *L. pullitarsis*.

galls can be recognized, the shoots were examined for galls.

Forty-four per cent of the shoots provided with a *L. lucens* egg carried a gall (Fig. 16.3a). The highest proportion of galled shoots was found in the diameter class between 3 and 4 mm (77 per cent). Thicker shoots produced proportionally fewer galls and on shoots thicker than 8 mm the larvae failed completely to induce a gall. The response curve is very similar for *L. rufitarsis* (Fig. 16.3b). Here again most galls were found on the thinnest shoots (diameter 2–4 mm). When the diameter increases, the proportion of galled shoots decreases. The thickest shoot carrying a gall was 6.3 mm wide. *L. pullitarsis* induced eggs on shoots from all diameter classes (Fig. 16.3c). The proportion of galled shoots is low for the thinnest shoots and rises on thicker shoots. The proportion of galls on shoots thicker than 6 mm is approximately equal.

It is clear that the diameter of a reed shoot definitely influences the survival of the *Lipara* larvae. For the species which form a nutrient rich gall chamber, gall formation and survival is highest on the thinnest shoots. Thicker shoots are more resistant to infestation. *Lipara pullitarsis*, the species which does not form a gall chamber, can cause galls on shoots of all diameter classes, but survival rate is highest on thicker shoots.

Mortality due to parasitization and predation

1. The Lipara-enemy community

When *Lipara* galls are opened and examined, a number of larvae turn out to be parasitized by Hymenoptera (Giraud 1863; Mook 1967; Chvála *et al.* 1974; Nartshuk 1977). During our study we found five hymenopteran parasitoids and one vertebrate predator attacking the *Lipara* species (De Bruyn 1987*b*).

Polemochartus liparae (Giraud) and *P. melas* (Giraud) (Hymenoptera: Braconidae) are egg-larva parasitoids. The female wasps oviposit in the *Lipara* eggs while they are attached to the outside of the reed shoot (Mook 1961). *Polemochartus* induces a premature pupation of its host. When the parasitized galls are opened during the winter, a dark (almost black) pupal case of the *Lipara* species, with a larva of the parasitoid inside, is found. The *Polemochartus* parasitoids emerge at the same time as their corresponding *Lipara* host, by clearing a way through the enwrapped leaves at the top of the gall.

Stenomalina liparae (Giraud) (Hymenoptera: Pteromalidae) is a larval parasitoid. The female wasps possess long ovipositors. Oviposition takes place after the *Lipara* larvae have entered the reed shoot, but before gall formation has been completed, by inserting the ovipositor through the reed shoot. During winter, only the shrivelled skin of the *Lipara* larva is left, in which the *S.liparae* larva overwinters. The adult wasps emerge a few days later than their host species by gnawing a little hole in the wall of the gall.

Tetrastichius legionarius Giraud (Hymenoptera: Eulophidae) are small, gregarious larval parasitoids. How the female wasps attack their hosts is unknown. During winter, one can find more than 40 *T. legionarius* specimens in a single *Lipara* larva. The adult parasitoids leave the gall through the enwrapped leaves of the gall 1–2 weeks after the *Lipara* flies have emerged.

The final parasitoid, *Scambus* sp. (Hymenoptera: Ichneumonidae) possesses a relatively long free ovipositor. The biology of this parasitoid is unknown. During winter, only the hibernating larvae, accompanied by a few remnants of the larval skin of the host can be found in the gall chamber. The adult *Scambus* wasps emerge through the enwrapped leaves approximately 1 week after the emergence of their host.

When *Lipara* galls are collected during winter, one can regularly find a gall with a large hole, probably hewn out by a bird. The gall chamber is empty. We have no personal observations of a bird searching for food in the reed beds. According to Mook (1967) a possible predator would be the Blue Tit, *Parus caeruleus* L. During winter, when there is a shortage of food, Blue Tits often forage in reed beds.

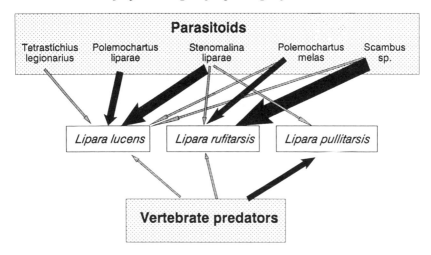

Fig. 16.4. Diagram illustrating the relationships and host specificity of the different parasitoids and predators. Arrow thickness corresponds to parasitoid impact.

To analyse the impact of the herbivore's enemies we carried out a small field experiment. During the winter season, when all gall inhabitants are in diapause, we collected *Lipara* galls in several reed beds. The galls were transported to the laboratory where they were dissected and the contents identified.

The parasitoids and predators mentioned above all show a more or less specific host spectrum (Fig. 16.4; De Bruyn 1987*b*). *Stenomalina liparae* attacks all three *Lipara* species although it is primarily a parasitoid of *L. lucens*. Of the specimens 93.28 per cent (194) were found in the galls of this species. It is also the most important parasitoid of *L. lucens* (16.4 per cent).

Polemochartus liparae only attacks *L. lucens* and *L. pullitarsis*. Most parasitoids were found in the galls of the first species. While the mortality of *L. lucens* due to *P. liparae* is rather low (6.85 per cent) it is practically negligible in *L. pullitarsis* (0.75 per cent). The other *Polemochartus* species, *P. melas* infests all three *Lipara* species, with a strong dominance for *L. rufitarsis*. However, the mortality rate here is still low (8.72 per cent).

The *Scambus* species is also primarily a parasitoid of *L. rufitarsis* (mortality rate 13.08 per cent). A few larvae were found in the galls of *L. lucens*. The latter only occurred when *L. rufitarsis* was also present in the same reed bed.

Finally, *T. legionarius* was exclusively found in the galls of *L. lucens*. This species is rather rare and the mortality rate only attains 3.61 per cent.

Most galls opened by the birds were those of *L. pullitarsis*, but the other two species were also attacked. Many galls were opened, but the *Lipara*

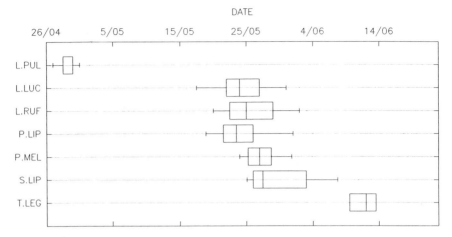

Fig. 16.5. Emergence sequence of the different *Lipara* species and their parasitoids. L.PUL, *L. pullitarsis*; L.LUC, *L. lucens*; L.RUF, *L. rufitarsis*; P.LIP, *P. liparae*; P.MEL, *P. melas*; S.LIP, *S. liparae*; T.LEG, *T. legionarius*.

larva was still present. This may indicate that the *Lipara* larvae were not the main target of the birds. Possibly the predators stopped searching when they did not find the inquilines which can co-occur in the *Lipara* galls in high numbers (De Bruyn 1985).

The mortality due to parasitoids and predators is highest for *L.rufitarsis* and *L. lucens*. However, in many localities, parasitoids are completely absent (*L. rufitarsis*, 53 per cent; *L. lucens*, 43 per cent; De Bruyn, unpublished data). For the remaining localities the overall mortality rate never exceeds 30 per cent (De Bruyn, 1985). *Lipara pullitarsis* is free of any parasitoid in 66 per cent of the localities visited and the overall mortality rate only attains 9.6 per cent (of which 6.6 per cent was caused by bird predation). These figures indicate the natural enemies of *Lipara* species might be of relative minor importance compared to the impact of mortality before gall formation.

An important factor in host–parasitoid relationships is temporal synchronization. To test this in the *Lipara*–parasitoid community, we collected galls in spring and transported them to the laboratory. Here they were stored in rearing vials at a constant temperature of 22°C. The galls were checked daily to count the emerged adults.

It is clear that *L. pullitarsis* can escape parasitism by emerging approximately 1 month earlier than the other *Lipara* species (Fig. 16.5). All parasitoid species are synchronized to the two other *Lipara* species.

According to the enemy hypothesis (Price *et al.* 1987*a*), gallers should benefit from enemy protection provided by the gall structure and should at least show reduced mortality in comparison with relatives feeding in other ways. Our data provide evidence for the opposite effect. Indeed, both *L. lucens* and *L. rufitarsis* which induce a clear gall on the reed shoots, suffer conspicuously more from enemy attack than *L. pullitarsis* which is more a shoot borer and does not produce a gall in the strict sense.

2. Shoot diameter dependent mortality

The probability of survival of *L. lucens* is not homogenously distributed for all shoot diameter classes (Fig. 16.6a). Mortality due to *S. liparae*, the most important parasitoid of *L. lucens*, is highest on the thinnest shoots. *Stenomalina liparae* inserts its eggs directly in its host by piercing through the reed shoot. To reach the host, the wasps need an ovipositor at least as long as the depth of the plant tissues. Ovipositor length of *S. liparae* was 1.91 ± 0.24 mm (mean $\pm$ SD, $n = 7$). When the wall of the shoot is too thick, the parasitoid cannot reach the host larva. Thus, the diameter of the shoot mediates accessibility of the host.

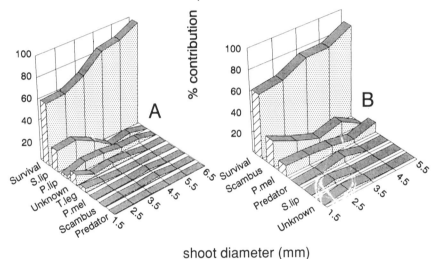

Fig. 16.6. Shoot diameter dependent mortality due to parasitization and predation. (A) *L. lucens*, (B) *L. rufitarsis*. (S.lip, *Sliparae*; P.lip, *P. liparae*; P.mel, *P. melas*; T.leg, *T. legionarius*; Scambus, *Scambus* sp.; Predator, bird predator; Survival, unparasitized *Lipara* larvae.)

The same mortality pattern is generated by *T. legionarius* and the unknown mortality factors. Some of the galls collected were fully developed but contained no larva or a depauperate *Lipara* larva. The latter may indicate *L. lucens* larvae can induce a gall on the very thin shoots, but the food resources are insufficient to support completion of the larval development. During the survival experiments, gall formation was also lower on the thinnest shoots (Fig. 16.3a) which supports this assumption.

The second important parasitoid of *L. lucens* is *P. liparae*. The infestation range is much broader than in *S. liparae*, but most galls are again found on thinner shoots, although the distribution is shifted to the right. The remaining parasitoids and the bird predation only contribute a negligible portion of the overall mortality.

The survival curve for *L. lucens* demonstrates mortality due to parasitization and predation is shoot diameter-dependent. All mortality factors caused by parasitization and predation of the larvae result in a higher survival chance of approximately 40 per cent on the thicker shoots were all galls carried fully developed larvae.

The most important parasitoid of *L. rufitarsis* is *Scambus* sp. (Fig. 16.6b). The highest infestation level was on thin shoots and decreases when shoot diameter increases. In the size class between 5 and 6 mm, no more *Scambus* larvae were found. At present, the biology of *Scambus* sp. is unknown. The fact that the female wasps possess a long, external ovipositor, may suggest that they are adapted to attack hosts concealed in the plant tissues and one can therefore assume the explanation used for the *S. lipara* parasitization of *L. lucens* may be valid for the *Scambus* species, although we have no data to confirm this. *Polemochartus melas* attacks the *L. rufitarsis* eggs while they are stuck on the surface of the reed shoot and is successful on all shoot diameters. The other parasitoids and the bird predator are noticeably less successful and do not show any shoot diameter-dependent pattern. Interestingly, the few *S. liparae* parasitoids found on *L. rufitarsis* were all confined to thin shoots.

The highest proportion of surviving larvae were found in the thinnest shoots and the overall survival rate decreases on thicker shoots, although less pronounced as in *L. lucens* (Fig. 16.6a). Even on the thickest shoots, 10 per cent of the galls were infested.

As mentioned above, the survival chance of *L. pullitarsis* is very high. The overall mortality rate is homogenously distributed over all diameter classes and never exceeds 15 per cent. The few parasitoids (*P. liparae*, *P. melas*, and *S. liparae*) were only found on thin shoots (diameter less than 6 mm). This disproportion is levelled out by a slightly higher mortality rate due to unknown factors.

Overall mortality rate

The results show that both the possibility of inducing galls and mortality due to natural enemies are shoot diameter-dependent and both these factors present a trade-off in mortality risk. In particular for *L. lucens*, a narrow shoot diameter makes the flies more vulnerable to parasitic attack (Fig. 16.6a) and exerts an upward pressure on the tri-trophic-level system. On the contrary, thicker shoots increase the difficulty of gall induction

(Fig. 16.3a) and exert a downward selective pressure in the opposite direction. However, the direction in which these interactions will evolve will depend upon the relative impact of the selective pressures and the existence of genetic variation in the species concerned for the selected trait.

The diameter-dependent failure to start gall formation is even more pronounced in *L. rufitarsis* (Fig. 16.3b), while the differences in parasitization rate for the different shoot diameter classes are less distinct (Fig. 16.6b). The resulting selection pressure may thus be in favour of thinner shoots.

Finally, survival rate up to gall formation in *L. pullitarsis* is higher on thicker shoots, although survival on thin shoots is also fairly high. Mortality due to parasitism and predation is practically negligible and diameter-independent. A possible explanation of the observed patterns is presented in the general discussion of the life cycle strategies.

Female oviposition preference

Phytophagous insects whose larvae develop at a restricted site are under strong selection pressure. Herbivores exploit resources that often vary in space and time. The larvae of galling insects have to use the available resources at the location where the female fly has oviposited the egg. Thus, a female selecting the correct position to deposit the eggs, can beneficially contribute to larval fitness. In the past, studies have reported some degree of correspondence between preference and performance (Craig *et al.* 1989; Fritz and Nobel 1989). Others have illustrated clear differences between the two traits (Thompson 1988; Roininen and Tahvanainen 1989).

Here we examine whether female oviposition preference of the *Lipara* species corresponds with performance patterns of the larvae, measured by offspring survival.

To test the female oviposition preference we set up a laboratory experiment (oviposition choice experiments). In a glass cage, 36 reed shoots, cut from the top 50 cm of a shoot, were offered to a single mated female fly. These shoots were spread over six diameter classes ranging from 3–4 to 8–9 mm. The shoots were arranged in six rows of six, each 5 cm apart. Diameter classes were randomized. Leaves were shortened to 5 cm so the adjacent shoots did not touch. For each replicate, new freshly cut shoots were used. The number of eggs and the location of these eggs on the shoots were recorded.

To define the number of eggs a single female can produce during her lifetime, an additional experiment was carried out (fecundity experiment). Galls were collected in the field in spring. Immediately after emergence,

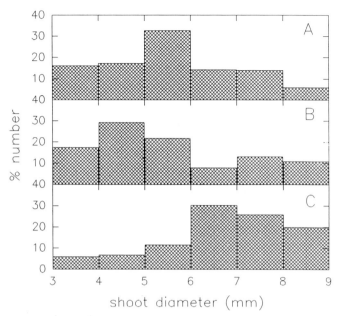

Fig. 16.7. Female oviposition preference; choice experiment (A) *L. lucens*, (B) *L. rufitarsis*,
(C) *L. pullitarsis*.

single females were mated and transferred to a small glass vial. Here
they were provided with fresh pieces of reed shoot of the appropriate
diameter class for oviposition and a few drops of a honey solution for
feeding. The vials were checked daily to count the eggs. In addition,
freshly emerged females were dissected to count the initial number off
eggs at emergence.

All species revealed a marked preference for a particular diameter
range in the oviposition choice experiments (Fig. 16.7,1). *Lipara lucens*
exhibits a clear peak preference for shoots of intermediate size and most
eggs oviposited were found on shoots between 5 and 6 mm (Fig. 16.7a).
Observations during the oviposition experiments demonstrated that the
female *L. lucens* wanders around on the surface of a shoot for a while
before ovipositing. An analogous result was obtained by Mook (1967)
who also showed that this searching time is shoot diameter-dependent.
After ovipositing one single egg, the female immediately leaves the shoot
to fly to another one. Under natural conditions in the field, most reed
shoots carry only one egg (De Bruyn, unpublished data), while practically
all eggs are situated in the top 20 cm of the reed shoot. The mean
number of eggs deposited by one female during the fecundity experiments
was 54.17 ($n = 22$) with a maximum of 84.

A similar preference pattern as in *L. lucens* was obtained for *L. rufitarsis*

(Fig. 16.7b). Here most eggs were found in the diameter class between 4 and 5 mm and *L. rufitarsis* females preferred thinner shoots. No behaviour data on oviposition nor sufficient field observations were available for *L. rufitarsis*. The mean number of eggs deposited by one female during the fecundity experiments was 28.86 ($n = 23$) with a maximum of 50.

Lipara pullitarsis oviposited most eggs on shoots between 6 and 7 mm (Fig. 16.7c), while thicker shoots also carried a reasonable number of eggs. In contrast to *L. lucens*, *L. pullitarsis* does not display any specific searching behaviour. Soon after alighting on a shoot, oviposition takes place. In many cases the female deposits one or more additional eggs without leaving the shoot. In the field, under natural conditions, reed shoots carry several *L. pullitarsis* eggs (De Bruyn, unpublished data). The maximum number found on a single shoot was 32. This multiple oviposition on a shoot implicates mortality will be high because only one single larva can develop in a shoot. The eggs are spread all over the shoot, on the culm as well as on the leaves and from the soil up to the top of the shoot. The mean number of eggs deposited by one female during the fecundity experiments was 92.00 ($n = 18$) with a maximum of 135.

If we compare the results obtained during the oviposition choice experiments (Fig. 16.7) with those from the survival experiments (Fig. 16.3) we find a relatively close correspondence for all three species. Both *L. lucens* and *L. rufitarsis* females prefer thinner shoots for oviposition, where survival of the larvae is also highest. The slight shift for oviposition preference to thicker shoots, especially apparent for *L. lucens*, may be accounted for by the upward selective pressure induced by the natural enemies (Fig. 16.6). Analogously, *L. pullitarsis* prefers mainly thicker shoots to deposit the eggs. Here also survival rate of the larvae up to gall formation is highest. A possible explanation for the observed patterns is presented in the general discussion of the life cycle strategies.

The life cycle strategies

For herbivores, a reed bed represents a rather adverse (low structural diversity, low nutritional value) habitat which is built up during a very short period (spring) and which declines slowly to disappear completely (above-ground parts die and dry up) at the end of the summer. During a long period of the *Lipara* life cycles, the habitat is not available for feeding and reproduction. The start of the reed-growing cycle however, is rather constant and therefore fairly predictable (Haslam 1970a; Vogel 1984). This study revealed that two adaptive exploitation strategies have evolved to overcome these problems. In both strategies, the diameter of the reed shoots plays a crucial role. Former studies on Cecidomyiid stem gallers (Diptera) and Lepidopteran shoot borers of the common reed also showed that the shoot diameter may play a crucial role in the survival of these herbivores (Mook and van der Toorn 1985; Tscharntke 1988, 1990).

Of course, it is not the shoot diameter in itself which influences the *Lipara* fitness, but the shoot diameter is merely a measure which can be used to quantify host plant resistance. Former studies showed that the diameter of a reed shoot is strongly correlated with final length, growth rate, and development of an ear (Mook 1967; Haslam 1971; Vogel 1984) and can, as such, be considered as measure of the strength of a shoot.

Based on the results of this study, the two life cycle strategies can be described as follows.

(a) The L. pullitarsis strategy   The adult flies emerge early in the season (April), approximately 1 month before the other two species (Fig. 16.5). At this time, the reed shoots have just emerged from the soil and are still small and even the thicker shoots are susceptible to infestation. Previous studies have already reported several times that host plant age may play an important role in herbivore resistance (for example, Coley 1980, 1988; Price *et al.* 1987*b*, *c*; Smith 1989). Thus, the early emergence of *L. pullitarsis* may explain why larval survival up to gall formation is approximately the same for all diameter classes (Fig. 16.3c). The presence of a *L. pullitarsis* larva does not cause the infested reed shoot to produce a distinct gall chamber as in the other species. The larvae have to feed upon the nutrient-poor leaves above the growing point throughout their development. The limited amount of food available in the thinner shoots may therefore explain the slightly higher mortality rate in these shoots.

A drawback of early emergence is the higher mortality risk due to adverse weather conditions. During this period of the year, low temperatures (even temperatures below zero) and heavy rain can still occur. Not only direct environmental effects may play a role, but former studies have also established that the reed itself is sensitive to these environmental effects (Haslam 1970*b*). Young tissues die rapidly when the temperature drops too low. During harsh winters shoot mortality can rise to 70–90 per cent. Frost kills the first, potential large shoots. These are replaced later by thinner shoots. All these factors make the habitat unpredictable for the herbivores.

To compensate, *L. pullitarsis* females produce a large number of eggs. Adult lifespan is relatively short. All eggs are oviposited during a short period, more or less randomly spread in the reed bed. As a consequence egg mortality is high, but shoot diameter-independent.

(b) The L. lucens strategy   *Lipara lucens* and *L. rufitarsis* adults emerge 1 month later than *L. pullitarsis* (Fig. 16.5). At this time, the mean temperature is

higher and the possibility of bad weather conditions is much lower. However, reed shoot development has continued over a longer period and host plant resistance is higher. Thick shoots can resist herbivore infestation and only the thinner, but resource limited, shoots are susceptible.

At some time during evolution, *L. lucens* and *L. rufitarsis* evolved the capability to manipulate their host plant. Due to the presence of a larva, the reed shoot starts to produce a gall, filled with a large amount of a nutrient-rich feeding tissue (parenchyma). In this way the food supply for the larvae is sufficient to survive on the thin reed shoots.

This study demonstrated that the female flies of both species are able to discriminate between shoots of different quality (Fig. 16.7). Before oviposition they carefully select a shoot of appropriate size and only a single egg is deposited on the top 20 cm of the shoot near the growing point. Total egg production is low. We can say the female flies have evolved a form of parental care. By carefully selecting the oviposition site, they increase the survival chance of their offspring. The habitat becomes favourable and predictable.

In conclusion, during the evolution of the genus *Lipara* the trade-off between obtaining the necessary food supply and growth-dependent host plant resistance has been a major force in structuring the life cycle strategies.

Acknowledgements

The author thanks M. De Meyer and J. Verwaerde for reading and making useful comments on this manuscript and providing computer assistance in drawing the figures. L.D.B. is a Senior Research Assistant of the Belgian National Fund for Scientific Research. This work was made within the framework of the Institute for the Study of Biological Evolution (University of Antwerp).

References

Abrahamson, W.G.., Armbruster, P.O., and Maddox, G.D. (1983). Numerical relationships of the *Solidago altissima* stem gall insect–parasitoid guild food chain. *Oecologia (Berlin)*, **58**, 351–7.

Abrahamson, W.G. and Weis, A.E. (1987). Nutritional ecology of Arthropod gall makers. In *Nutritional ecology of insects, mites, spiders, and related invertebrates*, (ed. F. Slanski and J.C. Rodriguez), pp. 235–58. Wiley, New York.

Alexander, H.M. (1991). Plant population heterogeneity and pathogen and herbivore levels: a field experiment. *Oecologia (Berlin)*, **86**(1), 125–31.

Allen, S.E. and Pearsall, W.H. (1963). Leaf analysis and shoot production in *Phragmites*. *Oikos*, **14**(2), 176–89.

Andow, D.A. and Prokrym, D.R. (1990). Plant structural complexity and host-finding by a parasitoid. *Oecologia (Berlin)*, **82**(2), 162–5.

Bernays, E.A. and Barbehenn, R. (1987). Nutritional ecology of grass foliage-chewing insects. In *Nutritional ecology of insects, mites, spiders, and related invertebrates*, (ed. F. Slanski and J.C. Rodriguez), pp. 235–58. Wiley, New York.

Beschovski, V.L. (1984). A zoogeographic review of endemic Palaearctic genera of Chloropidae (Diptera) in view of origin and formation. *Acta Zoologica Bulgarica*, **24**, 3–26.

Björk, S. (1967). Ecological investigations of *Phragmites communis*. Studies in theoretic and applied limnology. *Folia Limnologica Scandinavica*, **14**, 1–248.

Chvála, M., Doskočil, J., Mook, J.H., and Pokorny, V. (1974). The genus *Lipara* Meigen (Diptera, Chloropidae), systematics, morphology, behaviour and ecology. *Tijdschrift voor Entomologie*, **117**, 1–25.

Coley, P.D. (1980). Effects of leaf age and plant life history patterns on herbivory. *Nature*, **284**, 545–6.

Coley, P.D. (1988). Effects of plant growth rate and leaf lifetime on the amount and type of anti-herbivore defence. *Oecologia (Berlin)*, **74**, 531–6.

Coley, P.D., Bryant, J.P., and Chapin III, F.S. (1985). Resource availability and plant anti-herbivore defence. *Science*, **230**, 895–9.

Collinge, S.K. and Louda, S.M. (1988). Patterns of resource use by a Drosophilid (Diptera) leaf miner on a native crucifer. *Annals of the Entomological Society of America*, **81**(5), 733–41.

Confer, J.L. and Orloff, J. (1990). Spatial distribution of the goldenrod ball gall insects. *Great Lakes Entomologist* **23**(1), 33–7.

Craig, T.P., Itami, J.K., and Price, P.W. (1989). A strong relationship between oviposition preference and larval performance in a shoot-galling sawfly. *Ecology*, **70**(6), 1691–9.

Craig, T.P., Price, P.W., Clancy, K.M., Waring, G.L. and Sacchi, C.F. (1988). Forces preventing coevolution in the three-trophic-level system: willow, a gall-forming herbivore, and parasitoids. In *Chemical mediation of coevolution*, (ed. K.C. Spencer), pp. 57–79. Academic Press, New York.

Daniels, R.E. (1991). Variation in performance of *Phragmites australis* in experimental culture. *Aquatic Botany*, **42**, 41–8.

De Bruyn, L. (1985). The flies living in *Lipara* galls (Diptera: Chloropidae) on *Phragmites australis* (Cav.) Trin.ex Steud. *Bulletin et Annales de la Société Royale Belge d'Entomologie*, **121**, 485–8.

De Bruyn, L. (1987*a*). Habitat utilisation of three West-European *Lipara* species (Diptera: Chloropidae), a pest of the Common Reed, *Phragmites australis*. *Mededelingen van de Faculteit Landbouwwetenschappen, Rijksuniversiteit Gent*, **52**(2a), 267–71.

De Bruyn, L. (1987*b*). The parasite–predator community attacking *Lipara* species in Belgium. *Bulletin et Annales de la Société Royale Belge d'Entomologie*, **123**, 346–50.

De Bruyn, L. (1989*a*). Habitatselektie en levenscyclusstrategieën in het genus *Lipara* (Diptera: Chloropidae). Unpublished PhD thesis, University of Antwerp.

De Bruyn, L. (1989*b*). Influences of hostplant on larval survival and adult performance of *Lipara lucens* (Diptera: Chloropidae). *Mededelingen van de Faculteit Landbouwwetenschappen, Rijksuniversiteit Gent*, **54**/3a, 801–7.

Dykyjova, D. and Hradecká, A. (1973). Productivity of reedbelt stands in relation to the ecotype, microclimate and trophic conditions of the habitat. *Polskie Archiwum Hydrobiologii*, **20**, 111–19.

Faeth, S.H. (1985). Host leaf selection by leaf miners: interactions among three trophic levels. *Ecology*, **66**(3), 870–5.

Force, D.C. (1974). Ecology of insect host–parasitoid communities. *Science*, **184**, 624–32.

Fritz, R.S. and Nobel, J. (1989). Plant resistance, plant traits, and host plant choice of the leaf-folding sawfly on the arroyo willow. *Ecological Entomology*, **14**, 393–401.

Fritz, R.S. and Nobel, J. (1990). Host plant variation in mortality of the leaf-folding sawfly on the arroyo willow. *Ecological Entomology*, **15**, 25–35.

Fritz, R.S. and Price, P.W. (1988). Genetic variation among plants and insect community structure: willows and sawflies. *Ecology*, **69**(3), 845–56.

Giraud, J. (1863). Mémoire sur les insectes qui vivent sur le Roseau commun, *Phragmites communis* Trin. (*Arundo phragmites* L.) et plus spécialement sur seux de l'ordre des Hyménoptères. *Verhandlungen des Zoologisch-Botanischen Gesellschaft in Wien*, **13**, 1251–88.

Gould, F. (1991). Arthropod behavior and the efficacy of plant protectants. *Annual Review of Entomology*, **36**, 305–30.

Gross, P. and Price, P.W. (1988). Plant influences on parasitism of two leafminers: a test of enemy-free space. *Ecology*, **69**(5), 1506–16.

Hairston, N.G., Smith, F.E., and Slobodkin, L.B. (1960). Community structure population control and competition. *The American Naturalist*, **94**, 421–5.

Hartley, S.E. and Lawton, J.H. (1992). Host-plant manipulation by gall-insects: a test of the nutrition hypothesis. *Journal of Animal Ecology*, **61**(1), 113–19.

Haslam, S.M. (1970*a*). The development of annual population in *Phragmites communis* Trin. *Annals of Botany*, **34**, 571–91.

Haslam, S.M. (1970*b*). Variation of population type in *Phragmites communis*. *Annals of Botany*, **34**, 147–58.

Haslam, S.M. (1971). Shoot height and density in *Phragmites* stands. *Hidrobiologia*, **12**, 113–19.

Haslam, S.M. (1973). Some aspects of the history and autecology of *Phragmites communis* Trin. A review. *Polskie Archiwum Hydrobiologii*, **20**(1), 79–100.

Karban, R. and Myers, J.H. (1989). Induced plant responses to herbivory. *Annual Review of Ecology and Systematics*, **20**, 331–48.

Kimmerer, T.W. and Potter, D.A. (1987). Nutritional quality of specific leaf tissues and selective feeding by a specialist leafminer. *Oecologia (Berlin)*, **71**(4), 548–51.

Lawton, J.H. (1983). Plant architecture and the diversity of phytophagous insects. *Annual Review of Entomology*, **28**, 23–39.

Mook, J.H. (1961). Observations on the oviposition behaviour of *Polemon liparae* Giraud. *Archives Neérlandais de Zoologie*, **14**(3), 423–30.

Mook, J.H. (1967). Habitat selection by *Lipara lucens* Mg. (Diptera, Chloropidae) and its survival value. *Archives Néerlandais de Zoologie*, **17**, 469–549.

Mook, J.H. (1971). Influence of environment on some insects attacking Common Reed (*Phragmites communis* Trin.). *Hidrobiologia*, **12**, 305–12.

Mook, J.H. and van der Toorn, J. (1985). Delayed response of Common Reed *Phragmites australis* to herbivory as a cause of cyclic fluctuations in the density of the moth *Archanara geminipuncta*. *Oikos*, **44**(1), 142–8.

Nartshuk, E.P. (1977). Chloropidae of the genus *Lipara* Meigen (Diptera), their bionomics and parasites in Mongolia. (In Russian.) *Insects of Mongolia*, **5**, 711–15.

Pokorny, V. (1971). Flies of the genus *Lipara* Meigen on Common Reed. *Hidrobiologia*, **12**, 287–92.

Price, P.W. (1971). Niche breadth and dominance of parasitic insects sharing the same host species. *Ecology*, **52**, 587–96.

Price, P.W. (1972). Parasitoids utilizing the same host: adaptive nature of differences in size and form. *Ecology*, **53**, 190–5.

Price, P.W. (1973). Parasitoid strategies and community organization. *Environmental Entomology*, **2**, 623–6.

Price, P.W. (1985). *Evolutionary strategies of parasitic insects and mites*. Plenum, New York.

Price, P.W. (1989). Clonal development of coyote willow, *Salix exigua* (Salicaceae), and attack by the shoot-galling sawfly, *Euura exiguae* (Hymenoptera: Tenthredinidae). *Environmental Entomology*, **18**, 61–8.

Price, P.W. and Clancy, K.M. (1986). Interactions among three trophic levels: gall size and parasitoid attack. *Ecology*, **67**, 1593–1600.

Price, P.W., Bouton, C.E, Gross, P., McPheron, B.A., Thompson, J.N., and Weis, A.E. (1980). Interactions among three trophic levels: influence of plants on interactions between insect herbivores and natural enemies. *Annual Review of Ecology and Systematics*, **11**, 41–65.

Price, P.W., Fernandes, G.W., and Waring, G.L. (1987*a*). Adaptive nature of insect galls. *Environmental Entomology*, **16**, 15–24.

Price, P.W., Roininen, H., and Tahvanainen, J. (1987*b*). Why does the bud-galling sawfly, *Euura mucronata*, attack long shoots? *Oecologia (Berlin)*, **74**, 1–6.

Price, P.W., Roininen, H., and Tahvanainen, J. (1987*c*). Plant age and attack by the bud galler, *Euura mucronata*. *Oecologia (Berlin)*, **73**, 334–7.

Reavey, D. and Gaston, K.J. (1991). The importance of leaf structure in oviposition by leaf-mining microlepidoptera. *Oikos*, **61**(1), 19–28.

Roininen, H. and Tahvanainen, J. (1989). Host selection and larval performance of two willow-feeding sawflies. *Ecology*, **70**(1), 129–36.

Roininen, H., Price, P.W., and Tahvanainen, J. (1988). Field test of resource regulation by the bud-galling sawfly, *Euura mucronata*, on *Salix cinerea*. *Holarctic Ecology*, **11**, 136–9.

Ruppolt, W. (1957). Zur Biologie der Cecidogenen Diptere *Lipara lucens* Meigen (Chloropidae). *Wissenschaftliche Zeitschrift der Ernst-Moritz-Arndt-Universität Greifswald*, **6**, 279–91.

Skuhravý, V. (1981). *Invertebrates and vertebrates attacking common reed stands (Phragmites communis) in Czechoslovakia*. Československá Akademie Véd. Prague.

Slobodkin, L.B., Smith, F.E., and Hairston, N.G. (1967). Regulation in terrestrial ecosystems, and the implied balance of nature. *The American Naturalist*, **101**, 109–24.

Smith, C.M. (1989). *Plant resistance to insects. A fundamental approach.* Wiley, New York.

Smith, L. and Rutz, D.A. (1991). The influence of light and moisture gradients on the attack rate of parasitoids foraging for hosts in a laboratory arena (Hymenoptera: Pteromalidae). *Journal of Insect Behaviour,* **4**(2), 195–208.

Stiling, P. and Simberloff, D. (1989). Leaf abscission: induced defence against pests or response to damage. *Oikos,* **55**(1), 43–9.

Stinner, B.R. and Abrahamson, W.G. (1979). Energetics of the *Solidago canadensis*-stem gall insect-parasitoid guild interaction. *Ecology,* **60**(5), 918–26.

Strong Jr, D.R. and Levin, D.A. (1979). Species richness of plant parasites and growth form of their hosts. *The American Naturalist,* **114**, 1–22.

Thompson, J.N. (1988). Evolutionary ecology of the relationship between oviposition preference and performance of offspring in phytophagous insects. *Entomologia Experimentalis et Applicata,* **47**(1), 3–14.

Tscharntke, T. (1988). Variability of the grass *Phragmites australis* in relation to the behaviour and mortality of the gall-inducing midge *Giraudiella inclusa* (Diptera: Cecidomyiidae). *Oecologia (Berlin),* **76**(4), 504–12.

Tscharntke, T. (1989). Changes in shoot growth of *Phragmites australis* caused by the gallmaker *Giraudiella inclusa* (Diptera, Cecidomyiidae). *Oikos,* **54**(3), 370–7.

Tscharntke, T. (1990). Fluctuations in abundance of a stem-boring moth damaging shoots of *Phragmites australis*: causes and effects of overexploitation of food in a late-successional grass monoculture. *Journal of Applied Ecology,* **27**(2), 679–92.

van der Toorn, J. (1972). *Variability of Phragmites australis (Cav.) Trin. ex Steud. in relation to the environment.* Staatsuitgeverij s'Gravenhage.

Vogel, M. (1984). Ökologische Untersuchungen in einem *Phragmites*-Bestand. *Berliner ANL,* **8,** 130–66.

Wagner, W. (1907). Über die Gallen der *Lipara lucens. Verhandlungen des Vereins für Naturwissenschaften und Unterhaltungen zu Hamburg,* **13**, 120–35.

Waring, G.L. and Price, P.W. (1989). Parasitoid pressure and the radiation of a gallforming group (Cecidomyiidae, *Asphondylia* spp.) on Cresote Bush (*Larrea tridentata*). *Oecologia (Berlin),* **79**, 293–9.

Washburn, J.O. and Cornell, H.V. (1979). Chalcid parasitoid attack on a gall wasp population (*Acraspis hirta* (Hymenoptera: Cynipidae)) on *Quercus prinus* (Fagaceae). *The Canadian Entomologist,* **111,** 391–400.

Washburn, J.O. and Cornell, H.V. (1981). Parasitoids, patches, and phenology: their possible role in the local extinction of a cynipid gall wasp population. *Ecology,* **62,** 1597–1607.

Weis, A.E. (1982). Resource utilization patterns in a community of gall-attacking parasitoids. *Environmental Entomology,* **11,** 809–15.

Weis, A.E. and Abrahamson, W.G. (1985). Potential selective pressures by parasitoids on a plant–herbivore interaction. *Ecology,* **66**(4), 1261–9.

Weis, A.E., Abrahamson, W.G., and McRea, K.D. (1985). Host gall size and oviposition success by the parasitoid *Eurytoma gigantea. Ecological Entomology* **10**(3), 341–8.

Wetzel, T. (1967). Untersuchungen zum Auftreten und zur Schadwirkung der Larven von Fliegen (Diptera, Brachycera) an Gramineen. *Zeitschrift für angewandte Entomologie,* **59,** 260–8.

17. Induction and development of the bean gall caused by *Pontania proxima*

I.J. LEITCH

Jodrell Laboratory, Royal Botanic Gardens, Kew, Richmond, Surrey, UK

Abstract

The life cycle of the sawfly *Pontania proxima* (Hymenoptera: Tenthredinidae) which induces bean galls on the willow *Salix triandra* L. is comparatively simple when compared with many insect species that form galls. It has therefore been possible to establish an experimental laboratory system to induce galls and monitor their development.

Anatomical and ultrastructural changes, examined using light microscopy and both scanning and transmission electron microscopy, show that gall development is the result of cell division and cell expansion in all cell layers found in the uninfected leaf. Each layer, however, responds differently to the stimulus so that it remains more or less distinct in the mature gall. Once gall development is complete the larva hatches from the egg and begins eating the gall tissue. No specialized nutritive tissue is observed. The larva mainly relies on the greatly enlarged gall spongy mesophyll containing some vascular tissue.

The involvement of cytokinins was studied by measuring cytokinins extracted from:

(1) plant tissues (endogenous cytokinins);

(2) exudates collected from detached leaves (exported cytokinins).

Eight different types of cytokinins were measured using high performance liquid chromatography (HPLC) combined with a radioimmunoassay (RIA). The data show that the presence of the gall on a leaf leads to an accumulation of the cytokinin *iso*-pentenyladenine riboside (IPR) within the gall during the first 6 days of development and the elimination of IPR export, observed in uninfected leaves during the second week of leaf development.

A model of bean gall development based on these data and the significance of these observations in the manipulation of plant development by the insect is discussed.

Plant Galls (ed. Michèle A. J. Williams), Systematics Association Special Volume No. 49, pp. 283–300. Clarendon Press, Oxford, 1994. © The Systematics Association, 1994.

Introduction

Approximately 15 000 different insect galls have been recorded from around the world (Rohfritsch and Shorthouse 1982) and in each case they are thought to provide food and protection for the developing insect.

The majority of gall-inducing insects occur in the orders Thysanoptera, Heteroptera, Homoptera, Diptera, Lepidoptera, Hymenoptera, and Coleoptera. In general, galls induced by gall midges (Diptera: Cecidomyiidae) and cynipid wasps (Hymenoptera: Cynipidae) show the greatest structural complexity with well-developed nutritive and sclerenchymatous tissue. In contrast, galls produced by aphids (Homoptera: Aphididae) and allied groups are characterized by only limited disturbances to normal plant growth. Parenchymatous cells grow and surround the insect but typically the insect feeds from the underlying phloem tissue (Abrahamson and Weis 1987). Gall-forming insects are generally host-specific, some attack only one host species while others are restricted to a few closely related host species within the same genus and whilst galls can occur on all parts of the plant it is common to find that the gall-inducing insect is restricted to forming galls on a specific plant organ. In addition, galls are not equally abundant on different parts of the plant. For example, over 60 per cent of the galls induced by cynipid wasps on *Quercus* are found on the leaf (Mani 1964).

Studying the processes of gall induction and development

All insect galls that have been studied are initiated by a stimulus from the insect. In a few cases the gall is initiated during oviposition (for example, *Pontania proxima* Lepeltier, Hymenoptera: Tenthredinidae, described in detail below). In the majority of galls development does not proceed until the egg has hatched; here the gall-inducing stimulus usually comes from the presence and/or feeding activity of the larva. In each insect gall the presence of the egg and/or larva is essential for complete gall development. If at any stage during growth the egg or larva is removed then gall development will cease. In this way insect galls differ significantly from bacterial galls such as crown gall induced by *Agrobacterium tumefaciens* where the plant tissue becomes permanently transformed such that gall development continues even in the absence of the bacterium.

The processes that enable the insect to interact and direct plant development to produce species-specific gall structures are unknown. Since the gall is composed entirely of plant tissue, induction and development must involve some factor(s) that release the plant cells from the normal morphogenetic control. This is most likely to occur through the interaction of plant growth regulators and/or homeotic genes associated with normal plant development.

To enable a complete study of gall development, a biological system, suitable for laboratory experimentation and capable of providing material at all stages of gall development throughout the year, must be established. The system must be controllable so that material of a known age can be generated reproducibly. The life cycle of *P. proxima* which induces galls on the willow *Salix triandra* L. or *S. fragilis* L. (Fig. 17.1 and described below) has several features that make it ideally suited for studying gall development.

1. It is possible to grow willow saplings throughout the year in a growth cabinet.

2. Clonal material of *S. triandra* is readily available as it is grown commercially in Somerset for basket weaving. This means that genetically similar saplings can be used, eliminating one possible source of variation in the system.

3. The galls are induced by parthenogenetic females that do not feed. The rearing of the adult is therefore simple and parthenogenesis yields genetically uniform strains.

By growing the willows and rearing the insects it is possible to induce galls, as required, throughout the year. The methods are described in detail by Leitch (1990).

The life history of *Pontania proxima* Lep. (Fig. 17.1)

1. The process of oviposition

A single female is able to lay up to 35 eggs (Carleton 1939; Slepyan and Gabarayeva 1981) although Hovanitz (1959) quotes 50–100 eggs for a population he studied in California, USA. During oviposition, fluid from the accessory gland is injected into the mesophyll tissue. The effect of the fluid is rapid and a developing gall may be detected with the unaided eye within 24 h (Rey 1967). Oviposition is usually made in the leaf primordium. Magnus (1914) has shown that growth of the gall depends directly on the growth of the leaf such that small and large galls on the same leaf have been induced at different times. Galls induced in older, fully-grown leaves are undersized and insufficient for the larva to complete development; this is probably why ovipositing females select young and rapidly growing branches (Benes 1968).

The egg is white and oval when laid but undergoes changes in size and appearance during its incubation. Swammerdam (1758) noted that

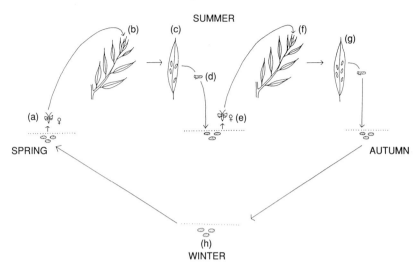

Fig. 17.1. Life cycle of *Pontania proxima* on *Salix triandra* or *S. fragilis*. (a) In late May parthenogenetic females emerge from the pupal cocoon in which they have overwintered. (b) Unfertilized eggs are laid in the leaf buds of *S. triandra* and give rise to 'bean' galls. (c) The galls mature June/July. (d) The fifth instar larva leaves the gall to pupate in the soil. (e) A second brood of parthenogenetic females emerge in August/September. (f) Unfertilized eggs are laid again in the leaf buds of *S. triandra* giving rise to further bean galls. (g) The galls mature September/October and the fifth instar larva leaves the gall and enters a prepupal stage. (h) In this form it overwinters within a cocoon until spring when it pupates and, soon after, the adult emerges.

just prior to hatching the egg was considerably swollen and that 'even the head and the two eyes of the caterpillar are generally observed through the integument'. The appearance of the egg prior to hatching is shown in Fig. 17.2a. In the laboratory the period of egg incubation was found to be less than 14 days which is similar to reports by Magnus (1914; 12–14 days) and Carleton (1939; 12–19 days). Hovanitz (1959) estimated a shorter incubation time of 6–7 days. These differences may reflect local variations between different populations of *P. proxima*.

2. *The larval stages*

There are five larval instars (the appearance of the first and third instar is shown in Fig. 17.2b and c, respectively and the second instar is shown in Fig. 17.3a). Benson (1950) noted that there was considerable variation in the time required to complete larval development in a single sawfly species. Carleton (1939) estimated that larval development was completed in 15 days whereas Hovanitz (1959) reported that 30–45 days were required.

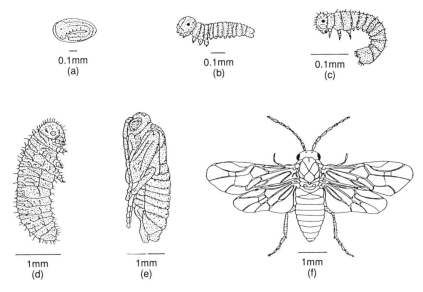

Fig. 17.2. Development of *Pontania proxima* (modified from Carleton 1939). (a) Egg just prior to hatching. (b) First instar larva. (c) Third instar larva. (d) Prepupa. (e) Pupa. (f) Adult.

When the egg hatches the gall is completely solid except for a small area around the egg. Gradually the tissue is eaten away by the developing larva. By the second or third larval instar a considerable hollow in the gall has been created (Fig. 17.3a). The larva then bites a small round aperture at one end of the gall on the underside of the leaf (Fig. 17.3c). By the fifth instar only the walls of the gall remain (Fig. 17.3d) and the larva stops feeding and leaves the gall through the hole to spin a cocoon and pupate.

3. The prepupa and pupal stages

Two to three days before spinning the cocoon the fifth instar larva becomes quiescent and its body shortens and fattens to become a prepupa (Fig. 17.2d); the prepupa is not a new larval instar as no ecdysis has taken place. The prepupa spins a cocoon of dark brown threads that are woven together to form a sealed capsule (Fig. 17.3b).

The duration of the prepupal stage for the first brood has been estimated by Carleton to be 10–12 days. In the second brood its duration is more variable being determined, at least in part, by temperature. If the autumn temperature is above a certain minimum (the exact temperature was not recorded by Carleton) then a proportion of the second

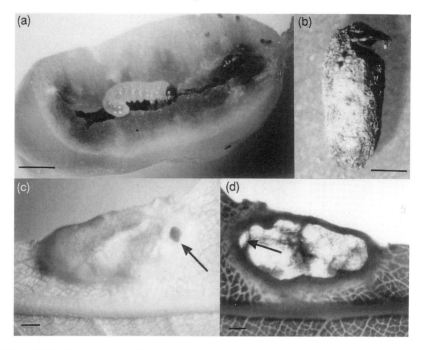

Fig. 17.3. Development of *Pontania proxima*. (a) Second instar larva inside gall. (b) Cocoon. (c) Abaxial epidermis of gall showing hole cut by larva (arrow). (d) Adaxial epidermis cut away to show that when the fifth larval instar has left the gall all but the epidermal tissue has been eaten. Arrow, exit hole cut by larva. Scale bar = 1 mm.

brood pupate within 10–12 days. Since the duration of the pupal stage is temperature-independent, sawflies that pupate in the autumn emerge 7–15 days later. These females are unable to lay eggs as no suitable juvenile leaves are present and so perish without producing a further brood. If the autumn temperature is sufficiently low to prevent pupation then the insect overwinters in the cocoon as a prepupa. Pupation is initiated the following spring as the temperature rises.

4. Eclosion and emergence of the adult

Eclosion takes place within the cocoon. With well-developed, pointed mandibles the insect cuts a round hole in the cocoon to emerge (Fig. 17.3b). The morphology of the adult sawfly is shown in Fig. 17.2f.

Host plants for *Pontania proxima*

Pontania proxima induces galls on the willows *S. triandra* and *S. fragilis*. However Carleton (1939) has noted that although no morphological differences exist between *P. proxima* developing on *S. triandra* and *S. fragilis* differences in behaviour suggest that two biological races do exist. This is supported by experimental evidence showing that adults reared from larvae collected from *S. triandra* fail to induce normal gall development on *S. fragilis*. The mortality rate of the larvae is high and the few adult sawflies that emerge are undersized and fail to lay eggs on *S. fragilis*. The mechanisms giving rise to this incompatibility are unknown but indicate that the process of gall development must involve specific interactions between the plant and insect.

The nature of the gall-inducing stimulus

The initial stimulus for gall development is thought to originate from the fluid injected by the adult sawfly; however, it is unclear as to the extent to which this stimulus is capable of completing gall development. Observations from the laboratory system showed that galls failing to complete development contained no egg. This can be interpreted in two ways, either these galls received less than the full dose of fluid required to develop fully or that a further stimulus is required. Carleton's (1939) work supports the first interpretation that complete gall development can only take place provided that the full dose of fluid is injected. In contrast, Hovanitz's (1959) work supports the second interpretation. He found that if the larva was removed from the gall, growth stopped within 2 days. He concluded that, while the fluid injected by the insect initiated gall development, a similar factor supplied by the larva was required to sustain it. In the laboratory system studied here gall development is usually complete before the egg hatches. It is therefore likely that if an additional stimulus is required then it comes from the egg rather than larva.

The fluid has been analysed by Hovanitz (1959) and McCalla *et al.* (1962) in an attempt to identify the chemical nature of the gall-inducing stimulus. Hovanitz (1959) found the fluid to be rich in both nucleic acids and protein. A more detailed analysis by McCalla *et al.* (1962) identified uric acid, two adenine derivatives, glutamic acid, and possibly uridine. The presence of adenine derivatives is of particular interest since most naturally occurring cytokinins are N6-substituted adenine derivatives and cytokinins are known plant growth regulators. The involvement of cytokinins in gall induction is discussed below.

Developmental morphology of the gall

To understand the processes involved in gall induction and development the chronological events during morphogenesis must be established. The major changes, studied by light microscopy and scanning and transmission electron microscopy, are outlined below and described in detail by Leitch (1990).

1. Changes in external appearance of the gall during development

The gall first appears as a translucent circle or oval (Fig. 17.4a and b). On the abaxial epidermis of the gall a lighter circular area marks the position of the egg (Fig. 17.4b). At this stage the gall is not visibly thicker than the leaf blade. By the end of the second day the gall has become opaque but in all other respects is unchanged (Fig. 17.4c and d). The first increases in gall dimensions are discernible on the third or fourth day (Figs 17.4e, 17.4f, and 17.5) and conspicuous changes in the gall's appearance are apparent by the fifth to sixth day (Fig. 17.4g and h). On the adaxial epidermis a band of pigmentation, which is red, is visible around the sides of the gall while the inner area remains pale. The

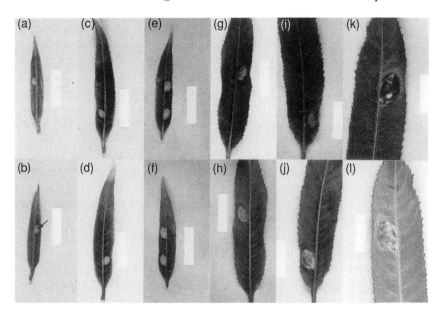

Fig. 17.4. Changes in the external appearance of the gall induced by *Pontania proxima*. (a), (c), (e), (g), (i), (k) Adaxial epidermis; (b), (d), (f), (h), (j), (l), abaxial epidermis. (a) and (b) One day old (arrow marks the position of the egg). (c) and (d) Two days old. (e) and (f) Three to four days old. (g) and (h) Five to six days old. (i) and (j) Nine days old. (k) and (l) Fourteen days old. Scale bar = 1 cm.

pigmentation is less extensive on the abaxial epidermis of the gall and the surface becomes pitted (Fig. 17.4h). As the gall continues to grow the pigmentation on the adaxial epidermis encroaches over the gall's surface (Fig. 17.4i) while the abaxial epidermis remains unpigmented (Fig. 17.4j).

Fourteen days after oviposition gall development is complete (Fig. 17.4k and l). The mature gall is oval in outline and measures approximately 9 mm long, 6 mm wide, and 4 mm thick protruding equally from

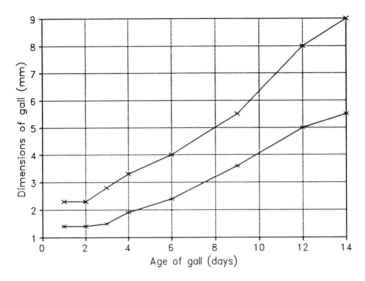

Fig. 17.5. Changes in the length (x) and width (*) of the gall induced by *Pontania proxima* during development. Data points represent mean values taken from 50 fresh galls.

both surfaces of the leaf. The adaxial epidermis has become deeply pigmented red and has a smooth and glossy cuticle (Fig. 17.4k) whereas the abaxial epidermis remains pale green and pitted. Occasionally, pigmented patches do occur as shown in Fig. 17.4l. The red gall pigment is not seen in mature uninfected leaves although the reddish/brown pigment found in young uninfected leaves may be the same. The gall pigment has been identified as cyanidin-3-monoglucoside by Blunden and Challen (1965).

Changes in the dimensions of the gall during its development (Fig. 17.5) indicate that although the gall is discernible 24 h after oviposition it is 3–4 days before steady, continuous growth is observed. The lag possibly represents the time required for the new pattern of tissue development to be established by the gall-inducing stimulus.

2. Structural changes during gall development

The young willow leaf into which *P. proxima* inserts its ovipositor is composed of five cell layers: the adaxial epidermis, palisade mesophyll, spongy mesophyll, hypodermis, and the abaxial epidermis. (Fig. 17.6a). Gall development results from cell division and cell expansion in all these cell layers and they have been termed gall adaxial and abaxial epidermis, gall palisade and gall spongy mesophyll, and gall hypodermis. It is possible that they are derived from the equivalent layers in the leaf but this has not been verified. However, each cell layer appears to respond differently to the gall stimulus and, thus, remains more or less distinct. Differences between the cell layers are as follows and are summarized in Fig. 17.6b.

(a) Gall adaxial and abaxial epidermis   The cells in each of the gall epidermal layers double in number. Cell division is predominantly periclinal and takes place during the first day of gall development. No cell expansion is observed in these cell layers.

(b) The gall palisade mesophyll and hypodermis   Cell division begins on the first day of gall development and is complete by the fifth day, to generate an eight-fold increase in cell number. The first few divisions are periclinal but the later ones may be both periclinal and anticlinal. Once cell division is complete, cell expansion, which in some cases may be considerable, is often observed.

(c) The gall spongy mesophyll   Cell division commences with the onset of gall development together with the other cell layers but continues for approximately 8 days to generate an eight-fold increase in cell number. As with the divisions in the gall palisade and hypodermis the early divisions are predominantly periclinal whereas the later divisions may be periclinal or anticlinal. Little or no cell expansion is observed in this layer.

Based on these observations gall development can be divided into 2 phases.

1. **Phase I (days 1–5)**: characterized by cell division alone (no cell expansion). By the end of this phase cell division in all but the gall spongy mesophyll layer is practically complete.

2. **Phase II (days 6–14)**: characterized by cell division in the gall spongy mesophyll and cell expansion in the gall palisade and hypodermis. Cell expansion continues until the end of gall development. The rate of growth during this phase is more rapid than Phase I and lasts until gall development is complete (Fig. 17.5).

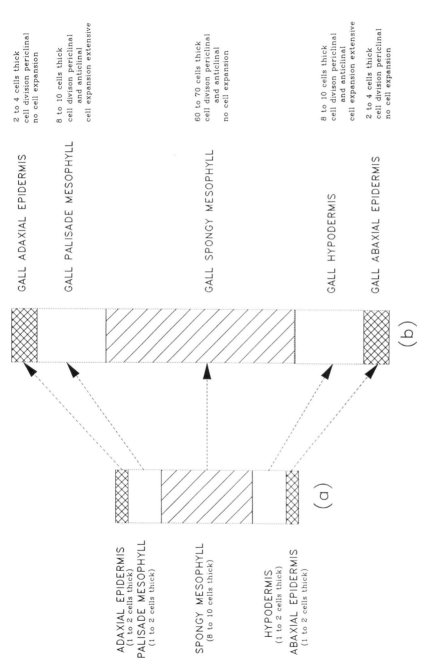

Fig. 17.6. Diagrammatic tissue map of (a) an uninfected leaf and (b) mature gall induced by *Pontania proxima*.

ADAXIAL EPIDERMIS
(1 to 2 cells thick)
PALISADE MESOPHYLL
(1 to 2 cells thick)

SPONGY MESOPHYLL
(8 to 10 cells thick)

HYPODERMIS
(1 to 2 cells thick)
ABAXIAL EPIDERMIS
(1 to 2 cells thick)

GALL ADAXIAL EPIDERMIS

2 to 4 cells thick
cell division periclinal
no cell expansion

GALL PALISADE MESOPHYLL

8 to 10 cells thick
cell divison periclinal
and anticlinal
cell expansion extensive

GALL SPONGY MESOPHYLL

60 to 70 cells thick
cell divison periclinal
and anticlinal
no cell expansion

GALL HYPODERMIS

8 to 10 cells thick
cell division periclinal
and anticlinal
cell expansion extensive

GALL ABAXIAL EPIDERMIS

2 to 4 cells thick
cell division periclinal
no cell expansion

(a)

(b)

The involvement of cytokinins in the gall development

1. A review of previous work

The involvement of cytokinins in insect galls was first suggested by McCalla *et al.* (1962) who observed that galls of *P. proxima* were always initiated adjacent to vascular tissue, a known source of cytokinins. However, although they also found two unidentified adenine derivatives in the ovipositional fluid of *P. proxima* that were active in a *Salix alba* L. bioassay, they dismissed the idea that these compounds could be cyto-kinins because they were inactive in a *Xanthium* leaf senescence bioassay. The data they presented on the chromatographic and chemical nature of the adenine derivatives in the ovipositional fluid are not inconsistent with their being cytokinins. The conclusions drawn from the *Xanthium* bioassay may be misleading as no recoveries of cytokinins from the fluid are presented. It is possible, for example, that the absence of activity in the assay was because all the cytokinins had been lost during extraction. It is also conceivable that the ovipositional fluid contains molecular species of cytokinins to which developing *Salix* leaves are sensitive but mature *Xanthium* leaf tissue is not. In addition, compounds might be present in the plant extract that interfere with the bioassay. These problems leave open the question as to whether cytokinins are present in the ovipositional fluid.

Other workers have also implicated cytokinin involvement in insect gall development (for example, Engelbrecht 1971; Van Staden 1975; Van Staden and Davey 1978; Abou-Mandour 1980). They all report the identification of higher levels of cytokinin in the gall tissue than the uninfected tissue when measured using a bioassay. However, as with the work of McCalla *et al.* (1962) none of the methods was adequately evaluated.

2. Measurements of cytokinins in galls induced by Pontania proxima

In the work reported here cytokinins were extracted from (i) plant tissues (endogenous cytokinins) and (ii) exudates collected from detached leaves (exported cytokinins). Carbohydrate export was also measured in leaf exudates to determine the effect of the gall on nutrient flow. Methods of cytokinin extraction and quantification using combined radioimmuoassay and high performance liquid chromatography (HPLC) were evaluated and are described in detail in Leitch (1990). Reliable differences between galled and uninfected tissues were found for four *iso*-pentenyl adenine type cytokinins (see Fig. 17.7 for cytokinin structures and abbreviations used).

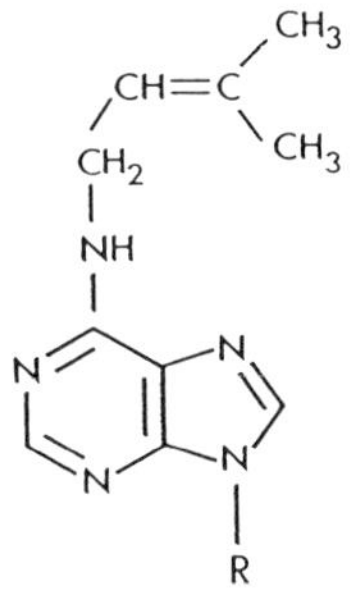

R	TRIVIAL NAME	ABBREVIATION
H	*iso*-pentenyl adenine	IP
Ribosyl	*iso*-pentenyl adenine riboside	IPR
Ribotide	*iso*-pentenyl adenine ribotide	IPRP
Glucosyl	*iso*-pentenyl adenine-9-glucoside	IP9G

Fig. 17.7. Structure, nomenclature, and abbreviations of the *iso*-pentenyl type cytokinins.

(a) Endogenous cytokinin measurements During the first 2 weeks of gall development the amounts of IPR, IP, IP9G, and IPRP extracted from gall tissue were always greater than those from the same weight of uninfected leaf tissue at the same stage of leaf development (Fig. 17.8). The predominant cytokinin was IPR which reached a peak value (approximately 50-fold greater than the uninfected leaf) when the gall was 6 days old (Fig. 17.8a). The other three *iso*-pentenyl adenine-type cytokinins reached peak values (approximately 20-fold greater than uninfected leaf) in gall tissue 4 days old but then declined rapidly approaching the amounts detected in uninfected leaves by the tenth to twelfth day of gall development (Fig. 17.8b,c, and d).

(b) Cytokinin and carbohydrate export The presence of the gall altered both IPR and carbohydrate export from the leaf. Peak values of IPR export (at 10 days) and carbohydrate export (at 4 weeks) by uninfected leaves were not detected for leaves bearing galls.

3. Origin of IPR in gall tissue

There are four mechanisms which could account for the high levels of IPR identified in the developing gall.

(a) Ovipositional fluid If this is the source of IPR it is unlikely that IPR *per se* is injected into the leaf as there is a time delay between the moment of oviposition (day 0) and the peak value of IPR within the gall tissue (day 6). If precursors rather than IPR itself were present in the ovipositional fluid, the delay of 6 days could represent the time taken for the precursors to be converted to IPR.

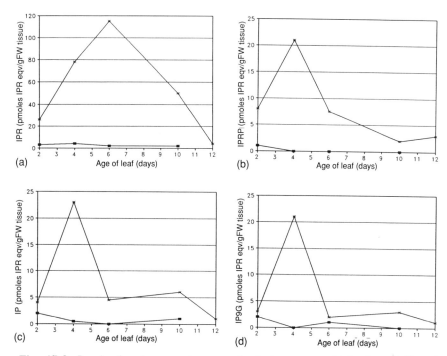

Fig. 17.8. Levels of endogenous *iso*-pentenyl adenine type cytokinins in gall (*) and uninfected leaf (■) tissue during the first 2 weeks of gall development. (a) IPR, (b) IPRP, (c) IP, (d) IP9G. Each data point is derived from duplicate assays on fractions from a single HPLC analysis.

(b) The egg Tsoupras *et al.* (1983) extracted IPR conjugated to phosphoecdysone from newly laid eggs of the locust *Locusta migratoria*. The conjugate was thought to be of maternal origin, being synthesized in the ovaries and then transferred to the oocytes where it was stored until oviposition. A similar conjugate may exist in the eggs of *P. proxima*. IPR conjugated in some form could be released from the egg following oviposition giving rise to the elevated levels of IPR observed.

(c) Changes in the rate of cytokinin biosynthesis and metabolism within the gall It has been shown that these can be influenced by many different physiological and environmental factors (Semdner *et al.* 1980). It is possible that the insect, directly or indirectly, alters the mechanisms controlling the level of IPR within the gall tissue.

(d) Disruption of IPR transport A mechanism could exist whereby the gall accumulates IPR from the surrounding tissue that would otherwise be exported. This idea is supported by the observation that it is exactly this

transport form of cytokinin, that is, IPR, that accumulates within the gall.

4. The role of IPR in directing nutrient flow in galled and uninfected leaves

The observation that carbohydrate export is reduced from galled leaves suggests that, in some way, the gall is affecting the normal flow of nutrients from the leaf. McCrea *et al.* (1985) also found an alteration in the flow of nutrients from insect-induced galls and discussed two mechanisms by which this might be achieved.

1. The gall could partially block normal nutrient translocation through the galled plant organ, thereby accumulating nutrients as the gall grows.

2. Alternatively the gall could actively redirect the translocation flow from other parts of the plant. This would draw in resources above those normally flowing through the gall.

Both mechanisms have been found to operate in gall tissue (Weis and Kapelinski 1984; McCrea *et al.* 1985).

The data presented here do not distinguish between these two alternatives although the enhanced level of IPR within the gall provides a mechanism by which reduced carbohydrate export might be achieved. Both Engelbrecht (1971) and Abou-Mandour (1980) have suggested that elevated levels of cytokinins might be responsible for redirecting nutrients required for gall development. It is suggested here that one role of IPR in the gall of *P. proxima* is to alter, directly or indirectly, the flow of nutrients in favour of gall development.

Hypothetical model for development of the gall induced by *Pontania proxima*

Although further research is necessary to fully understand the processes of gall development a hypothetical model of the interactive roles of the plant and insect in development of the gall induced by *P. proxima* is presented (Fig. 17.9).

1. **Phase I**: during oviposition a stimulus, either contained in the fluid or released as a result of wounding is capable of inducing cell division. It is possible that the cell divisions observed in the gall epidermal, palisade, and hypodermal layers are stimulated by one of the *iso-*

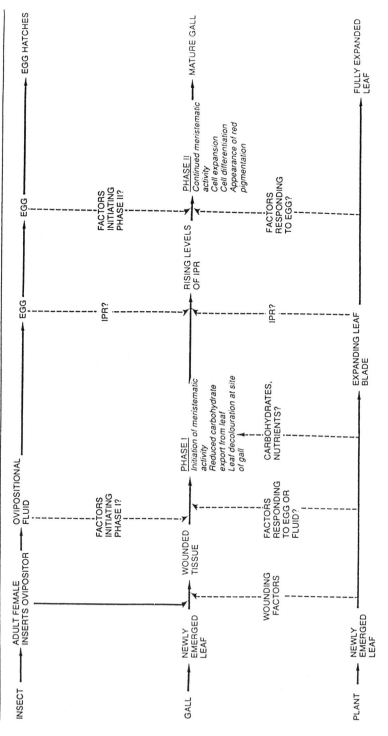

Fig.17.9 Schematic flow chart illustrating the possible factors that influence the species-specific development of the gall induced by *Pontania proxima*. (- - -) Hypothetical events, (——) known events.

pentenyl adenine-type cytokinins (that is, IP, IPRP, and IP9G) that reach a peak value on day 4 of gall growth.

2. **Phase II**: as the level of IPR rises a critical level is reached which triggers the onset of phase II of growth characterized by cell division in the gall spongy mesophyll and cell expansion in the gall palisade and hypodermis. IPR may have an additional effect by acting as a nutrient sink serving to draw in the resources required for gall development. The endogenous levels of IPR in the gall are observed to decline after day 6. This suggests that either the sensitivity of the cells to IPR has increased (for example, as a result of increased numbers of receptors and/or increased receptor affinity) so that IPR is capable of acting as a stimulus at a lower concentration or that other factors, possibly released from the egg, take over the role of inducing cell division and cell expansion.

Whatever the precise role of IPR it is likely that other factors are involved in producing the species-specific structure of the gall. The nature of these factors whether plant, insect or both is unknown.

Acknowledgements

I would like to thank Drs Colin Lazarus and David Hanke, for their helpful advice and discussion at all stages of this work and the Natural and Environmental Research Council for sponsoring the research.

References

Abou-Mandour, A.A. (1980). Investigations on cytokinins of parasitic origin III. Cytokinin-activities in *Melampyrum* and *Lathraea* and in galls on leaves of *Populus*, *Fagus* and *Quercus*. *Zeitschrift für Pflanzenphysiologie*, **97,** 59–66.

Abrahamson, W.G. and Weis, A.E. (1987). The nutritional ecology of arthropod gall-makers. In *Nutritional ecology of insects, mites and spiders*, (ed. F. Slansky and J.G. Rodriquez), pp. 235–58. Wiley, New York.

Benes, K. (1968). Galls and larvae of the European species of the genera *Phyllocolpa* and *Pontania* (Hymenoptera, Tenthredinidae). *Acta Entomologica Bohemoslovaca*, **65,** 112–37.

Benson, R.B. (1950). An introduction to the natural history of British sawflies. *Transactions of the Society for British Entomology*, **10,** 45–142.

Blunden, G. and Challen, S.B. (1965). Red pigment in leaf galls of *Salix fragilis* L. *Nature*, **208,** 388–9.

Carleton, M. (1939). The biology of *Pontania proxima* Lep., the bean gall sawfly of willows. *Journal of the Linnean Society of London, Zoology*, **40,** 575–624.

Engelbrecht, L. (1971). Cytokinin activity in larval infected leaves. *Biochemie und Physiologie der Pflanzen*, **162**, 9–27.

Hovanitz, W. (1959). Insects and plant galls. *Scientific American*, **201**, 151–62.

Leitch, I.J. (1990). Studies on the bean gall induced by *Pontania proxima* Lep. on *Salix triandra*. Unpublished D. Phil. thesis, University of Bristol.

Magnus, W. (1914). *Die Entstehung der Pflanzengallen verursacht durch Hymenopteran*. Fischer, Jena.

Mani, M.S. (1964). *Ecology of plant galls*. Junk Publishers, The Hague.

McCalla, D.R., Genthe, M.K., and Hovanitz, W. (1962). Chemical nature of an insect gall growth-factor. *Plant Physiology*, **37**, 98–103.

McCrea, K.D., Abrahamson, W.G., and Weis, A.E. (1985). Goldenrod ball gall effects on *Solidago altissima*: ^{14}C translocation and growth. *Ecology*, **66**, 1902–7.

Rey, L. (1967). Les premier stades de développement de la galle de *Pontania proxima* Lep. *Bulletin de la Société Botanique de France*, **115**, 413–24.

Rohfritsch, O. and Shorthouse, J.D. (1982). Insect galls. In *Molecular biology of plant tumours*, (ed. G. Kahl, and J.S. Schell), pp. 131–52. Academic Press, New York.

Semdner, G., Gross, D., Liebisch, H.W., and Schneider, G. (1980). Biosynthesis and metabolism of plant hormones. *Encyclopedia of Plant Physiology* (New Series), **9**, 281–444.

Slepyan, E.Z. and Gabarayeva, N.I. (1981). Structure and development of the gall caused by the larva of the sawfly *Pontania proxima* (Lepel.) (Hymenoptera, Tenthredinidae) on the leaves of the willow *Salix fragilis*. *Entomological Review*, **60**, 55–65.

Swammerdam, J. (1758). The history of the worms found in the tubercles and swellings of the leaves of the willow. In *The book of nature; or, the history of insects*, Vol. 2, (ed. J. Hill) (trans. T. Floyd), pp. 75–86. Seyfert, London.

Tsoupras, G., Luu, B., and Hoffmann, J.A. (1983). A cytokinin (isopentenyl-adenosyl-mononucleotide) linked to ecdysone in newly laid eggs of *Locusta migratoria*. *Science*, **220**, 507–9.

Van Staden, J. (1975). Cytokinins from larvae in *Erythrina latissima* galls. *Plant Science Letters*, **5**, 227–30.

Van Staden, J. and Davey, J.E. (1978). Endogenous cytokinins in the laminae and galls of *Erythrina latissima* leaves. *Botanical Gazette*, **139**, 36–41.

Weis, A.E. and Kapelinski, A. (1984). Manipulation of host plant development by the gall-midge *Rhabdophaga strobiloides*. *Ecological Entomology*, **9**, 457–65.

18. A willow gall from the galler's point of view

R.N. HIGTON* and D.J. MABBERLEY†

*Lord Williams's School, Thame, Oxon, UK † Department of Plant Sciences, University of Oxford, South Parks Road, Oxford, UK

Abstract

Investigations into gall induction in the leaves of *Salix fragilis* var. *russelliana* by *Pontania proxima* has demonstrated the following.

1. Methods of rearing both insect and host have resulted in an extension of the insect's flight period from 5 to 8 months, with three broods per year instead of the usual two.

2. A bioassay, based on microinjection techniques has identified the colleterial fluid, produced by the accessory glands of the insect, as containing the cecidogen. Preliminary studies have shown that the cecidogen has a molecular weight of less than 3 kDa. Thus, a single event, that is, the introduction of colleterial fluid during oviposition, is the initiator. In the plant it was observed that the gall effect was limited and the presence of an egg or larva was not required for the formation of a procecidium. Gall growth was mainly due to periclinal divisions of the provascular tissue.

Introduction

Most sawflies associated with the formation of plant galls are confined to the families Tenthredinidae and Xyelidae. In Britain there are 180 species referred to 16 genera in the tribe Nematini (Tenthredinidae) (Benson 1958). Of these, the three gall-making genera are *Phyllocolpa* Benson, species of which live in leaf-roll margins of species of *Salix* L. and *Populus* L. (Salicaceae) (Benson 1960), *Euura* E. Newman, species of which incite galls in the closed buds, petioles, and stems of *Salix* species, and *Pontania* O. Costa, species of which incite a wide variety of leaf galls in willows (Beneš 1968; Smith 1970). The structure of the gall and the

Plant Galls (ed. Michèle A. J. Williams), Systematics Association Special Volume No. 49, pp. 301–12. Clarendon Press, Oxford, 1994. © The Systematics Association, 1994.

life cycle, parasites, and inquilines of *Pontania proxima* (Lepeletier) and related species, together with their interactions, have been described by many authors, in this century including Magnus (1914), Bastin (1921), Carleton (1939), Benson (1943, 1950, 1954), Smith (1951), Caltagirone (1964), Rey (1967, 1968, 1972, 1973, 1974), Smith (1970), Sandlant (1979), Kay (1980), Kopelke (1985*a,b*, 1986, 1988), and Clancy *et al.* (1986).

However, literature reporting efforts to identify the cecidogen is much more limited (Hovanitz 1959; McCalla *et al.* 1962). Indeed, it would seem that little is known of the mechanisms controlling cecidogenesis of any insect gall.

The hypotheses developed over the latter half of this century have been based mainly on the involvement, either directly or indirectly, of plant growth regulators, notably auxins (Miles 1968*a,b*; Livingstone 1978) and cytokinins (Elzin 1983): however gibberellins (Byers *et al.* 1976) and abscisic acid (Bonga and Clark 1965) have also been implicated. Although ethylene has been shown to have a significant impact on all stages of plant development, it has received scant attention and other growth regulators, such as polyamines and brassinosteroids, have received none. In many ways, work on gall initiation has lagged behind the developments that have taken place in the study of plant growth regulators, both at experimental and theoretical levels. In reviewing the literature, what is clear is that there is no body of information favouring one hypothesis above all others. Indeed, it may be that different insect galls are initiated by different mechanisms and that behaviourial manipulations of the plant tissues, by the insect, play an important part in determining the resultant morphology of the gall.

Galling represents a highly specific link between organisms of different kingdoms, to such an extent that Cornell (1983) and Dawkins (1989) both view the gall as an extension of the insect's phenotype. Weis and Abrahamson (1986) proposed that, although the gall was plant tissue, the insect coded for the stimulus and the plant for the response. Thus, the gall has been considered a product of two genomes. Since the main beneficiary of the galling habit is the insect (Price *et al.* 1987; Fernandes and Price 1988), it would seem logical to concentrate on the insect in the search for the cecidogen. Until now biochemical anomalies found in gall tissue, compared with the rest of the host plant, have been heralded as cecidogens. To be that, they must be shown to be present in quantities physiologically significant in the insect itself.

In concentrating on one species pair, that is, *Pontania proxima–Salix fragilis* L. we have been able to develop techniques and methods that, in future studies, may be useful in leading to the identification of at least one such cecidogen.

Breeding host and galler

Salix fragilis var. *russelliana* (Smith) Koch, the Bedford willow, is a common lowland tree in England, frequently found by riversides. Following oviposition by *P. proxima* into the young leaves of the terminal bud, a cataplasmic procecidium forms within which, after eclosion, the larva develops. *Pontania proxima* demonstrates a high degree of host specificity. It has a bivoltine life cycle, with each cycle lasting approximately 36–56 days (Carleton 1939) and a flight period which extends from May to September (Magnus 1914, pp. 57 and 58). Unlike the prepupae of *Pontania triandrae* Benson, those of *P. proxima* undergo obligatory diapause (Carleton 1939). Thus, the seasonal nature of both host and sawfly represents limitations to work on gall induction. However, we have found it possible to extend these seasons.

1. Propagation of Salix fragilis

Fast-growing shoots of the type preferred by *P. proxima* for oviposition were supplied throughout the year from hardwood truncheons. The truncheons were 50 cm in length and 2–5 cm in thickness and were obtained in February from 2- to 3-year-old branches. After being dipped in a 0.05 per cent Benlate solution (ICI, Farnham, Surrey, UK) and wrapped in black plastic sheeting they were stored at 2°C. When required, truncheons were planted to two-thirds of their depth in a peat and sand rooting mixture and given a bottom heat of 22°C. A gradient between the rooting mixture and air temperature discouraged the premature emergence of shoots. Once rooted the truncheons were finally potted up in a free-draining compost. These truncheons could be kept throughout the winter in a heated greenhouse with supplementary lighting.

2. Rearing P. proxima

Heslop Harrison (1927), Carleton (1939), and Benson (1954) all provided details of the methods they used to rear sawflies but, in each case, there was no attempt to extend the flight period of the insect.

In our work, galls were collected during the latter stages of development of the larva. These stages were signalled by the appearance of an aperture on the abaxial surface of the gall and usually facing the leaf tip. The function of this aperture has been linked to either

(1) aeration of the gall;

(2) more feeding on areas of higher nutritional quality leading to the puncturing of the gall;

(3) as an aperture through which frass could be ejected (Carleton 1939; Lazareva *et al.* 1986).

Whatever the function of the aperture, a larva could sometimes be seen half-emerged from this hole and, as a prepupa, could use it to escape from the gall.

Once collected, leaves containing galls were placed on a 5 cm deep layer of a sand/peat mixture and kept at a temperature of 25°C in a relative humidity of 80 per cent and over and in a 16 h light/8 h dark regime).

The extension of the flight period involved, firstly, the continuation of emergence to produce a third brood and, secondly, early emergence in the spring. Complete abolition of diapause was not possible. A high relative humidity (80 per cent and over) was essential to the production of a third brood: without this the prepupa entered diapause. Early emergence could be encouraged by placing overwintered cocoons in an incubator, under the conditions detailed above.

Using these techniques it was possible to extend the flight season of these insects by 60 per cent, from 5 months to 8 months.

Development of a bioassay

Until now, experiments to induce gall formation in the leaves of *S. fragilis* have been hampered by the difficulty of injecting material into the young, delicate leaves of the type preferred by *P. proxima* for oviposition. Beyerinck (1887), Magnus (1914, p. 81), and Carleton (1939) all found the technical difficulties too great.

The use of insect extracts to induce gall formation has been attempted by several workers, for example, Boysen Jensen (1948), Leatherdale (1955), Schäller (1969), and Birch (1974), but no-one has produced, unequivocally, a gall. Indeed, McCalla *et al.* (1962) considered that the artificial production of normal galls would depend on precise amounts of cecidogen being placed in a precise place. This was something that they considered difficult, if not impossible. Although it may not, at present, be possible to create a completely artificial gall, a bioassay, based on the microinjection of insect extracts into willow leaves, has been developed.

1. Observations of oviposition

The mean dimensions of the saw of *P. proxima*, when measured 1 mm from the tip, are width, 0.17 mm and thickness, 0.022 mm ($n = 20$).

Oviposition lasts between 30 sec and 2 min per event. It begins when the sawfly arches her abdomen, a movement which brings her saw into perpendicular contact with the leaf. The two lancets of the saw then oscillate back and forth, driving it through a lateral vein into the provascular tissue of the leaf. Once the epidermis is pierced, the saw is then twisted through 90° so that its usually dorsal ventral axis is horizontal. The oscillating movement of the lancets continues until the saw is fully extended into the leaf. After a short pause, the ovipositor is withdrawn leaving a sickle-shaped wound, approximately 1.25 mm in length, running roughly parallel to the midrib. Oviposition usually disrupts only one or two layers of cells (between 0.09 mm and 0.11 mm in diameter) of provascular tissue.

2. Methods of injection

Using a tungsten microscalpel it is possible to imitate the dimensions and features of the wound caused during oviposition. Electron microscope filaments, approximately 1 mm in diameter, were etched thinner by immersion in heated and fused sodium nitrate. It was found that a needle with a diameter of 0.1 mm, 1 mm from the tip, caused very little disruption to the leaf tissues. A micropipette could then be inserted into the wound and its contents expelled into the leaf. It was found that the insertion of acid-washed beads (75–150 μm diameter) would approximate the size of an egg (640 μm—longest dimension). These glass beads functioned to hold the mesophyll layers apart, allowing injected liquids to enter. Sealing the wound, to prevent desiccation, was achieved by placing a ring of lanolin round the incision and covering the area with a fragment of coverslip.

The wound was always initiated through a lateral vein on the abaxial surface of the leaf running parallel to the midrib, in the direction of the petiole. This type of wound is similar to that caused by oviposition.

3. Identification of the source of the cecidogen

The internal reproductive organs and associated glands of *P. proxima* fill most of the abdominal cavity. In dorsal view the median oviduct, spermatheca, and lateral oviducts are obscured from view by a large colleterial sac (mean diameter 0.98 mm, $n = 93$) and its contents. At the distal end of the colleterial sac is a highly branched accessory gland, which empties through a single short tube into the colleterial sac. In Hymenoptera the accessory glands have a variety of functions; for example, many apocritans produce venoms which are capable of incapacitating their host organisms. It has been generally assumed that the colleterial fluid injected along with the egg during oviposition in *P. proxima* is responsible for gall formation or, at least, encouraging gall growth (Beyerinck 1887; Hovanitz 1959; McCalla *et al.* 1962). However, Magnus

(1914, p. 87) disputed this, maintaining that its only function was to seal the wound.

Hovanitz (1959) described how the eggs of *Pontania pacifica* Marlatt are laid and also the ontogeny of the gall, concluding that the egg is not necessary for the first phase of gall development, but that the colleterial fluid is: the first phase of growth of the gall lasts approximately 8 days, after which continued growth requires the presence and activity of a larva. Similar observations for *P. proxima* were made by Murphy (1929) and Lazereva *et al.* (1986). These reports contrast with those of other authors and those for other related species. Caltagirone (1964) studied the same species as Hovanitz and maintained that gall growth was directly related to the growth rate of the leaf, with slow-growing leaves producing the largest galls. He believed that the presence of a larva was not necessary for the growth of the gall. Smith (1970), who studied the bionomics of American species of *Phyllocolpa*, *Pontania*, and *Euura*, stated that, for galls of the type produced by *P. proxima*, hatching of the egg occurred after the gall had reached full size. Thus, although much of the literature has supplied strong evidence for the role of the colleterial fluid in gall formation, the latter has never been proved.

From our own observations, during which over 20 000 galls were dissected (Higton 1991) and direct observations of oviposition made and the ontogeny of the galls studied, it is clear that the presence of an egg or larva is not a prerequisite of gall development. Many fully-formed galls that contained no egg or larva have been observed. Moreover, the gall effect is limited: it does not spread (during the development of the gall) further than the original wound and a small surrounding area. This area is delineated by a ring of anthocyanin that forms around the site of the wound within 24 h of its occurrence. The rate of anticlinal divisions within the unaffected leaf and gall tissues remains similar, with the effect of the cecidogen dramatically increasing periclinal divisions. Thus, similar-sized leaves produce galls of comparable size. These observations agree with those made by Rey (1967, 1968).

At the start of gall development, the leaf is no more than seven to eight cells in thickness. After oviposition, material from the gall cavity is seen between the surrounding cells, but no colleterial fluid is observable in the cavity. Assuming that the cecidogen is contained in this fluid, as would be suggested by the eggless procecidia observed, then, during oviposition or soon after, the colleterial fluid moves into the surrounding tissues and the cecidogen exerts its effect—which is only on the rate and direction of cell division. In the gall of *P. proxima* there is no complex differentiation of cell layers, as in Cynipinae, to be considered. Nor, from electron microscopical studies of both insect and host, does it appear that the cecidogenic effect is the result of a mutualistic relationship with a virus as was suggested by Hovanitz (1959) and Cornell (1983).

To establish the cecidogen-containing organs of the insect, a series of over 100 dissections was performed and either the ovary or eggs or ducts of the reproductive system were implanted in wounds produced as explained above. Any effect on the leaf tissue was then noted over a period of time. Similarly, colleterial sacs were dissected out on the leaf and, using a microscalpel or micropipette, their contents were introduced into the wound. The same number of control experiments were carried out using distilled water. Removal of the fluid from the colleterial sac was complicated by its viscid nature: it had the appearance of granular albumen. If exposed to air at room temperature, it hardened quickly (in approximately 2 min), into a yellow crystalline mass. If it dried on the surface of the leaf it formed a transparent crust, in which were embedded the outlines of the epidermal cells it covered.

In all cases a response, by the tissues of the leaf, over and above that normally occurring at the site of a wound, was looked for. Three criteria were chosen to record positive results:

1. Callus formation was greater than that for a normal wound.

2. The callus should remain for the full life of the leaf; it was found that wound callus soon underwent necrosis.

3. There should be visual evidence of a red pigment in the callus.

Only those leaves which had been implanted with colleterial fluid developed callus material that met these criteria. Thus, it was a consistent observation that growth-promoting properties were contained in the colleterial fluid. The abundant nature and permanence of the callus material, initiated by the colleterial fluid, which contained cells that had undergone hyperplasy and hypertrophy, would seem to indicate that the colleterial fluid alone contains the cecidogen.

By contrast, in their study of the development of the horned oak gall which develops under the influence of the larva of *Callirhytis cornigera* (Osten Sacken) (Cynipidae) in twigs of *Quercus palustris* Muenchhausen (Fagaceae), Taft and Bissing (1988) observed that oviposition caused the development of wound-response phellogen and concluded that this arrangement of tissues formed a framework, from which the gall would develop after the hatching of the larva. The gall formed by *P. proxima* is much less complex in structure than any gall formed by members of Cynipinae. This does not, however, preclude a wound-response mechanism as having a role to play in the initial stages of formation of the *Pontania* gall.

Analytical techniques

McCalla *et al.* (1962) carried out a series of bioassays on material extracted from the colleterial sacs of *P. proxima*. The UV absorption spectrum of the extracted colleterial fluid demonstrated an increase in absorption from 300 nm downwards, with a peak at 260 nm. From this fraction they isolated six compounds including uric acid, two unidentified adenine derivatives, glutamic acid, and possibly uridine. They believed that these compounds were in sufficiently large quantities to have growth-promoting activity but, because they used *growing* galls for their work, this hypothesis may or may not be applicable to gall *induction*. Elzin (1983) commented on the work and added that, as cytokinins were adenine derivatives, the studies may suggest a role for these compounds in gall development; however, this view is not shared by all, for example, Osborne (1972).

1. Preliminary analysis of colleterial fluid

Preliminary analysis of the colleterial fluid extracted from *P. proxima* shows the following.

1. SDS–PAGE with a continuous buffer system demonstrates good separation of polypeptides. Twenty polypeptide units were consistently separated, the mass of these units ranging from 11.9 to 289 kDa. Separation using IEF (pH range 3–9) results in more than 37 bands being visible and covered a range of pH values from 3.5 to 9.3.

2. The uv absorption spectrum of both fresh and resuspended freeze-dried colleterial fluid demonstrates the same absorption pattern as that found by McCalla *et al.* (1962). From data obtained using the methods of Warburg and Christian (Dawson *et al.* 1986, pp. 541–2) protein concentration of the colleterial fluid ranged from 140 to 189 mg/ml.

3. The mean volume of colleterial fluid within each sac is 0.50 mm^3 ($n = 60$); when freeze-dried the mass of material from one sac is approximately 0.12 mg.

With the exception of the work using gel electrophoresis, it was not possible to collect enough specimens of *P. proxima* at any one time to allow fresh material to be used for all experimentation. Therefore, on emergence, adult sawflies were frozen and stored at $-75°C$. When injected into *Salix* leaves, this and resuspended freeze-dried colleterial material promoted callus formation similar to that produced by fresh material.

4. Separation by ultrafiltration and subsequent injection of the separated

material into *Salix* leaves suggests that the cecidogen is low molecular weight, that is, less than 3 kDa, with a UV absorption peak at 247 nm.

2. Future research

Preliminary analysis has shown that the colleterial fluid of *P. proxima* is amenable to further study. Using the techniques described in this paper, the broad aim of further research will be the separation, isolation, and identification of the cecidogen either directly or by a comparison with the colleterial fluids of closely related but non-galling species such as *Nematus ribesii* (Scopoli). However, the following remains as relevant today as it was 80 years ago:

'The galls arising in plant tissues through the presence of parasitic insects and fungi are of peculiar interest and significance, and offer a most attractive field of investigation, abounding in problems awaiting elucidation' (Swanton 1912, p. xi)

Acknowledgements

The research for this paper was carried out during the tenure of an SERC studentship awarded to R.N.H. whilst preparing a D.Phil thesis under the supervision of D.J.M. and Dr George McGavin, to whom both of us are very grateful for his entomological expertise. We are also indebted to the following for advice and encouragement: Professor F.R. Whatley at the Department of Plant Sciences, Dr R. Buxton and the Northmoor Trust who allowed the collection of galls at Little Wittenham Nature Reserve, Mr J. Baker, Dr A. Banham, Dr J. Coleman, Dr M. Fricker, Ms D.S. Higton, Mr C. Merriman, Mr P. Nichols, Mrs D. North, Dr D.J. Osborne, Ms A. Sing, and Dr J.M. Whatley.

References

Bastin, H. (1921). Vegetable galls. *Journal of Bath and West and Southern Counties Society*, 5th series, **15,** 30–56.

Beneš, K. (1968). Galls and larvae of the European species of genera *Phyllocolpa* and *Pontania* (Hymenoptera: Tenthredinidae). *Acta Entomologica Bohemoslovaca*, **65,** 112–37.

Benson, R.B. (1943). Collecting sawflies. *Amateur Entomologist*, **7**(40), 36–42.

Benson, R.B. (1950). An introduction to the natural history of British sawflies. *Transactions of the Society for British Entomology*, **10**(2), 45–142.

Benson, R.B. (1954). British sawfly galls of the genus *Nematus* [*Pontania*] on *Salix*.

(Hymenoptera: Tenthredinidae). *Journal of the Society for British Entomology*, **4**(9), 206–11.

Benson, R.B. (1958). *Handbook for the identification of British insects*. Vol. 6, part 2(c). *Hymenoptera. Symphyta*. Royal Entomological Society of London, London.

Benson, R.B. (1960). A new genus for the leaf-edge-rolling *Pontania* (Hymenoptera: Tenthredinidae). *Entomologist's Monthly Magazine* **96**, 59–60.

Beyerinck, M.W. (1887). De la cecide produite par le *Nematus capreae* sur le *Salix amygdalina*. *Archives Néerlandaises des Sciences Exactes et Naturelles*, **21**, 475–92.

Birch, M.L. (1974). Studies on gall formation in *Viola odorata* by *Dasyneura Uaffinis* Kieffer (Diptera: Cecidomyiidae), with emphasis on the experimental modification of this process. Unpublished PhD thesis, University of Reading.

Bonga, J.M. and Clark, J. (1965). The effect of β-inhibitor on histogenesis of balsam fir bark cultured *in vitro*. *Forest Science*, **11**(3), 271–8.

Boysen Jensen, P. (1948). Formation of galls by *Mikiola fagi*. *Physiologia Plantarum*, **1**, 95–108.

Byers, J.A., Brewer, J.W., and Denna, D.W. (1976). Plant growth hormones in pinyon insect galls. *Marcellia*, **39**, 125–34.

Caltagirone, L.E. (1964). Notes on the biology, parasites and inquilines of *Pontania pacifica*, a leaf gall incitant on *Salix lasiolepis*. *Annals of the Entomological Society of America*, **57**, 279–91.

Carleton, M. (1939). The biology of *Pontania proxima* (Lep.) the bean gall sawfly of willows. *Journal of the Linnean Society of London (Zoology)*, **40**, 575–624.

Clancy, K.M., Price, P.W., and Craig, T.P. (1986). Life history and natural enemies of an undescribed sawfly near *Pontania pacifica* (Hymenoptera: Tenthredinidae) that forms leaf galls on arroyo willow, *Salix lasiolepis*. *Annals of the Entomological Society of America*, **79**, 884–92.

Cornell, H.V. (1983). The secondary chemistry and complex morphology of galls formed by the Cynipinae (Hymenoptera): why and how? *American Midland Naturalist*, **110**(2), 225–32.

Dawkins, R. (1989). *The extended phenotype*. (Paperback edition with corrections.) Oxford University Press, Oxford.

Dawson, R.M., Elliot, D.C., Elliot, W.M., and Jones, K.M. (1986). *Data for biochemical research*. Oxford University Press, Oxford.

Elzin, G.W. (1983). Cytokinins and insect galls. *Comparative Biochemistry and Physiology*, **76A**(1), 17–19.

Fernandes, G.W. and Price, P.W. (1988). Biogeographical gradients in galling species richness. *Oecologia*, **76**, 161–7.

Heslop Harrison, J.W. (1927). Experiments on the egg-laying instincts of the sawfly *Pontania salicis* Christ., and their bearing on the inheritance of acquired characteristics; with some remarks on a new principle in evolution. *Proceedings of the Royal Society of London. Series B: Biological Sciences*, **101**, 115–26.

Higton, R.N. (1991). Studies in gall induction with special reference to the *Pontania-Salix* system. Unpublished D. Phil thesis, Bodleian Library, University of Oxford.

Hovanitz, W. (1959). Insects and plant galls. *Scientific American*, **201**(5), 151–62.

Kay, M.K. (1980). *Pontania proxima* (Lepeletier) (Hymenoptera: Tenthredinidae). *Willow gall sawfly*, Forest and Timber Insects in New Zealand, No. 45. Forest Research Institute, New Zealand Forest Service, New Zealand.

Kopelke, J.-P. (1985*a*). Biologie und Parasiten der gallenbildenden Blattwespe *Pontania proxima* (Lepeletier 1823). (Insecta: Hymenoptera: Tenthredinidae). *Senckenbergiana Biologica*, **65**(3/6), 215–39.

Kopelke, J.-P. (1985*b*). Über die Biologie und Parasiten der gallenbildenden Blattwespenarten *Pontania dolichura* (Thoms. 1871), *P vesicator* (Bremi, 1849) und *P viminalis* (L. 1758) (Hymenmoptera: Tenthredinidae). *Faunistisch-Ökologische Mitteilungen*, **5**, 331–44.

Kopelke, J.-P. (1986). Zur Taxonomie und Biologie neuer *Pontania*-Arten der *dolichura*-Gruppe. (Insecta: Hymenoptera: Tenthredinidae). *Senckenbergiana Biologica*, **67**, 51–71.

Kopelke, J.-P. (1988). Zur Biologie und Ökologie der Arten des Brutparasiten–Parasitoiden-Komplexes von gallenbildenden Blattwespen der Gattung *Pontania* (Hymenoptera: Tenthredinidae: Nematinae). *Mitteilungen der Deutschen Gesellschaft für Allgemeine und Angewandte Entomologie*, **6**, 150–5.

Lazareva, A.I., Ogorodnikova, V.I., and Trusevich, A.G. (1986). The willow bean-gall sawfly. *Zashchita Rastenii*, **9**, 33–4.

Leatherdale, D. (1955). Plant hyperplasia induced with a cell-free insect extract. *Nature (London)*, **175**, 553–4.

Livingstone, D. (1978). Phytosuccivorous bugs and cecidogenesis. *Journal of the Indian Academy of Wood Science*, **9**(1), 39–45.

Magnus, W. (1914). *Die Entstehung der Pflanzengallen verursacht durch Hymenopteran*. Fischer, Jena.

McCalla, D.R., Genthe, M.K., and Hovanitz, W. (1962). Chemical nature of an insect gall growth-factor. *Plant Physiology*, **37**, 98–103.

Miles, P.W. (1968*a*). Insect secretions in plants. *Annual Review of Phytopathology*, **6**, 137–64.

Miles, P.W. (1968*b*). Studies on the salivary physiology of plant-bugs: experimental induction of galls. *Journal of Insect Physiology*, **14**, 97–106.

Murphy, I.S. (1929). The oviposition of *Pontania gallicola* Steph. (the bean gall of willows). *Entomologist's Monthly Magazine*, **65**(787), 270–2.

Osborne, D.J. (1972). Mutual regulations of growth and development in plants and insects. In *Insect/plant relations*, Symposia of the Royal Entomological Society of London, No. 6, (ed. H.F. van Emden), pp. 33–42. Blackwell Scientific Publications, London.

Price, P.W., Fernandes, G.W., and Waring, G.L. (1987). Adaptive nature of insect galls. *Environmental Entomology*, **16**, 15–24.

Rey, L. (1967). Les premiers stades de développement de la galle de *Pontania proxima* Lep. *Bulletin. Société Botanique de France*, **114**, 80–95.

Rey, L. (1968). La galle de *Pontania proxima* Lep.: stades ultérieurs du développement. *Bulletin. Société Botanique de France*, **115**, 413–24.

Rey, L. (1972). Sur l'évolution particulière des chloroplastes dans la galle de *Pontania proxima* Lep. sur la feuille de *Salix triandra* L. *Journal de Microscopie (Paris)*, **14**(3), 87a.

Rey, L. (1973). Ultrastructure des chloroplastes au cours de leur évolution pathologique dans le tissue central de la jeune galle de *Pontania proxima* Lep. *Comptes rendus hebdomadaires des Séances de l'Académie des Sciences. Série D*, **276**(7), 1157–60.

Rey, L. (1974). Modifications ultrastructurales pathologiques présentées par les chloroplastes de la galle de *Pontania proxima* Lep. en fin de croissance. *Comptes rendus hebdomadaires des Séances de l'Académie des Sciences. Série D*, **278**(10), 1345–8.

Sandlant, G.R. (1979). Arthropod successori inhabiting willow galls during Autumn in Christchurch, New Zealand. *Mauri Ora*, **7**, 83–93.

Schäller, G. (1969). Untersuchungen zur Erzeugung Künstlicher Pflanzengallen. *Marcellia*, **35**, 131–53.

Smith, E.L. (1970). Biosystematics and morphology of Symphyta II. Biology of the gall-making Nematine sawflies in the California region. *Annals of the Entomological Society of America*, **63**, 36–51.

Smith, K.M. (1951). *A textbook of agricultural entomology*. Cambridge University Press, Cambridge.

Swanton, E.W. (1912). *British plant galls*. Methuen, London.

Taft, J.B. and Bissing, D.R. (1988). Developmental anatomy of the horned oak gall induced by *Callirhytis cornigera* on *Quercus palustris* (pin oak). *American Journal of Zoology*, **75**, 26–36.

Weis, A.E. and Abrahamson, W.G. (1986). Evolution of host–gall manipulation by gall-makers: ecological and genetic factors in the *Solidago–Eurosta* system. *American Naturalist*, **127**(5), 681–95.

19. Assemblages of herbivorous chalcid wasps and their parasitoids associated with grasses—problems of species and specificity

MICHAEL F. CLARIDGE and HASSAN ALI DAWAH

School of Pure and Applied Biology, University of Wales, Cardiff, UK

'the green myriads in the peopled grass'
F. Walker (1835)

Abstract

Most chalcid wasps are parasitoids, but the family Eurytomidae includes many herbivores. Of the numerous species of *Tetramesa* in temperate regions, all are phytophagous and, so far as known, use only grasses as their host plants. Many of these form distinctive galls in which their larvae feed. Others feed within developing flowering stems without accompanying gall formation. *Tetramesa* larvae are attacked by a series of characteristic parasitoids including species of *Pediobius* (Eulophidae), *Chlorocytus* and *Homoporus* (Pteromalidae), and *Eurytoma* and *Sycophila* (Eurytomidae).

Taxonomy of chalcid wasps is generally very difficult and poorly known. Biparental sexual reproduction is usual for *Tetramesa* and its parasitoids so that the biological species concept is appropriate. Morphological differentiation of species is usually very slight and thus other techniques, including enzyme electrophoresis have been used to recognize and identify groups of sibling species.

Details are given of food webs associated with *Tetramesa* and other phytophagous Eurytomidae which in Britain attack the grasses *Calamagrostis epigejos* (L.) Roth, *Deschampsia cespitosa* (L.) Beauv., *Elymus repens* (L.) Gould, *E. farctus* (Viv.) Runemark ex Meldris, *Festuca rubra* L., and *F. ovina* L. The *Tetramesa* show very high levels of specificity. Each *Tetramesa* is attacked by mostly monophagous parasitoids. Many previously supposed polyphagous parasitoids are in reality groups of monophagous sibling species. The food webs associated with grasses show extreme compartmentation with few species interlinking each herbivore association.

Plant Galls (ed. Michèle A. J. Williams), Systematics Association Special Volume No. 49, pp. 313–29. Clarendon Press, Oxford, 1994. © The Systematics Association, 1994.

Introduction

Chalcid wasps are mostly parasitoids that attack other insects. Noyes (1990) estimates that almost 2000 genera including more than 18 000 valid species have so far been described. Since tropical faunas have been little studied the real numbers must be very much higher. Though most of these insects are parasitoids, some phytophagous species have been recorded from many of the 21 families generally recognized (Gauld and Bolton 1988). Apart from the Torymidae, which are now often taken to include the fig-wasps (Agaonidae), the Eurytomidae are best known for the inclusion of herbivorous species.

The evolution of the phytophagous habit within the Chalcidoidea and, indeed, within the Hymenoptera generally, has been a major area of speculation and controversy. It is generally assumed that the order was primitively herbivorous (Gauld and Bolton 1988) with modern groups of Symphyta, such as the Xyelidae, resembling the earliest forms. The earliest Apocrita are usually thought to have been parasitoids and, therefore, all phytophagous Apocrita must be secondarily herbivorous. Within the Chalcidoidea the phytophagous habit has clearly evolved separately in several different groups.

The Eurytomidae is a morphologically conservative family, but includes species of very diverse habits. Many are true herbivores, many are parasitoids, and some are parasitic in their early larval instars and herbivorous in later ones. Noyes (1990) estimated that 1200 species have so far been described world-wide in 79 genera. The relatively small number of genera compared to other comparable chalcid families is undoubtedly due to the slight morphological diversity of the adult insects, on which taxonomy is generally based (Claridge 1961*a,b*). The supposedly most primitive living Euryomidae are usually considered to be among the Buresiinae, Rileyinae, and Heimbrinae (Zerova 1992) which seem to be exclusively parasitic. Thus, phytophagous forms appear to have evolved independently in several different lines. Among the groups that Zerova believes to be most recently evolved are the very large subfamilies, Eurytominae and Harmolitinae (considered as part of the Eurytominae by Boucek 1988). It is these that include most phytophagous species, many of which cause distinctive galls on their host plants. A well-known gall-forming species is the pest of citrus in parts of eastern Australia, *Bruchophagus fellis* (Girault) (Noble 1936). In Sweden, *Nikanoria metallica* (Erdös) has been recorded as galling stems of *Astragalus glycyphyllus* (Anderson 1967).

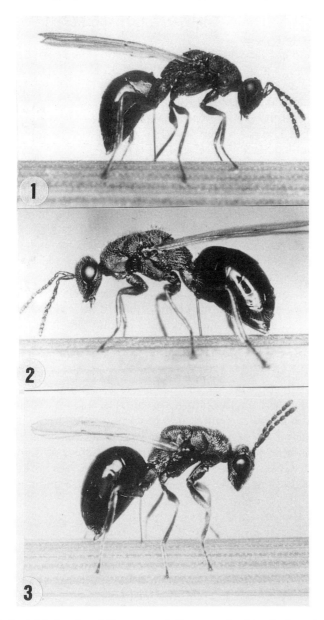

Figs 19.1–3. Adult females of *Tetramesa* and *Eurytoma* ovipositing in flowering stems of grass hosts. **Fig. 19.1.** *Tetramesa petiolata* (Walk.) on *Deschampsia cespitosa*. **Fig. 19.2.** *Tetramesa calamagrostidis* (von. Schlechtendal) on *Calamagrostis epigejos*. **Fig. 19.3.** *Eurytoma pollux* Claridge, a parasite of *T. calamagrostidis*, on *C. epigejos*. Photographs by D. E. Windsor. Magnification approx. × 15.

Grass-associated Eurytomidae

Many Eurytomids are associated with grasses. The biggest group is the genus *Tetramesa* (= *Harmolita*) (Figs 19.1 and 19.2). All species, so far as known, are exclusively phytophagous in their larval stages, usually in flowering stems of different grasses, to which they show high levels of specificity (Claridge 1961*b*; Zerova 1976, 1978). They are abundant in the temperate regions of the world—Zerova (1976) records 61 species from the former Soviet Union and Peck (1963) 63 from America, north of Mexico. Bouček (1988) regards *Tetramesa* as primarily a north temperate genus, with few known species from the southern hemisphere.

Most species of *Tetramesa* live during their larval stages in the developing flowering stems of their hosts where many form distinct galls. Others live usually singly at or just above the nodes of flowering stems with little external evidence of their presence and no obvious galling. *Tetramesa* species may thus be divided broadly into galling and non-galling species.

Of the gall-forming species of *Tetramesa*, most cause stunting and distortion of growth (Figs 19.6 and 19.7) and even complete abortion (Figs 19.4 and 19.5) of the flowering stems of their hosts. In some, the galls include a single chamber occupied by only one larva (Fig.19.8). In others, the galls include a sequence of several successive chambers, each occupied by a single *Tetramesa* larva (Figs 19.6 and 19.7). A few unusual species form distinctive galls on the developing flowers of their grass hosts (Fig.19.9). Perhaps the most unusual *Tetramesa* galls are caused by the numerous and poorly known species related to *T. linearis* (Walker), including the so-called joint-worm pests of cereals in N. America. In these the larvae are generally gregarious and live in separate hardened cells in the stem wall or its surrounding leaf sheaths, but not in the central stem cavity. When, as is usual, a number of larval cells are present together, the flowering stem is generally badly stunted and the developing flower head much reduced (Phillips 1920).

The joint-worms and straw-worms, larvae of *Tetramesa*, have been regarded as major pests of cereals in North America (Phillips 1920) and the former Soviet Union (Zerova 1976). With the advent of newer cultivars of major cereal crops and the widespread use of insecticides, these insects now seem only rarely to cause major problems. However, Spears and Barr (1985) recently showed significant detrimental effects of attack by *Tetramesa* on the productivity of four important pasture grasses in Idaho, USA.

In addition to *Tetramesa*, a few species of *Eurytoma* and *Ahtola* also attack grasses where they live and feed like the non-galling *Tetramesa* species at the nodes of flowering stems. Such a species is *Eurytoma suecica* von Rosen, known as a pest of cereals in Sweden (von Rosen 1956).

The herbivorous Eurytomidae associated with grasses are attacked by

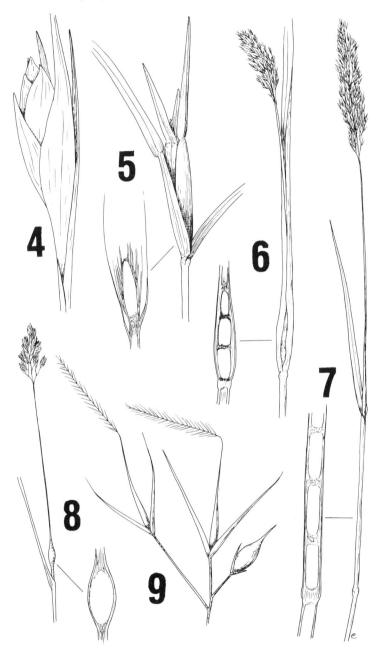

Figs 19.4–9. Sketches of galls caused by *Tetramesa* species (not to same scales). **Fig. 19.4.** *Tetramesa hyalipennis* (Walk.) on *Elymus farctus*. **Fig. 19.5.** *Tetramesa hyalipennis* on *E. repens*. **Fig. 19.6.** *Tetramesa calamagrostidis* on *Calamagrostis epigejos*. **Fig. 19.7.** *Tetramesa exima* (Giraud) on *C. epigejos*. **Fig. 19.8.** *Tetramesa brevicollis* (Walk.) on *Festuca rubra*. **Fig. 19.9.** *Tetramesa stipae (De Steff.) on Stipa tortilis*. After Claridge (1961).

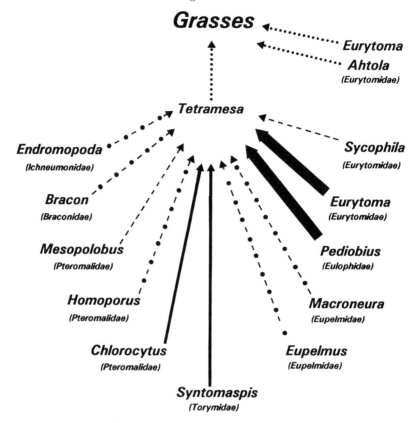

Fig. 19.10. Generalized food web of phytophagous Eurytomidae and their parasitoids associated with grasses. Herbivore links indicated by simple dotted lines. Parasitoid connections are shown by different line conventions to indicate frequency of attack: thick solid lines indicate relationships observed on more than 60 separate occasions, thin solid lines between 30 and 50, dashed lines between 20 and 30, and dot-dashed lines between 5 and 10.

a very characteristic assemblage of hymenopteran parasitoids (Fig. 19.10). Most of these are groups of species or genera that are highly specific to *Tetramesa* and other Eurytomid hosts in grasses, including species of the Eulophid genus, *Pediobius*, the Pteromalid genera, *Chlorocytus* and *Homoporus*, and the Eurytomid genera, *Eurytoma* (Fig. 19.3) and *Sycophila* (= *Eudecatoma*). Also more polyphagous species of the Eupelmid genera, *Eupelmus* and *Macroneura* and of the Ichneumonid genus, *Endromopoda* (= *Ephialtes*), regularly attack the larvae. Some of the parasitic species of *Eurytoma* are very similar to and difficult to distinguish from their phytophagous relatives.

The taxonomy of *Tetramesa* and of the major parasitoid genera of Eulophidae, Pteromalidae, and Eurytomidae is extremely difficult. Each of the genera concerned shows very little morphological differentiation and particular species tend to vary greatly in adult size. Species taxonomy has thus been based generally on very small and often variable characters. Claridge (1961*b*) stressed the importance of biological characters, particularly host specificity and mode of larval life, in the discrimination of species of *Tetramesa*. Species have been described from N. America (Phillips and Emery 1919; Phillips 1920, 1936) and from the former USSR (Zerova 1976, 1978) on a basis primarily of host specificity. Rearing experiments, particularly those by Phillips and his co-workers in the USA, established without doubt the reality of high levels of host specificity. Similar levels of specificity have also been suggested for many of the parasitoids of *Tetramesa*.

There is no doubt that the major problem confronting the exploration of the ecological nature of assemblages of these insects and their specificity is the difficulty of recognizing and characterizing different species of both the herbivores and their parasitoids.

Species concepts in chalcid wasps

Biparental sexual reproduction is the normal mode of reproduction for most chalcid wasps. However, in many genera parthenogenetic all-female clones, often recognized as distinct species, have frequently evolved. The haplo-diploid system of sex determination and the normal arrhenotokous system of reproduction of the Hymenoptera generally appear to preadapt the group to developing thelytokous parthenogenetic forms. Frequently such all-female clones produce among their usual all-female offspring small numbers of sexually inactive males; this occurs particularly under situations of environmental stress. This may lead the unsuspecting observer to assume that such a population is a biparentally reproducing one. The only certain way of determining whether arrhenotoky or thelytoky is the normal mode for any particular population is to allow virgin females to lay eggs. If the subsequent generation of adults is all female then thelytoky must be involved. If arrhenotoky is normal, then such unmated females will produce only males.

The importance of determining the mode of reproduction of any organism is vital when considering the nature of the species concept that may be employed. For biparental sexually reproducing organisms the biological species concept, as developed by Mayr (1942), Cain (1954), and others, is widely accepted as the most useful (Claridge 1988, 1991). This is based critically on reproductive isolation between species. The simplest and probably still the best definition is one of the original ones of Mayr (1942, p. 120):

'Species are groups of actually or potentially interbreeding populations which are reproductively isolated from other such groups.'

The critical criterion for the application of the biological species concept is the observation of reproductive isolation. This is only very rarely observed in practice so that genetic markers for reproductive isolation are normally used. The most obvious of such markers are morphological ones. Thus, the idea that species are primarily morphologically recognizable entities is based on the use of such markers. However, we know that reproductively isolated biological species exist in many groups of organisms that show no obvious morphological differentiation, so-called sibling or cryptic species. Parasitic organisms generally and herbivorous insects and their insect parasitoids, in particular, are rich in sibling species (Claridge 1988). The problem of finding appropriate markers for the differentiation of genetically isolated host-specific species from host-adapted populations of one biological species are difficult, but critical for such organisms. Micromorphological and cytological features, including particularly chromosome form and number, may provide such characters. However, more frequently in recent years, biochemical and molecular techniques have revealed a whole new series of sibling species. Particularly valuable is the electrophoretic analysis of enzyme differences between different populations (see Loxdale and den Hollander 1989). Differences between mobility patterns of isozymes may indicate unequivocally the presence of reproductive isolation. The application of such techniques to species of *Tetramesa* (Dawah 1987) and their parasitoids of the genera *Pediobius* (Dawah 1988*a*) (Figs 19.11 and 19.12), *Eurytoma* (Dawah 1988*b*) and *Chlorocytus* (Dawah 1989) has demonstrated the existence of more host-specific biological species than was predicted by purely morphological analyses.

A different approach to determining biological species status is to determine the specific mate recognition systems which must characterize each species and result in reproductive isolation. In all biparental insect species these systems consist of behavioural interactions between males and females with the exchange of signals which may be visual, olfactory, acoustic, tactile, etc. The distinctive patterns of these signals provide us with the best evidence for biological species status (Claridge 1988, 1991).

In the chalcid wasps considered here, the specific mate recognition signals are poorly understood but are probably primarily chemical—both olfactory and gustatory. However, mate recognition tests, where females are given a choice of mates from their own and other populations, provide powerful supplementary evidence on species status. Dawah (1987, 1988) has used these methods with both *Tetramesa* and *Pediobius* species to confirm species status of otherwise doubtful forms. Most of the known species of *Tetramesa* and their parasitoids appear to be biparental

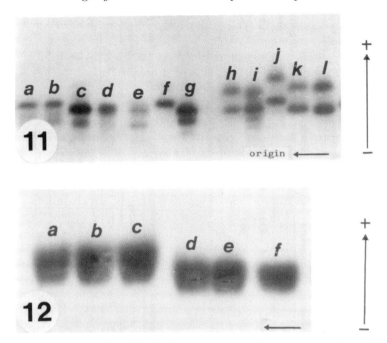

Figs 19.11 and 19.12. Fig. 19.11. Esterase representative zymograms for individuals of two sibling species: (a)–(g) *Pediobius calamagrostidis* Dawah, (h)–(l) *Pediobius deschampsiae* Dawah. **Fig. 19.12.** Lactic dehydrogenase representative zymograms: (a)–(c) *P. calamagrostidis*, (d)–(f) *P. deschampsiae*. After Dawah (1988a).

arrhenotokous species so that the biological species criterion of reproductive isolation is a real one. However, any species that reproduces by thelytokous parthenogenesis poses the same problems for the application of a species concept as do truly asexual organisms. In these, reproductive isolation has no meaning and they exist as isolated clones. No satisfactory species concept exists for such organisms, but they can be characterized by useful markers which may be morphological or biochemical and may be associated with distinct and important biological characteristics. In the chalcid wasps under consideration here, *T. linearis* (Walker), a gall former on stems of *Elymus repens* and its common parasite, *Eurytoma flavimana* (Boheman), are thelytokous. We have reared only very small numbers of males among many hundreds of females.

Assemblages of *Tetramesa* and their parasitoids

Detailed studies, made by sampling populations of a number of different grasses in Britain consistently over many years, have enabled us to build up patterns of ecological relationships and food webs based on a more certain biological taxonomy. Here we shall select a limited set of grasses to illustrate some of the generalities that we can make from a wider study (in preparation). For the nomenclature of grasses we follow *Flora Europaea* (Tutin *et al.* 1980).

Assemblages associated with *Deschampsia cespitosa* and *Calamagrostis epigejos*

Deschampsia cespitosa and *Calamagrostis epigejos* are large perennial tussock grasses found growing in close association in woodlands in central and southern England. *Deschampsia cespitosa* is much more widely distributed and common throughout Britain than *C. epigejos*. However, *C. epigejos* is also found frequently in moist maritime sand dune areas where it may occur with the related marram grass, is *Ammophila arenaria* (L.) Link.

The food web associated with *D. cespitosa* (Fig. 19.13) commonly includes the gall-forming species of *Tetramesa*, *T. airae* (von Schlechtendal), and the non-gall-forming species, *T. petiolata* (Walker) (Fig. 19.1). Also an undescribed phytophagous species of *Eurytoma* near to *suecica* commonly feeds at the flowering stem nodes. The web associated with *C. epigejos* (Fig. 19.14) is similar, with two species of *Tetramesa*, *T. eximia* (Giraud) (Fig. 19.7) and *T. calamagrostidis* (von Schlechtendal) (Figs 19.2 and 19.6) which are both gallformers and a species of *Eurytoma* near to *suecica*, possibly the same as that from *D. cespitosa*.

The general patterns of parasitoids associated with the two grasses are similar, but all of the species of more specific feeders are different, but very closely related. The specific status of the *Tetramesa*, *Eurytoma* (Fig. 19.3), *Chlorocytus*, and *Pediobius* (Figs 19.11 and 19.12) have mostly been established with certainty only by enzyme electrophoresis (Dawah 1987, 1988*a,b*, 1989). Thus, the links between the two webs in Britain are limited to some of the rarer polyphagous species including *Macroneura vesicularis* (Retzius) and *Mesopolobus graminum* (Hårdh). Interestingly, *Tetramesa eximia*, in addition to attacking *C. epigejos*, is also common in *Ammophila arenaria* in coastal sand dunes. This is the only species of *Tetramesa* known to us to attack species of more than one genus of grasses. The biological species status of insects from both grasses has been confirmed by electrophoresis and by mate choice experiments (Dawah 1987). Interestingly, in *Ammophila* the associated parasitoid complex shows many differences to that in *Calamagrostis* (Fig. 19.15).

13

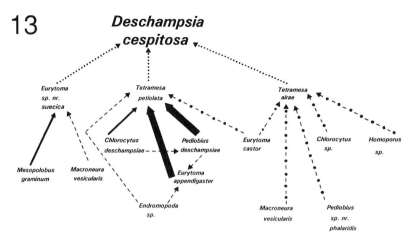

14

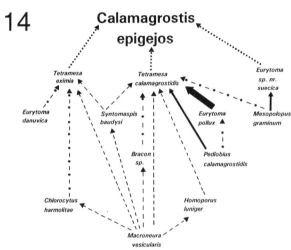

15

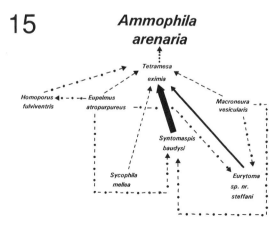

Figs 19.13–15. Food webs of phytophagous Eurytomidae and their parasitoids associated with the grasses. **Fig. 19.13.** *Deschampsia cespitosa.* **Fig. 19.14.** *Calamagrostis epigejos.* **Fig. 19.15.** *Ammophila arenaria* (L.) (Conventions as in Fig. 19.10.)

Assemblages associated with *Elymus* species and *Festuca rubra/ovina*

In Britain there are three species of *Elymus* (= Agropyron), each of generally similar growth form. We have mainly studied *E. repens* and *E. farctus* (= *E. junceiforme*). Three distinct species of *Tetramesa* are commonly associated with these grasses (Fig. 19.16).

1. *Tetramesa cornuta* (Walker) is a non-galling species that lives singly at flowering stem nodes.

2. *Tetramesa hyalipennis* (Walker) normally forms a very distinct apical gall causing complete abortion of the flowering shoot (Figs 19.4 and 19.5). The gall encloses a single larva.

3. *Tetramesa linearis* (Walker) is a very characteristic stem gall former, as described above (p. 310).

The parasitoid complexes of these species are quite distinct (Fig. 19.16) with no recorded overlap between them. Each species of *Tetramesa* has two or more associated monophagous parasitoids.

Of the large grass genus *Festuca* the two taxonomically difficult species groups of *F. ovina* and *F. rubra* share a common assemblage of two *Tetramesa* species (Fig. 19.17).

1. *Tetramesa brevicornis* (Walker) is closely related to *T. linearis* from *Elymus* and forms the same type of characteristic stem wall galls.

2. *Tetramesa brevicollis* (Walker) makes distinct swollen stem galls each occupied by only a single larva (Fig. 19.8). The parasitoid assemblages from each of these *Tetramesa* are quite distinct and characterized by monophagous species. They are linked only by the polyphagous *Eupelmus atropurpureus* and *Macroneura vesicularis*. The *Eurytoma* species associated with each have not yet been described but are certainly different.

These two sets of assemblages of *Tetramesa* and their parasitoids show that even grasses with similar growth forms and habits are colonized by quite different species of *Tetramesa*, including gall formers and non-gall formers. Even more remarkably the associated parasitoids tend to be almost completely monophagous. Indeed, as more detailed biosystematic studies are made, so more and more supposed polyphagous species are revealed as complexes of genetically distinct sibling species. For example, recent electrophoretic and morphometric work in Cardiff has shown that *Sycophila mellea* (Curtis) consists of several host specific biological species. It was previously regarded as one species that attacked only *Tetramesa* galls, but showed no specificity among them (Claridge 1959).

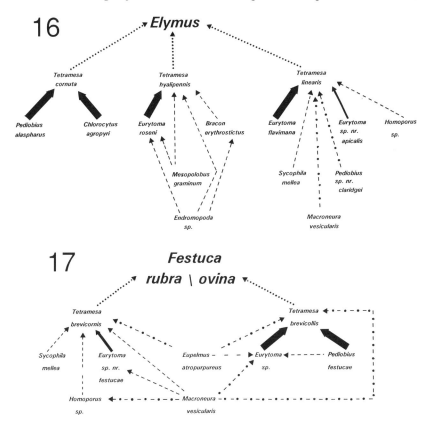

Figs 19.16 and 19.17. Food webs of phytophagous Eurytomidae and their parasitoids associated with **Fig. 19.16.** *Elymus repens* and *E. farctus* and **Fig. 19.17.** *Festuca rubra* and *F. ovina* L. (Conventions as in Fig. 19.10.).

Discussion

Our knowledge of the community structure of assemblages of insect herbivores and their parasitoids is remarkably meagre. This is the more remarkable considering the enormous numbers of species of these organisms that contribute to the total of terrestrial biological diversity.

Most of what we know about the detailed food webs associated with insect herbivores is derived from studies on species that live within the tissues of their host plants in their larval stages. These endophytic larvae are relatively easily sampled and, since they are confined within plant material, it is a simple matter to obtain precise records of associated parasitoids. Thus, gall formers, stem-borers, leaf-miners, and flower head-

and fruit-feeders have been most widely studied (Claridge 1987).

In a major series of studies Askew (1984) has documented more than 50 species of gall wasps and their inquilines (Cynipidae) and parasitoids associated with oaks, *Quercus* species, and with roses, *Rosa* species, in Britain. He found that each gall is typically attacked by a large number of different relatively polyphagous parasitoids and very few monophagous ones. However, the assemblages associated with galls on oaks are almost totally different to those on roses. One of the few species originally thought to link the webs on the different host plants was *Eurytoma rosae* Nees. However some of the earliest biosystematic studies on these insects showed it to be a complex of sibling species. The true *E. rosae* was associated only with rose galls and *E. brunniventris* Ratzburg only with oak galls (Claridge and Askew 1960). Similar studies to those of Askew on oak and rose galls have been made by Zwölfer (1968, 1979, 1980) on the herbivorous insects inhabiting the flower heads of thistles, knapweeds, and related species of Compositae in Europe. Here he found a diverse but characteristic set of herbivore taxa, including gall flies (Tephritidae), and gall wasps (Cynipidae). Unlike the studies of Askew, Zwölfer typically found that each herbivore species within each flower head assemblage had its own specific group of associated parasitoids.

The phytophagous Eurytomidae associated with grasses have been studied only sporadically and almost exclusively in northern temperate regions of the world. All authors are agreed that despite taxonomic difficulties, species of *Tetramesa* show great specificity, almost always to the generic or major species groups level of their grass hosts. Large scale rearings of insects were made in North America earlier in this century (Phillips and Emery 1919; Phillips 1920, 1936; Phillips and Poos 1922) because a number of species were pests of various important cereals and forage grasses—the so-called straw- and joint-worms. These rearings confirmed high levels of host specificity, but unfortunately relatively little was published on the assemblages of parasitoids associated with the insects.

Our own studies over many years show extreme specificity and parallel those of Zwölfer on flower heads in showing extreme compartmentation in each web. This is due to the high levels of specificity shown by most of the associated parasitoids even when their *Tetramesa* hosts inhabit the same host grass.

It is clear that the real nature of interactions in assemblages such as those described here can only be discovered if modern biosystematic techniques are used to determine biological species boundaries. In this way more and more polyphagous species are being shown to be groups of more specific biological species. Only after such assemblages have been carefully documented and diversity and status of host-associated populations clarified will it be possible to ask meaningful questions concerning the ways in which these extremely diverse assemblages may have evolved. The long-standing arguments over the importance or otherwise of sympatric speciation by host race formation without spatial separation has still not been resolved (Claridge 1988, 1993). The insects discussed here will provide excellent material for the application of biochemical and molecular techniques to test critically the various hypotheses.

Acknowledgements

Over the years that we have been interested in *Tetramesa* species and their parasitoids many individuals around the world, too many to name, have helped us by generously loaning material and sending reprints of their works. We are deeply grateful to them all.

We have been particularly fortunate to know and to discuss taxonomic problems frequently with two of the most outstanding chalcid taxomists that the world has yet produced: Dr Zdenek Bouček and Dr Marcus Graham. M.F.C. was greatly influenced by them in his early days and we have both benefited continuously from their wisdom and knowledge. We thank them and dedicate this chapter to them without permission.

We are also indebted to David Windsor who took the photographs used in Figs 19.1–19.3.

References

Anderson, H. (1967). Three insect galls of species new to Sweden. *Opuscula Entomologica*, **32**, 282–3.

Askew, R.R. (1984). The biology of gall wasps. In *Biology of gall insects*, (ed. T.N. Ananthakrishnan), pp. 223–71. Edward Arnold, London.

Bouček, Z. (1988). *Australasian Chalcidoidea (Hymenoptera). A biosystematic revision of genera of fourteen families, with a reclassification of species*. CAB International, Wallingford, UK.

Cain, A.J. (1954). *Animal species and their evolution*. Hutchinson, London.

Claridge M.F. (1959). A contribution to the biology and taxonomy of the British species of the genus *Eudecatoma* Ashmead (= *Decatoma* aucct. nec Spinola) (Hymenoptera: Eurytomidae). *Transactions of the Society for British Entomology*, **13**, 149–68.

Claridge, M.F. (1961a). An advance towards a natural classification of eurytomid genera (Hymenoptera: Chalcidoidea), with particular reference to British forms. *Transactions of the Society for British Entomology*, **14**, 167–85.

Claridge, M.F. (1961b). A contribution to the biology and taxonomy of some Palaearctic species of *Tetramesa* Walker (= *Isosoma* Walk., = *Harmolita* Motsch.) (Hymenoptera: Eurytomidae), with particular reference to the British fauna.

Transactions of the Royal Entomological Society of London, **113**, 175–216.

Claridge, M.F. (1987). Insect assemblages—diversity, organization and evolution. In *Organization of communities: past and present*, (ed. P. Giller and J. Gee), pp. 140–61. Blackwell, Oxford.

Claridge, M.F. (1988). Species concepts and speciation in parasites. In *Prospects in systematics*, (ed. D.L. Hawksworth), pp. 92–111. Clarendon Press, Oxford.

Claridge, M.F. (1991). Genetic and biological diversity of insect pests and their natural enemies. In *The biodiversity of microorganisms and invertebrates: its role in sustainable agriculture*, (ed. D. L. Hawksworth), pp. 183–94. CAB International, Wallingford, UK.

Claridge, M.F. (1993). Speciation in herbivores—the role of acoustic signals in leafhoppers and planthoppers. In *Evolutionary patterns and processes*, (ed. D. Edwards and D. Lees), pp. 285–97. Academic Press, London.

Claridge, M.F. and Askew, R.R. (1960). Sibling species in the *Eurytoma rosae* group (Hymenoptera: Eurytomidae). *Entomophaga*, **5**, 141–53.

Dawah, H.A. (1987). Biological species problems in some *Tetramesa* (Hymenoptera: Eurytomidae). *Biological Journal of the Linnean Society*, **32**, 237–45.

Dawah, H.A. (1988*a*). Taxonomic studies on the *Pediobius eubius* complex (Hymenoptera: Chalcidoidea: Eulophidae) in Britain, parasitoids of Eurytomidae in Gramineae. *Journal of Natural History*, **22**, 1147–71.

Dawah, H.A. (1988*b*). Differentiation between *Eurytoma appendigaster* group (Hymenoptera: Eurytomidae) using electrophoretic esterase patterns. *Journal of Applied Entomology*, **105**, 144–8.

Dawah, H.A. (1989). Separation of four *Chlorocytus* species (Hymenoptera: Pteromalidae), parasitoids of stem-boring Hymenoptera (Eurytomidae and Cephidae) using enzyme electrophoresis. *The Entomologist*, **108**, 216–22.

Gauld, I.D. and Bolton, B. (ed.) (1988). *The Hymenoptera*, pp. 1–310. Oxford University Press, London.

Loxdale, H.D. and den Hollander, J. (ed.) (1989). *Electrophoretic studies on agricultural pests*, pp. 1–477. Clarendon Press, Oxford.

Mayr, E. (1942). *Systematics and the origin of species from the viewpoint of a zoologist*, pp. 1–334. Columbia University Press, New York.

Noble, N.S. (1936). The citrus gall wasp (*Eurytoma fellis* Girault). *Department of Agriculture, New South Wales, Science Bulletin*, **53**, 1–41.

Noyes, J.S. (1990). The number of described chalcidoid taxa in the world that are currently regarded as valid. *Chalcid Forum*, **13**, 9–10.

Peck, O. (1963). A catalogue of the Nearctic Chalcidoidea (Insecta: Hymenoptera). *Canadian Entomologist, Supplement*, **30**, 1–1092.

Phillips, W.J. (1920). Studies on the life history and habits of the jointworm flies of the genus *Harmolita* (*Isosoma*), with recommendations for control. *United States Department of Agriculture Bulletin*, **808**, 1–27.

Phillips, W.J. (1936). A second revision of chalcid flies of the genus *Harmolita* (*Isosoma*) of America north of Mexico, with descriptions of 20 new species. *United States Department of Agriculture Technical Bulletin*, **518**, 1–25.

Phillips, W.J. and Emery, W.T. (1919). A revision of the chalcid flies of the genus *Harmolita* of America north of Mexico. *Proceedings of the United States National Museum*, **55**, 433–71.

Phillips, W.J. and Poos, F.W. (1922). Five new species belonging to the genus *Harmolita* Motschulsky (*Isosoma* Walker). *Kansas University Science*, **14**, 350–3.

Rosen, H. von (1956). Eine phytophage *Eurytoma* in Mittel—und Nordschweden. *Opuscula Entomologica*, **21**, 16–20.

Spears, B.M. and Barr, W.F. (1985). Effects of jointworms on the growth and reproduction of four native range grasses of Idaho. *Journal for Range Management*, **38**, 44–6.

Tutin, T.G., Heywood, V.H., Burges, N.A., Moore, D.M., Valentine, D.H., Walters, S.M. *et al.* (1980). *Flora Europaea*, Vd. 5, pp. 118–267. Cambridge University Press, Cambridge.

Walker, F. (1835). Monographia Chalciditum. *The Entomological Magazine*, **2**, 13.

Zerova M.D. (1976). Hymenoptera, Chalcidoidea family Eurytomidae, subfamilies Rileyinae and Harmolitinae. (In Russian.) *Fauna SSSR*, (n.s.), **7**, 1–23, Leningrad.

Zerova, M.D. (1978). Parasitic Hymenoptera, Chalcidoidea, Eurytomidae. (In Ukrainian.) *Fauna Ukraini*, **11**, 1–465.

Zerova, M.D. (1992). Problems of phylogeny and developmental trends in the family Eurytomidae. *Chalcid Forum*, **15, 1 –6.**

Zwölfer, H. (1968). Untersuchungen zur biologischen Bekämpfung von *Centaurea solstitialis* L.—Strukturmerkmale der Wirtspflanze als Auslöser des Eiablageverhaltens bei *Urophora siruna-seva* (Hg.) (Diptera: Trypetidae). *Zeitschrift für angewandte Entomologie*, **61**, 119–30.

Zwölfer, H. (1979). Strategies and counterstrategies in insect population systems competing for space and food in flower heads and plant galls. *Fortschritte der Zoologie*, **25**, 331–53.

Zwölfer, H. (1980). Distelblütenkopfe als ökologische Kleinsysteme: Konkurrenz und Koexistenz in phytophagenkomplexen. *Mitteilungen der deutschen Gesellschaft für allgemeine und angewantdte Entomologie*, **2**, 21–37.

20. The ecology of the pea galls of *Cynips divisa*

FRANCIS GILBERT, CHARLOTTE ASTBURY,
JENNY BEDINGFIELD, BRUCE ENNIS,
SAMANTHA LAWSON, and TRACEY SITCH

Department of Life Science, University of Nottingham, Nottingham, UK

Abstract

The results of a long-term study into the population ecology of pea galls is presented. Pea galls are produced by the agamic generation of the cynipid gall wasp *Cynips divisa* Hartig and they were studied over the period 1985–1991. Population biology is the sum effect of variations in fitness between individuals, variations which are determined by life history parameters. Components of fitness were measured for a large sample, calculating survival, size, fecundity, and maturity at eclosion: the fitness consequences of ovipositional decisions made in positioning the gall on different trees, leaves, and positions within leaves are analysed. Evidence for a 6–7 year population cycle is discussed.

Introduction

Understanding the structure and dynamics of communities of species is a central problem in ecology, particularly in view of new ideas about food webs (Pimm and Lawton 1979; Lawton 1989) and population dynamics (Hassell *et al.* 1991). Marrying up these contrasting disciplines in an integrated view of populations embedded within communities will be a difficult task, since ecologists have different ideas about the importance of the various forces involved (for example, top-downs vs bottom-up controls on population and community dynamics; Matson and Hunter (1991)). Recently there has been a lot of interest in interactions between more than two trophic levels, usually concentrating upon the nature of

Plant Galls (ed. Michèle A. J. Williams), Systematics Association Special Volume No. 49, pp. 331–49. Clarendon Press, Oxford, 1994. © The Systematics Association, 1994.

the interaction between plants, herbivores, and parasitoids (Price *et al.* 1980; Zwölfer 1987; Tscharntke 1992). Herbivores are subject to selection pressures from both the defensive mechanisms of the plant and attack by parasitoids. What determines their fitness under these conditions?

We have been studying this question since 1985 using the pea galls of the agamic generation of the cynipine gall wasp *Cynips divisa* Hartig, a relatively well-studied system (see Askew 1960, 1961*a,b*, 1965, 1975, 1985; Sitch *et al.* 1988). Gall systems are particularly useful for studying this since their position is fixed at the outset; the consequences of the decision-making process of oviposition or settlement by the female can then be assessed (for example Whitham 1980; Sitch *et al.* 1988). The relationship between the gall maker and the plant is a fascinating but elusive problem (Askew 1985; Crawley 1985; Abrahamson and Weis 1987; Weis *et al.* 1988; see reviews in Shorthouse and Rohfritsch 1992), as is the exact evolutionary advantage of creating the gall (Abrahamson and Weis 1987; Price *et al.* 1987; Cornell 1990; Hartley and Lawton 1992). Our studies on pea galls have been hampered by great year to year fluctuations in gall densities, including some years of very high mortality, but these changes in abundance have themselves proved very interesting. Long-term data on non-economic insect populations are becoming more available (for example Wolda 1983; Owen and Gilbert 1989), including some galling insects (Miyashita *et al.* 1965; Redfern and Cameron 1978; Washburn and Cornell 1981; Moran and Whitham 1988; Wool 1990).

We report here on data gathered between 1985 and 1991 to investigate two questions: first, is there any long-term pattern to population densities? and, second, is there any evidence for positional effects in components of fitness?

Materials and methods

Cynips divisa has a heterogonic life cycle with a sexual generation from red wart galls on oak buds, followed by an agamic generation from pea galls on the leaves (Askew 1985). The species has a wide distribution in the British Isles, from the Isle of Wight (D.E. Biggs, personal communication) to Scotland (C.K. Leach, personal communication). Mapping schemes are currently being undertaken (for example Bowdrey 1987; Griffiths *et al.* 1990), but it will be some time before we have an accurate indication of the distribution. Most records appear to be of a few scattered galls, quite unlike the situation in Oxford in 1958, where Askew (1961*a*) records them as very common.

We collected galled leaves from young oaks in Clumber Park, Nottingham during August and September, each year when they were available between 1985 and 1991. In 1991 galls were abundant on several trees, as in 1985 and samples were taken from eight different trees: a large sample was taken from one single tree with a high proportion of leaves galled. Galls were concentrated on the leaves of young oaks growing under the canopy of older larger oaks (Askew (1961*a*) notes that his pea galls were collected from oaks only 9 years old): because of the extreme patchiness of galled leaves between and within trees, only leaves with galls were collected. The position of galls (aborted or otherwise) was recorded according to the scheme reported in Sitch *et al.* (1988); briefly, the vein number, side of the leaf (left or right of the midrib) were all recorded for individual leaves. In addition, the length of the midrib and in most cases the maximum width of each leaf was measured. The width of individual galls was measured under the microscope using an eyepiece graticule and their weights measured using a Cahn 23 electrobalance: the weight of the wasp in the gall was later removed to leave the true weight of the gall alone. Galls were then dissected. At this stage the rare thin-walled galls of the related *Cynips agama* Hartig were obvious. The gall contents were identified to generic level using the key developed by Askew (1985; and personal communication) (see Sitch *et al.* 1988).

If the gall maker had survived, adult female wasps were in the gall awaiting emergence: females were removed and four measurements taken (wing length in mm, fresh body weight to the nearest 0.1 mg, fresh weight of the abdomen alone to the nearest 0.1 mg, and on dissection the number of eggs in the abdomen). No weights were taken in 1985; since survival was only again high in 1991, all analyses of fitness components refer only to this year. Parasitic inhabitants of the gall were identified, removed, and weighed (fresh weight) to the nearest 0.1 mg in 1988, but not in other years.

Females eclose from the pupa with an abdomen full of stored material from the pupal fat body, which they then use to mature their eggs. On natural emergence from the gall the eggs are completely or almost completely developed (personal observation). We counted the number of matured eggs in the ovarioles by dissecting the abdomen in Griffith's Ringer under a binocular microscope.

We then calculated several fitness components from these data: survival (the percentage of all galls that produce female wasps), fecundity (the number of eggs in the abdomen), size (body weight in mg), relative weight (standardized deviation from the double-log regression line that predicts body mass from wing length), allocation to reproduction (percentage of the total body mass represented by the abdomen), and degree of maturity (standardized deviation from the double-log regression line that predicts the number of eggs from wing length).

Calculations were performed using the SPSS statistical package implemented on an ICL mainframe computer. All mean values in the text and on graphs are quoted ± 1 standard error. Standard errors of proportions and percentages were calculated from the equation

$$SE = \sqrt{((p(p - 1/(n - 1)))}$$

where p = proportion and n = sample size.

Results

1. Overall patterns in fitness components

Gall weight (all, 31.7 ± 0.6 mg, range 1–84; with surviving wasps, 47.4 ± 0.7 mg, range 19.5–77.0) was, not surprisingly, strongly related to gall diameter[3], an index of gall volume ($r^2 = 0.76$, $n = 849$). Gall weight was also related ($r^2 = 0.23$, $n = 397$, $p \ll 0.001$) to the weight of the female wasp, where she survived (slope = 0.040 ± 0.004, intercept = 1.72 ± 0.18). Female body weight (mean, 3.81 ± 0.04 mg, range 2.0–7.0) was slightly better fitted to an allometric rather than a simple linear function of wing length, since the slope of the double-log regression is only 0.84 ± 0.07 ($r^2 = 0.24$, $n = 496$, $p \ll 0.001$); deviations from this regression were saved, to give a measure of size-adjusted or relative weight. The number of eggs in the abdomen (mean 78 ± 2, range 3–243) was also better fitted as an allometric function of wing length ($r^2 = 0.18$, $n = 484$, $p \ll 0.001$, slope = 2.24 ± 0.22, intercept = 0.19 ± 0.16); deviations from this line were also saved as a measure of maturity. Allocation to the abdomen (mean 60.1 ± 0.3 per cent, range 21–84) was weakly positively related to wing length ($r^2 = 0.01$, $n = 496$, $p < 0.01$) with a slope of 0.011 ± 0.004.

　Galls with surviving wasps were a mean of 5.73 ± 0.03 (range 3.2–9.4) mm in diameter. Galls containing parasitoids or inquilines were always on average significantly smaller whatever the species: the smallest galls were those where fungal attack appeared to have been the cause of death (mean 3.62 ± 0.07 mm).

2. Differences between years

Galls were nearly always on the same set of young trees; despite intensive search, other small oaks in the immediate vicinity were never galled. Figure 20.1 shows the variation in gall density and mortality on these trees during the study period. The data suggests that one cycle of a long-term population cycle occurred at this site and that a decline in gall-maker populations followed increased mortality rates by parasitoids.

　There was no effect of leaf side (left/right) in any year. The distribution of galls on the different veins of the leaf showed highly significant

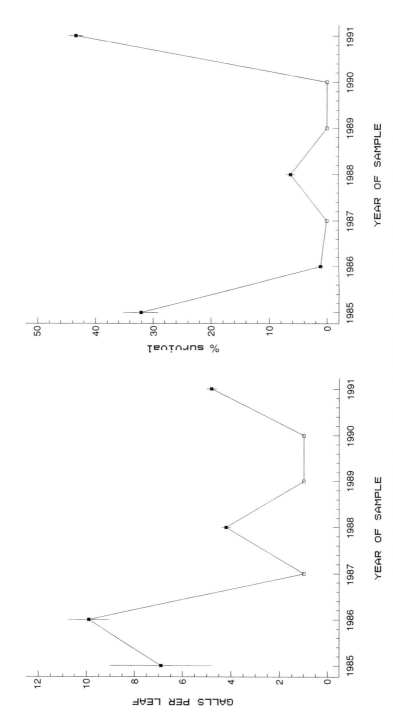

Fig. 20.1. (a) Gall density and (b) survival of the gall maker during the years of the study.

differences between the years: most galls were placed on veins 3–5 of the oak leaf in 1986 and 1988, but on veins 4–7 in 1991. Approximately 90 per cent were positioned on the half of the vein nearest the midrib, but there were also differences between years. Considering only the gall nearest the midrib, in 1991 this gall was on average approximately 12.4 ± 0.3 mm from the midrib, but less than 9.7 mm on average in 1986 and 1988 ($F_{2,1886} = 35.8$, $p \ll 0.001$). Restricting the analysis to single-gall veins, this difference still remained ($F_{2,1454} = 15.7$, $p \ll 0.001$). In 1991 the first gall was on average placed 38 per cent along the length of the vein, but only 26 per cent in 1988 ($F_{1,1104} = 125.5$, $p \ll 0.001$), with again a significant difference remaining for just the single-gall veins (35 per cent vs, 26 per cent; $F_{1,904} = 48.4$, $p \ll 0.001$).

There was always a positive correlation between leaf size and gall number (1986, $r = 0.45$, $n = 43$, $p = 0.001$; 1988, $r = 0.37$, $n = 257$, $p \ll 0.001$; 1991, $r = 0.33$, $n = 353$, $p \ll 0.001$).

Survival rates decreased with gall density on leaves in years when survival was low to very low (in 1986, $F_{1,41} = 3.55$, $n = 42$, $p = 0.066$, slope $= -0.007 \pm 0.004$; and in 1988, $F_{1,266} = 23.6$, $p \ll 0.001$, slope $= -0.028 \pm 0.006$), but were unrelated to gall density in 1991 when survival was high ($F_{1,351} = 0.001$, $p > 0.05$). In 1988 a specific sample of low (1–3) and high (> 5) gall density leaves was taken, in which 37 survivors were found amongst a total of 535 galls: there was a highly significant difference in survival rates between low (17.5 per cent) and high (4.4 per cent) density leaves ($\chi_1^2 = 20.1$, $p < 0.001$).

In 1991, survival was not related to the number of galls on veins, but overall survival decreased with order on the vein (nearest the midrib 45 per cent, second position 38 per cent, third 29 per cent, fourth 20 per cent: $\chi_5^2 = 15.1$, $p < 0.01$).

Failure rates (galls aborted at an early stage of development) on each vein increased with the number of galls on the vein in 1986 ($\chi_4^2 = 10.8$, $p = 0.03$), but there was no detectable effect in 1988 ($\chi_3^2 = 6.5$, $p = 0.09$) or 1991 ($\chi_5^2 = 1.7$, $p = 0.9$).

3. Fitness component differences in 1991

(a) Tree differences   Genetic, age, and environmental differences should lead to variation between trees in resistance; thus we predict that we should find significant differences in fitness components between trees. Across the eight trees sampled, there were strong differences in survival (from 0 to 53 per cent: $\chi_7^2 = 173$, $p \ll 0.001$). On average there were no differences between trees in female weights ($F_{6,693} = 1.45$, $p > 0.05$) or wing lengths ($F_{1,563} = 1.5$, $p > 0.05$). There were differences in the mean

relative weights (-1.2 ± 0.4 to $+0.4 \pm 0.3$ standard deviations; $F_{6,489} = 2.98$, $p > 0.01$), allocation (58 per cent ± 0.4 to 64 per cent ± 1; $F_{6,682} = 3.8$, $p = 0.001$), number of eggs (52 ± 11 to 106 ± 8; $F_{6,479} = 4.03$, $p < 0.001$), and maturity (-0.6 ± 0.1 to $+0.6 \pm 0.1$ standard deviations; $F_{6,479} = 4.4$, $p < 0.001$) for females on different trees. Because of these differences, we restricted most analyses of within-leaf positional effects to the large sample taken from one tree, although results from other trees match those presented here.

(b) Leaf size We predicted that larger leaves can provide more nutrients, which lead to larger heavier galls and heavier wasps that can allocate more to reproduction and mature faster. We had no specific prediction about how gall density should vary with leaf size, and therefore performed a two-tailed test. Gall density ($r = 0.41$, $n = 166$, $p < 0.001$) and the means per leaf for gall diameter ($r = 0.21$, $n = 166$, $p < 0.01$), gall weight ($r = 0.18$, $n = 96$, $p < 0.05$), female weight ($r = 0.15$, $n = 96$, $p = 0.07$), allocation ($r = 0.14$, $n = 91$, $p = 0.09$), egg number ($r = 0.20$, $n = 88$, $p < 0.05$), relative size ($r = 0.16$, $n = 91$, $p = 0.05$), and maturity ($r = 0.18$, $n = 88$, $p < 0.05$) were all siginificantly related to leaf area. These results are mirrored for all but gall diameter for the large sample from one tree.

(c) Gall number on the leaf We performed multiple regression using the mean values per leaf of the fitness components with leaf area and gall number as independent variables. The prediction is that fitness components should increase with leaf area (see above), but decrease with the independent effect of gall number. We used only the large sample data from one tree to avoid intertree differences. In all cases except female weight (where it was not significantly different from zero), the partial regression coefficient for gall number was negative, significantly so for survival (slope $= -0.034 \pm 0.009$, $p < 0.001$). This coefficient was also negative for gall weight (slope $= -0.5 \pm 0.38$, $p = 0.10$). All partial regression coefficients for leaf area were positive, most significantly different from zero.

(d) Overall effects of vein number For the large sample for one tree, we could analyse fitness differences on the different veins. We predicted that since most galls were on veins 4–7 in 1991, which are the longest veins, there would be differences in fitness components. Eight hundred and fifty-five galls were analysed from the midrib and veins 2 to 11. There were no differences in survival probability ($\chi^2_{10} = 14.8$, $p > 0.05$), female weight ($F_{9,469} = 0.63$, $p > 0.05$), relative weight ($F_{9,399} = 0.9$, $p > 0.05$), allocation ($F_{9,465} = 0.9$, $p > 0.05$), number of eggs ($F_{9,393} = 0.9$, $p > 0.05$), or maturity ($F_{9,393} = 0.9$, $p > 0.05$).

(e) Overall effect of distance along a vein   Using the large sample, we tested the prediction that components of fitness would decrease in galls further away from the midrib. The reason for making this prediction arises from two hypotheses concerning the source of the nutrients flowing from the plant to the gall; either from the leaf down the veins or from the branch, up the veins. Either gives the prediction that it is better to be near the midrib, but if the former explanation is correct, the hypothesis is framed in terms of distances relative to vein length. The results might be biased by including all galls, since where there are several on a vein, distal galls are inevitably further away from the midrib than distal ones and any differences in fitness components with order on the vein (see below) will confound interpretation. We therefore used only data for veins with a single gall on them. There was no difference in distances ($F_{1,365} = 0.22$, $p > 0.05$) or relative distances ($F_{1,239} = 1.4$, $p > 0.05$) between survivors and non-survivors, although in both cases the mean values are in the expected direction (non-survivors further away from the midrib). Size, allocation, and the number of eggs were not affected by distance. Relative weight ($r = -0.25$, $n = 125$, $p < 0.01$) and maturity ($r = -0.15$, $n = 125$, $p < 0.05$) decreased with distance, as expected.

(f) Overall effect of the number of galls on a vein   Again for the large sample from one tree, we analysed fitness differences resulting from different numbers of galls on a single vein. We predicted that competition for nutrients will lead to a decrease in fitness components. Survival did in fact decrease with increasing gall density ($\chi^2_5 = 19.8$, $p < 0.01$). There were significant effects on female weight ($F_{3,392} = 7.7$, $p \ll 0.001$) and relative weight ($F_{3,331} = 2.9$, $p < 0.05$), but none on allocation ($F_{3,388} = 1.0$, $p > 0.05$) or the number of eggs ($F_{5,323} = 0.95$, $p > 0.05$). The index of maturity showed significant differences between mean values ($F_{3,323} = 2.4$, $p = 0.06$), but these were not in the predicted order.

We also predicted that, if a significant proportion of nutrients come down the vein from the leaf during larval development, the effects of competition should be most severe on the gall nearest the midrib. We assigned galls to their sequence order along veins, the first being the nearest to the midrib, and tested whether the number of galls on a vein affects the fitness components of the first gall. Survival was not affected ($\chi^2_5 = 6.4$, $p > 0.05$). Female weights ($F_{2,490} = 5.5$, $p < 0.01$) and relative weights ($F_{2,326} = 5.3$, $p < 0.01$) were significantly different in the expected direction, but allocation, egg number, and maturity were not affected.

(g) Overall order effects   Applying a similar argument to the effects of distance along veins, we predicted that there should be an effect of sequence order on a vein on components of fitness. The exact form this takes, unlike before, depends on which hypothesis concerning the source

of nutrients is thought to be correct. There were substantial differences in survival with order, and also in female weights, but not with other fitness components. However, these results are confounded by the impact of the number of galls on a vein.

A more sensitive test uses the cases where all females survived on a vein. There were 46 two-gall and two three-gall veins where this was true, not all with full data for each female. For two-gall veins, gall weights ($t_{17} = 3.2$, $p < 0.003$) and allocations ($t_{31} = 2.05$, $p < 0.05$) were significantly lower for the distal gall, but gall diameter, female weight, relative weight, egg number, and maturity were not different ($t < 0.83$, $p > 0.21$). In 1986 with rather few data, we suggested that having a companion on a vein disrupts the relationship between size and fecundity (Sitch *et al.* 1988). This is also apparent in these more extensive data, since there is a good correlation between weight and wing length for the proximal galls of the pair ($r^2 = 0.59$, $n = 26$, $p \ll 0.001$), but this disappears for the distal galls ($r^2 = 0.05$, $p > 0.05$).

(h) Interactions between gall number on veins and order Figure 20.2 shows fitness components split into categories of gall number on veins and order on veins. There are significant differences in survival, and the pattern is clear: the more galls and the further down in the sequence, the lower the survival.

The effect of other fitness componets was assessed by two-way ANOVA with leaf length as a covariate: only in the case of allocation was the effect of leaf length significant ($F_{1,379} = 12.9$, $p < 0.001$). There were no significant effects of gall number or order in the cases of wing length, allocation, number of eggs, or maturity. There was a significant effect of gall number on female weight ($F_{2,383} = 5.5$, $p < 0.01$) and a significant gall number × order interaction ($F_{1,383} = 4.64$, $p < 0.05$).

Gall weight and diameter for galls where female wasps survived were also analysed in this two-way design with leaf length as a covariate. Gall diameter was strongly affected by leaf length ($F_{1,383} = 13.0$, $p < 0.001$), but of the main effects only order is significant ($F_{2,383} = 4.8$, $p < 0.01$). Gall weight is significantly affected by gall number ($F_{2,136} = 4.9$, $p < 0.01$ and order ($F_{2,136} = 14.6$, $p < 0.001$), but there is no interaction ($F_{1,136} = 0.5$, $p > 0.05$) and no apparent effect of the covariate ($F_{1,136} = 2.9$, $0.1 < p < 0.05$).

4. Food web relationships

In 1988 there were few surviving female wasps, but many parasitoids. We therefore weighed the parasitoid larvae to look for evidence of the interactions between them.

Synergus nervosus Hartig was found in the main chamber of 36 per cent of the galls, whilst *S. pallicornis* Hartig was found in the walls of 49 per

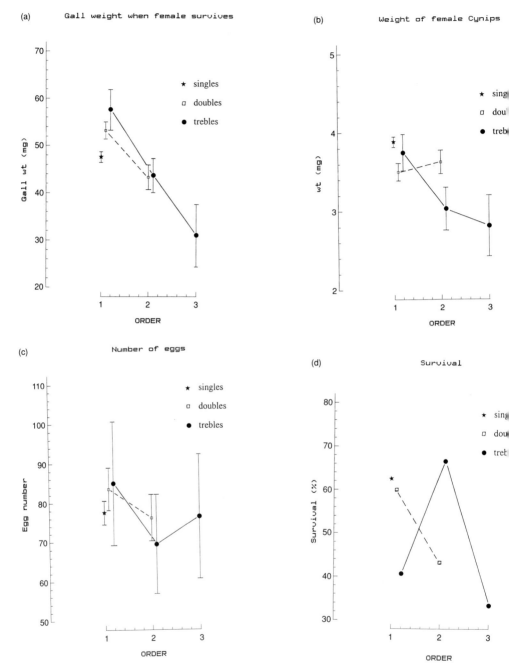

Fig. 20.2. Plot of fitness components split into categories of the number of galls on a single vein and the order on the vein: (a) gall weight for surviving gall makers, (b) weight of female wasp, (c) number of eggs carried by dissected females, (d) survival.

Table 20.1. Multiple regression equations assessing the effect of gall diameter and the number or weight of *S. pallicornis* on the weight of the other inhabitants

| Dependent variable | Independent variables | | Intercept | R^2 | F |
	Gall diameter	Other independent variable			
S. pallicornis wt $n = 97$	0.13 ± 0.06	*S. nervosus* wt nsd from 0.0	nsd from 0.0	0.45	6.6**
S. pallicornis wt $n = 49$	0.28 ± 0.06	*Eurytoma* wt -0.17 ± 0.07	-0.84 ± 0.30	0.44	10.6***
S. pallicornis wt $n = 97$	0.15 ± 0.04	Number of *S. pallicornis* nsd from 0.0	-0.38 ± 0.16	0.50	15.6***
S. nervosus wt $n = 60$	0.41 ± 0.07	Number of *S. pallicornis* 0.06 ± 0.03	-0.95 ± 0.32	0.59	40.3***
Eurytoma wt $n = 49$	0.88 ± 0.17	Number of *S. pallicornis* -0.35 ± 0.08	-2.78 ± 0.80	0.42	16.4***

All figures cited are significantly different from zero at the 5 per cent level, except where indicated by 'nsd from 0.0'.
** $= p < 0.01$; *** $= p < 0.001$.

cent of the galls. Unlike in 1986, *S. pallicornis* were not solely confined to galls already containing *S. nervosus*, even though there was a strong positive association ($\chi_1^2 = 42.2$, $p < 0.001$). Both *Synergus* species were found in relatively large galls ($F_{1,413} > 11.5$, $p < 0.001$). *Eurytoma brunniventris* Ratzeburg were found in 24 per cent of the galls, virtually always in the main chamber.

The mean weights (mg) of the different larval gall inhabitants were *Eurytoma*, 1.09 ± 0.11 (range, 0.27–4.46), $n = 49$; *S. nervosus*, 1.27 ± 0.06 (range, 0.22–1.95), $n = 60$; and *S. pallicornis*, 0.40 ± 0.02 (range 0.04–1.00), $n = 97$. There were between one and six *S. pallicornis* in an individual gall ($n = 138$). Both the combined weights of *S. pallicornis* ($F_{1,58} = 16.9$, $p < 0.001$) and individual weights ($F_{1,95} = 30.9$, $p \ll 0.001$) are strongly related to gall diameter. Similar positive regressions are found for *S. nervosus* ($F_{1,59} = 61.12$, $p \ll 0.001$) and *Eurytoma* ($F_{1,48} = 10.9$, $p < 0.05$).

We investigated the effect of competition in *S. pallicornis* by removing the effect of gall diameter via multiple regression (Table 20.1). We looked for the effect of *S. pallicornis* on the other inhabitants by regressing gall diameter and the number of *S. pallicornis* on the weights of *S. pallicornis*, *S. nervosus*, and *Eurytoma*. Regressing gall diameter and weights of *S. nervosus* or *Eurytoma* on the weight of *S. pallicornis* assessed the effect of competition on *S. pallicornis*. Significant regressions are detected by testing whether the slope is different from zero via a *t*-test.

Greater numbers of *S. pallicornis* had no effect on the weights of individual *S. pallicornis* ($t = 0.73$, $p > 0.05$). However, there was a significant negative effect on *Eurytoma* weights ($t = 4.12$, $p < 0.001$) and a significant positive effect on *S. nervosus* weight ($t = 2.25$, $p < 0.05$). The effect of the other inhabitants on *S. pallicornis* was very different. No effect of *S. nervosus* could be detected ($t = 0.35$, $p > 0.05$), but there was a negative effect of *Eurytoma* ($t = 2.5$, $p < 0.05$). This latter negative effect is presumably a competitive effect via the food supply, since *Eurytoma* was virtually never found as a direct parasitoid of *S. pallicornis* (although Askew (1961*a*) notes that it frequently is).

Discussion

There have been a number of recent reviews of the biology of gall systems, each from a slightly different perspective (Askew 1975, 1985; Abrahamson and Weis 1987; Weis *et al.* 1988; various papers in Shorthouse and Rohrfritsch 1992). Because of the general lack of long-term data, few of these reviews are able to assess the population biology of gall systems. However, more long-term studies are becoming available and will perhaps lead to new syntheses in the near future. Within populations, individual 'fitness' is determined through individual variation in life history parameters such as developmental rate, fecundity, and probability of survival. Population biology can be addressed through considering the life history consequences of oviposition site selection (Weiss *et al.* 1988). These consequences are particularly transparent in the case of gall systems because the gall is sessile.

The timing of oviposition and habitat selection during oviposition are key components of the life history of gall makers (Weis *et al.* 1988). We have no data on the consequences of variation in the timing of oviposition in *Cynips divisa*. In *C. divisa* oviposition occurs during May (Askew 1985) when oak leaves are expanding; presumably the precise timing of gall initiation relative to bud burst is important, if not critical, like many other gall systems (for example, Whitham 1980).

Whitham's (1980) elegant study found that increasing numbers of galls on a leaf affected various fitness components such that mean fitnesses per leaf were equalized whatever the gall density. The influences on fitness in *Cynips divisa* are best summarized by a path diagram (Fig. 20.3) of the inferred relationships between components, based on the results of several stepwise multiple regressions. We have not tested any particular path model since this would involve some circularity. The major components are the negative effects of gall density on leaves and on veins,

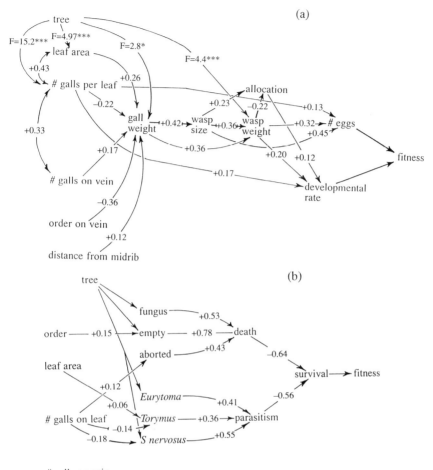

Fig. 20.3. Path diagram detailing influences upon components of fitness in *Cynips divisa*, as inferred from 1991 data. Double-headed arrows indicate correlated variables, single-headed arrows indicate inferred causal relationships. Figures are path coefficients (standardized partial regression coefficients) or correlation coefficients. Arrows drawn from 'Tree' indicate a significant variance component or chi-squared for the effect of different trees after all other influences have been taken into account. (a) Components of reproduction, (b) components of survival.

order effects, the positive influence of leaf area on fitness, and strong differences between trees. Which vein the female chooses has no effect.

As can be seen from the results for the only fitness component we can assess over several years (survival), the way in which these influences affect survival differs from year to year. The existence of this variability makes us cautious about interpreting too definitively the results of the analyses on other fitness components; the way in which gall order on a vein, for example, affects female weight might change from year to year. Why should there be this variability? We suggest that this is mainly a host plant factor and, in particular, the timing of oviposition relative to bud burst. We plan manipulative experiments to test this in future years.

Our results from a year when mortality was very high can be used to show the way in which other components of the food web interact. Askew (1961*a*) drew a very detailed food web for the gall of *Cynips divisa* on the basis of his extensive rearing programme involving more than 2000 galls (see Fig. 20.4a). We are unable to match his superbly detailed study, but we are able to suggest that there are some extra and subtle interactions between the web species (Fig. 20.4b). We base our conclusions on the analysis of larval weights. As would be expected on a simple resource availability argument, larval weights increase with an increasing gall size, evident in both species of *Synergus* and *Eurytoma*. One would expect therefore that where two individuals occur together in the same gall, there might be competition for resources. *Eurytoma* is a parasitoid, feeding directly on either the gall maker *Cynips* or on the inquiline *S. nervosus*; Askew (1961*a*) notes that it also can feed on plant tissue of the gall. The *Synergus* species are inquilines, feeding on gall tissue in the wall (*S. pallicornis*) or the inside (*S. nervosus*) and are unable to create the gall themselves, but may modify it (Askew, 1961*a*, 1985).

After removing the effect of gall diameter, the weight of *S. nervosus* **increases** when there are more *S. pallicornis* present in the walls, but there is no reciprocal effect. This positive facilitative influence on *S. nervosus* suggests that *S. pallicornis* may be able to induce the plant to produce more gall tissue. There is therefore here a positive feedback of density on resource availability (see Bianchi *et al.* 1989), as is the case with the gall makers proper. The form of the relationship implies that each additional *S. pallicornis* in the wall increases the mean weight of *S. nervosus* by 5 per cent. Thus, at least some *Synergus* species are not solely parasitic on the gall, but can induce the plant to provide them with food; they may merely be unable to initiate the process of gall formation. Presumably there is a spectrum of abilities, from true gall makers that need no assistance at all in creating a gall, to true inquilines that cannot play any part in inducing plant growth. Such a spectrum might perhaps be expected, since the inquilines appear to have evolved from gall makers (Askew 1985).

In contrast, the weight of *Eurytoma* decreases as the numbers of *S. pallicornis* rise, and the weights of *S. pallicornis* decrease as the weight of *Eurytoma* increases. We interpret this as supporting the idea that part of the food of *E. brunniventris* consists of plant material, for which it is in direct competition with *S. pallicornis*. Each additional individual *S. pallicornis* in

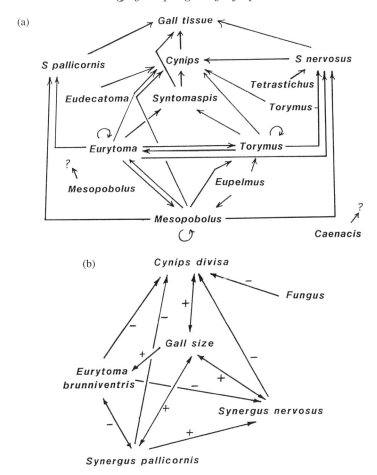

Fig. 20.4. (a) Food web based on *Cynips divisa* galls, redrawn from Askew (1961), (b) Food web of the commoner inhabitants of *Cynips divisa* galls. Signs show the nature of the interaction, either positive/facilitation (+) or negative/competition (−). Arrows point to a confirmed significant interaction; absence of an arrow means that no interaction has been detected.

the gall reduces the mean size of *Eurytoma* by 32 per cent, but a 10 per cent increase in the weight of *Eurytoma* decreases the mean weight of *S. pallicornis* by only 5 per cent. This interaction is therefore very potent and asymmetric.

How do these components combine to produce fluctuations in population size? In a widely cited paper, Washburn and Cornell (1981) interpreted their 3-year data from a site in New York State (USA) as implying that populations of the cynipid *Xanthoteras* were evanescent at one site, starting at high population density in one year, to be followed by declining populations to extinction 2 years later. They attributed this decline to the action of natural enemies, particularly the inquiline *Synergus*. A similar situation could pertain in Clumber Park. The entire park is a patchwork of young oaks of different genotypes and ages, and *Cynips* populations could shift patchily from year to year (cf. Huffaker 1958), the small-scale equivalent to the Rothamsted moth and aphid data (see Taylor 1986). However, unless the scale of this population patchiness is large, we do not feel that this is a correct interpretation since other trees in the vicinity were searched intensively and unsuccessfully for galls.

Alternatively, gall maker populations could really be responding to age-related long-term changes in host plant chemistry or other factors related to resistance or they could be undergoing population cycles that crash almost to extinction in particular years. The very high mortalities of 1986 and 1988 are clearly not unusual in *Cynips*, since Askew (1961*a*) also recorded a very low rate of survival (2–3 per cent) in *C. divisa* galls in Whytham Wood (Oxford) in 1958. Survival is improved at low gall densities on leaves and this suggests a possible mechanism for cycling. Although there are only 7 years of data, we suggest that there might well be a 6–7-year cycle of population density. In fact, many longer-term population studies on gall systems are consistent with long populations cycles. Washburn and Cornell (1981) could have been looking only at the decline phase of a 6–7-year cycle in the cynipid *Xanthoteras*; a 2-year or longer cycle may be pertinent for the cynipid *Dryocosmus* in Japan (Miyashita *et al.* 1965). The very interesting long-term data for the cynipid *Disholcaspis* (Frankie *et al.* 1992) on urban live oaks show great variation from tree to tree in any one year (for example, 0.6–88.6 per cent survival) and from year to year; again, the data suggest an overall long-term cycle. A cycle has been inferred also from a 9-year study of *Cynips quercusfolii* in The Netherlands (Wiebes-Rijks and Shorthouse 1992). In other non-cynipid gall systems the story is similar. In pemphigid aphids on *Pistacia* there is evidence for a 2- or 4-year cycle (Wool 1990); although the authors interpret population fluctuations in terms of rainfall, there is evidence for a 4-year cycle in *Pemphigus betae* on *Populus* in Utah (see Fig. 3 in Moran and Whitham 1988). Finally, a 12–14-year cycle is suggested by the data for populations of yew gall midge *Taxomyia* (Redfern and Cameron 1978; Cameron and Redfern 1978). We suggest that the interaction between plants, gall makers, and their parasitoids and inquilines will often generate long-term population cycles. Most authors (for example, Washburn and Cornell 1981; Frankie *et al.* 1992; Wiebes-Rijks and Shorthouse 1992) attribute changes in gall density to the action of natural enemies; while this appears to be similar to the situation in *Cynips divisa* in Clumber Park, experimental studies will probably be necessary to assess the relative impact of natural enemies as against host plant factors.

Acknowledgements

We thank Colin Hartley for advice and encouragement, the National Trust for permission to work in Clumber Park, and Nottingham University Insect Ecology Class of 1987 and Animal Ecology class of 1992 for help with some of the data collection. We thank also Drs Chris Leach, David Biggs, and Jerry Bowdrey, who were very helpful with information about pea galls elsewhere in the country.

References

Abrahamson W.G. and Weis, A.E. (1987). Nutritional ecology of arthropod gall makers. In *Nutritional ecology of insects, mites and spiders*, (ed. F. Slansky and G. Rodriguez), pp. 235–58. Wiley, New York.

Askew, R.R. (1960). On the Palaearctic species of *Syntomaspis* Forster (Hym., Chalcidoidea, Torymidae). *Entomologists' Monthly Magazine*, **96,** 184–91.

Askew, R.R. (1961*a*). On the biology of the inhabitants of oak galls of Cynipidae (Hymenoptera) in Britain. *Transactions of the Society for British Entomology,***14,** 237–68.

Askew, R.R. (1961*b*). A study of the biology of the species of *Mesopobolus* Westwood (Hymenoptera: Pteromalidae) associated with cynipid galls on oak. *Transactions of the Royal Entomological Society of London*, **113,** 155–76.

Askew, R.R. (1965). The biology of the British species of the genus *Torymus* Dalman (Hymenoptera: Torymidae) associated with galls of Cynipidae (Hymenoptera) on oak, with special reference to alternation of forms. *Transactions of the Society for British Entomology*, **16,** 215–32.

Askew, R.R. (1975). The organisation of chalcid-dominated parasitoid communities centred upon endophytic hosts. In *Evolutionary strategies of parasitic insects and mites*, (ed. P.W. Price), pp. 130–53. Plenum, New York.

Askew, R.R. (1985). The biology of gall wasps. In *The biology of gall insects*, (ed. T.N. Ananthakrishnan), pp. 223–71. Edward Arnold, London.

Bianchi, T.S., Jones, C.G. and Shachak, G. (1989). Positive feedback on consumer population density on resource supply. *Trends in Ecology and Evolution*, **4**(8), 234–8.

Bowdrey, J.P. (1987). Oak galls of Essex. *Essex Biological Record Centres Publication* **6.** Colchester Museum Service, Colchester.

Cameron, R.A.D. and Redfern, M. (1978). Population dynamics of two hymenopteran parasites of the yew gall midge *Taxomyia taxi* (Inchbald). *Ecological Entomology*, **3,** 265–72.

Cornell, H.V. (1990). Survivorship, life history, and concealment: a comparison of leaf miners and gall formers. *American Naturalist*, **136,** 581–97.

Crawley, M.J. (1985). Reduction of oak fecundity by low-density herbivore populations. *Nature*, **314,** 163–4.

Frankie, G.W., Morgan, D.L., and Grissell, E.E. (1992). Effects of urbanization on the distribution and abundance of the cynipid gall wasp, *Disholcaspis cinerosa*, on ornamental live oak in Texas, USA. In *Biology of insect-induced galls*, (ed. J.D. Shorthouse and O. Rohfritsch), pp. 258–79, Oxford University Press, Oxford.

Griffiths, R., Jones, L., and Leach, C.K. (1990). *An introduction to the study of plant galls in Leicestershire*. Leicester Polytechnic, Leicester.

Hartley, S.E. and Lawton J.H. (1992). Host-plant manipulation by gall-insects: a test of the nutrition hypothesis. *Journal of animal Ecology*, **61,** 113–19.

Hassell, M.P., Comins, H.N., and May, R.M. (1991). Spatial structure and chaos in insect population dynamics. *Nature*, **353,** 255–8.

Huffaker, C.B, (1958). Experimental studies on predation: dispersion factors and predator–prey oscillations. *Hilgardia*, **27,** 343–83.

Lawton, J.H. (1989). Food webs. In *Ecological concepts*, (ed. J.M. Cherrett), pp. 43–78. Blackwell, Oxford.

Matson, P.A. and Hunter, M.D. (1992). The relative contributions of top-down and bottom-up forces in population and community ecology. *Ecology*, **73,** 723–66.

Miyashita, K., Ito, Y., Nakamura, K., Nakamura, M., and Kondo, M. (1965). Population dynamics of the Chestnut Gall-wasp, *Dryocosmus kuriphilus* Yasumatsu (Hymenoptera; Cynipidae). III. Five year observation on population fluctuation. *Japanese Journal of Applied Entomology and Zoology*, **9,** 42–52.

Moran, N.A. and Whitham, T.G. (1988). Population fluctuations in complex life cycles: an example from *Pemphigus* aphids. *Ecology*, **69,** 1214–18.

Owen, J. and Gilbert, F.S. (1989). On the abundance of hoverflies. *Oikos*, **55,** 183–93.

Pimm, S.L. and Lawton, J.H. (1979). Are food webs divided into compartments? *Journal of Animal Ecology*, **49,** 879–98.

Price, P.W., Bouton, C.E., Gross, P., McPheron, B.A., Thompson, J.N., and Weiss, A.E. (1980). Interactions among three trophic levels: influence of plants on interactions between insect herbivores and natural enemies. *Annual Review of Ecology and Systematics*, **11,** 41–65.

Price, P.W., Fernandes, G.W., and Waring, G.L. (1987). The adaptive nature of insect galls. *Environmental Entomology*, **16,** 15–24.

Redfern, M. and Cameron, R.A.D. (1978). Population dynamics of the yew gall midge *Taxomyia taxi* (Inchbald) (Diptera: Cecidomyiidae). *Ecological Entomology*, **3,** 251–63.

Shorthouse, J.D., and Rohfritsch, O. (1992). *Biology of insect-induced galls*. Oxford University Press, Oxford.

Sitch, T.A., Grewcock, D.A., and Gilbert, F.S., (1988). Factors affecting components of fitness in a gall making wasp (*Cynips divisa* Hartig). *Oecologia*, **76,** 371 –5.

Taylor, L.R. (1986). Synoptic dynamics, migration and the Rothamsted insect survey. *Journal of Animal Ecology*, **55,** 1–38.

Tscharntke, T. (1992). Coexistence, tritrophic interactions and density dependence in a species-rich parasitoid community. *Journal of Animal Ecology*, **61,** 59–67.

Washburn, J.O., and Cornell, H.V. (1981). Parasitoids, patches and phenology:

their probable role in the local extinction of a cynipid gall wasp population. *Ecology*, **62**, 1597–607.

Weis, A.E., Walton, R., and Crego, C.L. (1988). Reactive plant tissue sites and the population biology of gall makers. *Annual Review of Entomology*, **33**, 467–86.

Whitham, T.G. (1980). The theory of habitat selection examined and extended using *Pemphigus* aphids. *American Naturalist*, **115**, 449–66.

Wiebes-Rijks, A.A., and Shorthouse, J.D. (1992). Ecological relationships of insects inhabiting cynipid galls. In *Biology of insect-induced galls*, (ed. J.D. Shorthouse and O. Rohfritsch), pp. 238–57, Oxford University Press, Oxford.

Wolda, H. (1983). 'Long-term' stability of tropical insect populations. *Research in Population Ecology* (suppl.), **3**, 112–26.

Wool, D. (1990). Regular alternation of high and low population size of gall-forming aphids: analysis of ten years of data. *Oikos*, **57**, 73–9.

Zwölfer, H. (1987). Species richness, species packing, and evolution in insect–plant systems. *Ecological Studies*, **61**, 301–19 (ed. E.D. Schulze and H. Zwölfer). Springer-Verlag, Berlin.

21.

The biogeography and population genetics of the invading gall wasp *Andricus quercuscalicis* (Hymenoptera: Cynipidae)

PAUL J. SUNNUCKS*, G.N. STONE†,
K. SCHÖNROGGE†, and G. CSÓKA‡

Institute of Zoology, Regent's Park, London, UK † Imperial College at Silwood Park, Ascot, Berkshire, UK ‡ Department of Forest Protection, Forest Research Institute, P.O. Box 49, Gödöllő, 2100 Hungary

Abstract

Andricus quercuscalicis is one of a group of five European cynipid gall wasps which have a life cycle involving alternation of generations between English oak, *Quercus robur* and Turkey oak, *Q. cerris*. The natural range of this species is limited to areas where the two oaks occur together, in Europe south of the Alps, Tatras, and Carpathians and east of the Carpathians in a small region of the Ukraine. While *Q. robur* is widespread and abundant in much of Europe north and west of these regions, *Q. cerris* is only found where it has been planted by man in parks and gardens. Although *Q. cerris* can self-seed in much of its new range, over the time scale involved (200–300 years) this species has probably yet to spread significantly away from introduction sites.

The anthropogenic distribution of *Q. cerris* has generated a patchy distribution of areas containing both oak species. *Andricus quercuscalicis* and other cynipids requiring both oak species have spread northwards and westwards as far as Ireland and Denmark. Our prediction has been that the widely-spaced and patchy distribution of *Q. cerris* should result in new patches being colonized by only a small fraction of source populations in the gall wasp's native range. As *A. quercuscalicis* has continued to spread northwards and westwards, this process should have been repeated many times. Each founding event is assumed to be associated with a genetic bottleneck, a series of which should result in a dramatic loss of genetic variability along the invasion pathway. A simple model of this invasion mechanism with supporting evidence from analyses of allozyme variation from almost 1000 individuals in over 40 populations from sample sites between Hungary

Plant Galls (ed. Michèle A. J. Williams), Systematics Association Special Volume No. 49, pp. 351–68. Clarendon Press, Oxford, 1994. © The Systematics Association, 1994.

and the British Isles is presented. Hypotheses resulting from this model of the invasion process, including investigation of genetic variability in other cynipid species with the same pattern of host alternation and genetic patterns associated with the eastward spread of *A. quercuscalicis* through the Commonwealth of Independent States are discussed. The use of DNA techniques, including multilocus fingerprinting (and the use of single locus probes) are reviewed as tools in finer resolution analyses of the colonization process.

Introduction

The distributions of many organisms are determined neither by climate nor by physiological limitation, but by other factors such as natural barriers to dispersal or mortality inflicted by enemies (Lawton 1986). When dispersed artificially beyond such limitations, an organism may be capable of rapid range expansion. A fundamental limiting factor for gall formers is the distribution of host plants; changes in geographic range of hosts can have profound effects on the geographic range of the animals. Cynipid gallwasps (Hymenoptera: Cynipidae) have at least three dispersal mechanisms. Most species can fly and so disperse themselves. Cynipids may also be dispersed as galls on plants—the probable cause of invasion of North America by *Diplolepis mayri* (Schlectendal) and *Diplolepis rosae* (L.) (Ritchie and Peters 1981) and of Japan and America by *Dryocosmus kurriphilus* (Payne 1978; Moriya *et al.* 1989). A third mechanism is through human trade in cynipid galls (Larew 1987).

Perhaps the best known current cynipid invader is *Andricus quercuscalicis* (Burgsdorf), one of a group of four species which have spread from southeastern Europe to reach Britain in the last 200 years (Stone and Sunnucks 1993). These invasions were made possible by human dispersal of an obligate host plant, the Turkey oak, (*Quercus cerris* L.). This chapter concerns the biogeography and population genetics of *A. quercuscalicis*. First, we discuss historical oak distributions in Europe and biogeographical patterns which made the invasions possible. We ask which factors have been important in the invasion process, and explore how existing population genetic models may be useful in understanding the spread of this gall wasp and other species. Finally, we consider other cynipid invasions which may follow human dispersal of oaks.

The palaeobotany of oaks in Europe

In Europe there are seven gall wasp species with a life cycle involving oaks in the two taxonomic sections *Cerris* and *Quercus*—six in the genus *Andricus* (*A. kollari* (Hartig), *A. quercuscalicis*, *A. lignicola* (Hartig), *A. corruptrix*

(Schlectendal), *A. gemmea* Giraud, and *A. tinctoriusnostrus* Stefan) and one in the genus *Fiorella*. For all these species, a sexual generation develops in galls only on *Q. cerris* and an asexual (agamic) generation develops on oaks in the section *Quercus*, particularly *Quercus petraea* (Mattuschka) Liebl. and *Quercus robur* L. *Andricus quercuscalicis* has an asexual generation which develops on the acorns of the English oak (*Q. robur*) in the summer and autumn, and an alternating sexual generation which develops in spring on the male flowers of the Turkey oak (*Q. cerris*). *Andricus quercuscalicis* is unusual within the group in that the agamic generation can develop only on *Q. robur*. All of these cynipid species can only survive where the necessary species in both oak taxa exist.

The current geographic overlap between *Q. cerris* and *Q. robur* is a product of both palaeoclimatic patterns and human intervention. Fossils show that the ancestors of both oak species occurred together in northern Europe as long as 20 million years ago (mya). Fossil members of the genus *Andricus* are also known from 20 mya (Larew 1987) and cynipids with life cycles involving alternation between these two oak groups may therefore have existed for this long. During each of at least nine Pleistocene ice ages (1 million–18 000 years ago) oaks retreated to southern refugia. In the last ice age there were three such areas: the Iberian peninsula, Italy, and the Balkans (Huntley and Birks 1983; Bennet 1986; Roberts 1989). While *Q. robur* and *Q. petraea* apparently survived in all three refuges, *Q. cerris* persisted only in the Balkan refuge and perhaps also the Italian one. After the retreat of the last ice sheets (13 000–10 000 years ago) the warming effect of the sea in the west and the persistence of extremely cold conditions in the Alps, Tartras, and Carpathians meant that the advance north from the Iberian refuge was far more rapid than expansion from the other refugia (Huntley and Birks 1983). *Quercus robur* and *A. petraea* advanced rapidly and are now found throughout Europe, including the southeast (Fig. 21.1a). *Quercus cerris* remained restricted to southern refuges and before human intervention began, *Q. cerris* was native only to areas south and east of the Alps, Tatras, and Carpathian ranges, extending eastwards into Turkey (Fig. 21.1b). It is assumed that *Q. cerris*-associated fauna would have been similarly restricted. *Andricus quercuscalicis* is now found throughout the overlapping native ranges of the two oaks (Fig. 21.1). From these regions expansion to the west was limited by the absence of *Q. cerris* (Jalas and Suominen 1987) and to the east through Asia Minor by rarity of *Q. robur*.

Human dispersal of *Quercus cerris*

In the last 300–400 years *Q. cerris* has been planted extensively outside its native range, creating a highly clumped and patchy distribution across

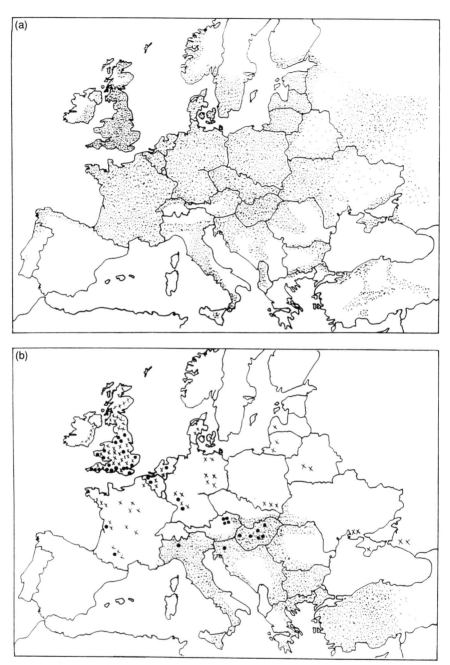

Fig. 21.1. (a) The current distribution of *Quercus robur* in Europe (after Jalas and Suominen 1987) (b) The native (close hatching (after Jalas and Suominen 1987) and introduced (marked with xs) range of *Quercus cerris*. It is assumed that the native range of *A. quercuscalicis* followed that of *Q. cerris*. The filled circles indicate sample sites for populations used in the allozyme investigation.

northwestern Europe. *Quercus cerris* has reached Ireland in the west, Scotland and Denmark in the north, and Yalta, Georgia, in the east. Thus, gall wasps requiring this oak have had the potential to expand their ranges. *Andricus quercuscalicis* reached eastern Germany as early as 1631 (Gauss 1977) and The Netherlands by 1882 (Beijerinck 1897), but the English Channel apparently prevented invasion by *A. quercuscalicis* for over a century. Although Turkey Oak was introduced to Britain in 1735 (Askew 1984), *A. quercuscalicis* only reached Britain in the 1950s and has since spread rapidly throughout England and Wales (Hails and Crawley 1991). Three other gall wasps in the genus *Andricus* with the same pattern of host alternation have also invaded Britain. *Andricus kollari* is thought to have been intentionally introduced to Britain in the nineteenth century, while *A. lignicola* and *A. corruptrix* have recently become established in Britain without known introductions (Askew 1984).

Population genetic factors in the invasion of *Andricus quercuscalicis*

As *Q. robur* is widespread throughout the invaded range of *A. quercuscalicis*, the pattern of spread of this gall wasp has probably been determined by availability of *Q. cerris* and the dispersal abilities of the agamic generation. Invasion by *A. quercuscalicis* has depended on colonization and survival in widely spaced patches of suitable habitat (Carter and Prince 1981). This is likely to have resulted in population subdivision and the generation of regional genetic differentiation (McCauley 1991). Analysis of these genetic differences may be used to determine the phylogenetic relationships between the subpopulations and, thus, the pattern of the invasion.

Genetic variation is quantified by determining the occurrence and frequencies of different forms (alleles) of the same genes (loci) in different populations. Observed differences in allele frequencies in subpopulations will depend on a number of factors including

(1) number of founders of a population;

(2) opportunity for genetic drift—stochastic loss of alleles in small populations (Hartl 1980);

(3) species-specific levels of variability. If the species has little genetic variability, the changes occurring during colonization may contain too little information to assess phylogeographic patterns.

Population differentiation as a result of the invasion process depends on three factors outlined by McCauley (1991).

1. *The number of founders and population size.* Where the number of

founders of a population is small, the colony may be said to have passed through a population and genetic *bottleneck*. If the number of colonists is small and their genes are a randomly drawn subset of the source gene pool, initial allele frequencies in the colony may be markedly different from those in the source population due to stochastic processes. Alleles rare in the source population are likely to be absent in the founders and the colony. Alternatively, rare alleles which occur by chance in the founders may be established at unusually high frequencies. Thus small founder number will tend to greatly increase the variance in allele frequencies between colonies. If colonies remain small, rare alleles may be lost with high probability through genetic drift.

2. *Origin of colonists.* If uninhabited patches of *Q. cerris* are relatively distant from existing populations of *A. quercuscalicis*, only agamic females from the nearest population will be able to reach them (a 'stepping stone' model; Maynard Smith 1989). Genetic variability in the new colony can then only be a subset of the variation in the source population and variation may decline steeply along a line of invasion. In contrast, if the uninhabited *Q. cerris* patch is close to many inhabited patches, founding individuals of the colony can come from many source populations (an 'island' model; Maynard Smith 1989) and may then have more variation than each contributing source population individually (McCauley 1991).

3. *Subsequent rate of migrant exchange between the colony and other populations.* Even very low rates of immigration reduce genetic differences between populations (Maynard Smith 1989). Marked genetic subdivision thus suggests very low rates of gene exchange between populations. The few data on cynipid dispersal abilities suggest that numbers reaching hosts from a source decrease sharply with distance (Docters van Leuwen 1959; Notton 1990). This pattern has been seen in other migrating insects (for example, Florence *et al.* 1982; Harrison 1989) and we assume that the same applies for asexual *A. quercuscalicis*. As *A. quercuscalicis* disperses from the endemic range of *Q. cerris* and the distribution of the tree becomes patchier, colonies are more likely to be founded by fewer individuals from fewer neighbouring colonies (stepping stone model). Although the *potential* rate of population increase of *A. quercuscalicis* is very high (Hails and Crawley 1991), this may not be fully realized (Hails 1988) and genetic drift may affect colonies during the early phase of population growth. Our prediction is that colonies should become increasingly different from the original source population with distance from it.

Finally, there may be genetic intrapopulation variation in dispersal ability. In at least one insect, southern pine beetle, *Dendroctonus frontalis* Zimm. (Florence *et al.* 1982), some genotypes predictably spread further than others and the genotypes founding populations at different distances from a source are not random. Insects can also respond rapidly to selection for dispersal ability (Dingle 1985). The possibility that selection for dispersal ability may have occurred must be considered in interpreting genetic patterns in *A. quercuscalicis*.

The biochemical population genetics of *Andricus quercuscalicis*

Methods

Allozyme studies can be rapid, inexpensive, and very informative and have been used for over two decades to investigate population processes in invertebrates (Pamilo *et al.* 1978; Hebert *et al.* 1991). Cellulose acetate electrophoresis requires small amounts of material so is suitable for study of even small insects. Even so, individuals of the agamic generation of *A. quercuscalicis* weigh approximately 10 times as much as the sexual generation, so agamic females were used in genetic investigations. Samples were prepared and gels run and stained following standard methods (Harris and Hopkinson 1976; Richardson *et al.* 1986: abbreviations used here follow the latter). Population genetic analyses were carried out using BIOSYS-1 (Swofford and Selander 1981). Phylogenetic trees (phylograms) were constructed using six common measures of genetic distance (Nei's Minimum distance, Nei's Genetic distance, Rogers' Genetic distance, Modified Rogers' Genetic distance, Edwards' E distance and Cavalli-Sforza and Edwards' Arc distance) and two common tree-building procedures (UPGMA and Wagner procedure). These phylograms show mean genetic similarity between populations and are generated using differences in gene frequencies.

2. Levels of variation observed

Investigations into more than 30 protein loci identified 13 which could be reliably scored from each small individual. Electrophoretic variation in these 13 loci was analysed in 823 gall wasps from 39 populations along possible invasion pathways across Europe (Fig. 21.1b). Eight loci were polymorphic (frequency of the commonest allele < 0.99 in a given population), with up to five alleles at a locus, totalling 24 alleles (Table 21.1). The allozyme variation (mean expected heterozygosity) detected in *A. quercuscalicis* was considerably higher than that reported for even the most variable Hymenoptera (Graur 1985). In subsequent analyses, the genetic markers used in this study are assumed to be unaffected by geographic change in selection pressure. This assumption is supported by the following evidence:

Table 21.1 Populations sampled, mean sample size per locus per site, mean number of alleles per locus (all loci), total number of alleles per locus (polymorphic loci), mean observed heterozygosity, and gene frequencies of six alleles lost in western Europe

Population	Mean values Sample /locus	Alleles /locus /total	Heterozygosity (SE)	Allele frequencies MDHs A	PGM A	GOTs C	GPD2 A	PEPb A	GOTm C
SW England									
Franchis	14.9	1.2 (11)	0.080 (0.049)	0.00	0.07	0.00	0.00	0.00	0.00
Lostwithiel	14.9	1.2 (10)	0.070 (0.052)	0.00	0.00	0.00	0.00	0.00	0.00
Plymouth	11.9	1.2 (11)	0.065 (0.039)	0.00	0.14	0.00	0.00	0.00	0.00
B Tracey	11.8	1.2 (11)	0.052 (0.040)	0.00	0.00	0.00	0.00	0.17	0.00
Taunton	12.0	1.4 (14)	0.077 (0.042)	0.00	0.13	0.00	0.00	0.04	0.04
Winterbourne	12.5	1.3 (12)	0.105 (0.059)	0.00	0.13	0.00	0.00	0.12	0.00
New Forest	22.0	1.3 (12)	0.101 (0.060)	0.00	0.09	0.00	0.00	0.09	0.00
Wales									
St Clears	12.8	1.2 (11)	0.074 (0.047)	0.00	0.04	0.00	0.00	0.00	0.00
Orielton	8.0	1.2 (11)	0.106 (0.060)	0.00	0.44	0.00	0.00	0.00	0.00
Aberystwyth	18.0	1.3 (12)	0.111 (0.062)	0.00	0.22	0.00	0.00	0.25	0.00
Central, SE, and N England									
Darlington	12.0	1.4 (13)	0.103 (0.047)	0.00	0.46	0.00	0.00	0.65	0.08
E Anglia	23.9	1.4 (13)	0.094 (0.047)	0.00	0.51	0.00	0.00	0.28	0.17
Ascot	12.0	1.3 (12)	0.122 (0.066)	0.00	0.33	0.00	0.00	0.21	0.00
Mere Sands	11.9	1.4 (13)	0.145 (0.072)	0.00	0.63	0.00	0.00	0.54	0.04
Sherwood Forest	12.0	1.4 (13)	0.103 (0.050)	0.00	0.46	0.00	0.00	0.33	0.04
Wkye	13.8	1.3 (12)	0.101 (0.058)	0.00	0.29	0.00	0.00	0.54	0.00
Caton	13.3	1.4 (13)	0.086 (0.050)	0.00	0.29	0.00	0.00	0.38	0.08
London 1	79.3	1.4 (13)	0.152 (0.073)	0.00	0.50	0.00	0.00	0.34	0.09
London 2	52.7	1.4 (13)	0.132 (0.063)	0.00	0.40	0.00	0.00	0.36	0.10
Hinks Hill	23.9	1.4 (13)	0.146 (0.072)	0.00	0.48	0.00	0.00	0.28	0.06
Danbury	24.0	1.4 (13)	0.103 (0.050)	0.00	0.46	0.00	0.00	0.46	0.10

Belgium									
Bruges	30.0	1.4 (13)	0.092 (0.045)	0.23	0.98	0.00	0.00	0.00	0.03
Tielt	13.0	1.3 (12)	0.124 (0.064)	0.35	0.89	0.00	0.00	0.00	0.00
Holland									
Wageningen	20.5	1.5 (14)	0.122 (0.049)	0.24	0.83	0.00	0.00	0.32	0.18
France									
C De Loire	21.7	1.5 (14)	0.158 (0.066)	0.16	0.57	0.00	0.00	0.10	0.07
Agen	8.0	1.4 (13)	0.125 (0.062)	0.69	0.56	0.19	0.00	0.00	0.00
Germany									
Ludwigsberg	31.9	1.7 (17)	0.184 (0.065)	0.23	0.27	0.28	0.39	0.31	0.11
Worms	18.9	1.5 (16)	0.123 (0.047)	0.16	0.66	0.63	0.53	0.26	0.00
Italy									
Lago Maggiore	8.0	1.5 (14)	0.163 (0.075)	0.13	0.38	0.38	0.19	0.00	0.00
Austria									
Rottenbach	27.0	1.8 (17)	0.234 (0.076)	0.33	0.37	0.44	0.22	0.15	0.20
Rosenau	19.9	1.8 (17)	0.217 (0.071)	0.30	0.40	0.30	0.17	0.18	0.18
Kirchberg	19.7	1.7 (16)	0.188 (0.066)	0.20	0.35	0.53	0.08	0.18	0.20
Weitra	19.5	1.6 (15)	0.240 (0.083)	0.38	0.33	0.35	0.15	0.38	0.17
Slovenia									
Ljubyana	50.7	1.7 (17)	0.198 (0.068)	0.33	0.32	0.24	0.08	0.28	0.24
Hungary									
Izsakfa	51.0	1.9 (20)	0.213 (0.069)	0.35	0.36	0.28	0.16	0.27	0.16
Miskolc	11.0	1.7 (17)	0.245 (0.083)	0.32	0.27	0.36	0.09	0.18	0.18
Szambathely	11.0	1.7 (17)	0.189 (0.065)	0.27	0.17	0.46	0.18	0.18	0.32
Szarvas	6.8	1.7 (17)	0.202 (0.077)	0.29	0.21	0.29	0.10	0.14	0.14
Tiszakurt	13.9	1.8 (19)	0.207 (0.079)	0.29	0.25	0.39	0.07	0.15	0.11

1) allele frequencies recorded were consistent with Hardy Weinberg equilibrium, with no evidence for selection either of heterozygotes or homozygotes;

2) there was no geographic pattern in Wright's F_{is} statistic, deviation from expection of which may indicate selection;

3) although there were alleles with a geographic trend in frequency (for example GOT-S C), this was associated with a general loss of variation.

Thus, to explain this by selection, there would have to be homozygous advantage at a number of loci with diverse function.

3. Process and pattern of variation

The highest genetic variability was found in Hungary where 17–20 of the 24 known alleles at polymorphic loci were represented in each population, followed closely by Slovenia and Austria (15–17 alleles). Several rare alleles at GOT-S, GPI, and PGM were detected only in Hungary and Austria. Alleles were lost from populations westwards with only 10 alleles remaining in southwest Britain, for example, MDH-S A allele, GOT-S C, and αGPD2 A were present in central and eastern Europe, but absent from most of coastal Europe and all of Britain (Table 21.1). No new alleles were found outside the native range of *Q. cerris* and populations generally contain a subset of alleles present further east. The lowest genetic diversity was detected in the southwest of Britain, with progressive loss of PGM A, PEP-B A, and GOT-M B (Table 21.1).

When analysing the association between genetic measures and geographic location it is necessary to identify a point of reference against which comparisons can be made; we have taken Miskolc in Hungary, being in the native range of *Q. cerris* and the site with the highest observed mean heterozygosity (the probability that any locus in any individual is heterozygous). Observed mean heterozygosity and mean number of alleles per locus both declined steeply with distance from Miskolc (Fig. 21.2). In addition, the six common measures of genetic distance all increase sharply with physical distance of populations from Hungary, indicating increase in genetic difference. The most likely explanation for the observed pattern of variation is that gall wasps colonized western Europe and Britain from central Europe, variation being lost through founding events involving small numbers of colonizers and perhaps later through genetic drift. All general measures of genetic variation (total number of alleles, mean number of alleles per locus, mean heterozygosity per locus) declined in an approximately linear fashion with distance from

(a)

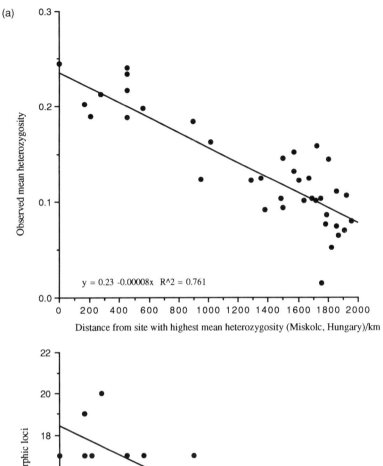

Fig. 21.2. Mean observed heterozygosity (a) and the total number of alleles at polymorphic loci per population (b) as a function of distance from Miskolc.

Miskolc, suggesting that there have been no abrupt discontinuities in invasion by *A. quercuscalicis*.

Genetic differentiation between populations allows examination of likely patterns of migration between them. Phylogenetic relationships between populations were computed using the methods and measures of distance given above. Figure 21.3 shows a Wagner network based on Modified Rogers distance (Wright 1978), rooted at Miskolc. The pattern of nodes in the tree agrees well with the geographic distribution of sample sites. Trees generated by both UPGMA and distance Wagner analyses of various measures of genetic distance were broadly similar, with certain groupings being produced by all methods. This consistency suggests that Figure 21.3 can be regarded as a reasonable summary of the relationships between sites, a fact also borne out by high cophenetic correlation (statistic of tree stability).

Sites in central and eastern Europe share three alleles absent from Britain and coastal Europe (see above). It is extremely unlikely that had Britain been colonized from these areas, all three alleles would be lost. Coastal European sites have only one allele, MDH-S *A*, which is absent from Britain and are therefore more probable source populations. This is supported by the fact that Modified Rogers' genetic distance between central France and three widely spaced British coastal sites was 0.057–0.065, compared with 0.097–0.116 between the same British sites and Miskolc. Within Britain there was a clear division between two clusters of sites. Sites in south-eastern, eastern and northern England form one group, while sites in southern, central and western England and Wales form a second. It has been suggested previously that *A. quercuscalicis* arrived in Britain from the Channel Islands, invading Cornwall and Devon first (Hayhow 1983). The allozyme data suggest that this is unlikely as these areas lack an allele (PGM *A*) found in all other British populations and so could not have given rise to them unless they lost the allele later. It is probable that Britain has been invaded at least twice, once in the southwest and once in the east or perhaps the first invaders came into eastern Britain and lost variation on the way into the southwest.

4. Number of colonists of Britain

The probability of chance loss of the MDH-S *A* allele (absent in 418 British gall wasps) can be estimated as a function of the number of colonists arriving in Britain. The frequency of the MDH-*A* allele in the rest of Europe is approximately 0.30, with a minimum of 0.16 in western France. Even assuming that British alleles were drawn at random from this lowest detected frequency (and that no genetic drift occurred after colonization) nine founders would carry at least one copy of the allele with a probability of 0.95, and 13 founders with a probability of more

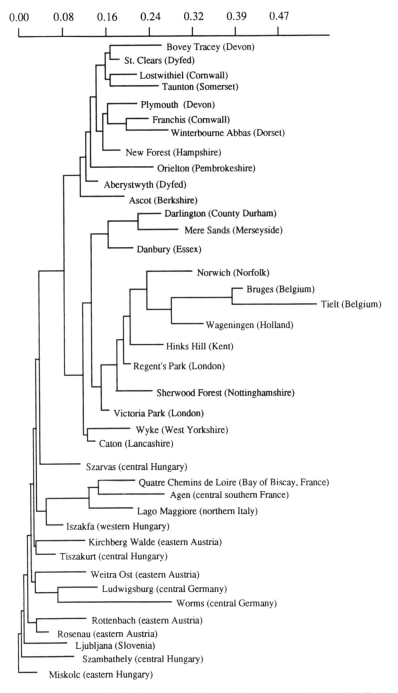

Fig. 21.3. A phylogram of *Andricus quercuscalicis* populations generated using the distance Wagner procedure on modified Rogers distance data.

than 0.99. This points to very low (<15) numbers of colonists reaching Britain.

5. Patchiness and geographic barriers to gene flow in A. quercuscalicis

Andricus quercuscalicis has been known in The Netherlands since the 1880s but in Britain only since the 1950s, so the English Channel may have acted as a substantial barrier to migration. However, genetic variation in French, Belgian, and Dutch sites is very low compared to Germany and central Europe. This suggests that there are considerable terrestrial and/or habitat barriers to the spread of the insect in western Europe and Britain. In the present study, values of Wright's statistic of genetic subdivision (F_{st}) were relatively high over the whole sample (mean over all loci $= 0.18$). An inverse relationship between F_{st} and dispersal ability of a species has been demonstrated theoretically and is suggested by empirical studies (Hebert *et al.* 1991). The result from the whole range disguises differences within the sample: F_{st} was recalculated with data just from within the native range of *Q. cerris* (mean $= 0.025$) and with data from an approximately equal land area in Britain (mean $= 0.158$). Comparison of these values with those in Hebert *et al.* 1991) suggests that *A. quercuscalicis* in its native range is as subdivided as an aphid 'known for its migration ability', whilst within Britain it is equivalent to a species of 'restricted dispersal capabilities'. This could be due to the patchiness of the environment in Britain, but also invading populations may have had too little time to exchange migrants.

Very high variance in allele frequencies (for example, frequency of PGM *A* in the Low Countries was 0.83–0.98 compared to 0.20–0.30 at most other sites; the German sites had *a* GPD2 *A* frequencies of 0.39 and 0.53, compared to 0.10–0.20 in most of their neighbours) suggests that small numbers of individuals often found populations and that invaded areas tend to remain isolated for considerable periods. Although there are inequalities in allele distributions in the natural range of *Q. cerris* (some sites in Hungary had alleles absent which reached frequencies of 0.07 elsewhere in the country (GOT-S *D*, PGM *C*) and there were large ranges even of common alleles (for example GOT-S *A*, 0.09–0.36)) these differences are small compared to those in the invaded range.

6. Future work

The results described suggest that application of molecular techniques with greater resolution may help to understand the invasion and dispersal of *A. quercuscalicis*. We have already produced minisatellite DNA 'fingerprints' from individual *A. quercuscalicis* and *A. lignicola* and preliminary comparisons between populations of the former are encouraging. DNA techniques have been applied to other Hymenoptera and a search is underway for suitable genetic markers. We wish to address the following specific predictions.

1. If the decline in variability in *A. quercuscalicis* really is imposed by the distribution of *Q. cerris*, then we should expect to find similar patterns in other gall wasps with the same life cycle. We will test this in *A. corruptrix* and *A. lignicola*.

2. We expect to find no such decline for species over the same range which attack only *Q. robur* and *Q. petraea*. These species should not have experienced population bottlenecks imposed by the distribution of their host trees. We will collect two suitable species, *A. fecundator* (Hartig) and *A. inflator* Hartig, to test this prediction.

3. We expect narrower bottlenecks in species which attack only *Q. cerris*. These species need to disperse the entire distance between *Q. cerris* patches in order to colonize, so populations should arise from smaller numbers of founders than those of host-alternating species. Bottlenecks in these species may also be more extreme because these cynipids do not have host alternation and should not have faced such strong selection for dispersal ability in their endemic range. We intend to work in future on the only known invading obligate *Q. cerris* feeders, *A. grossulariae* Giraud and *Chilaspis nitida* Giraud.

The future of cynipid invasions in Europe

In *A. quercuscalicis* and cynipids with a similar pattern of host alternation, both generations must have evolved the capacity to disperse to their required host. Flightless or short-winged generations are found only in cynipids in which both generations are able to exploit the same oak (Askew 1984). If this generalization is valid, it is surprising that *A. gemmea* and *A. tinctoriusnostrus*, two cynipids with host alternation between *A. cerris* and oaks in the section *Quercus*, have yet to increase their geographic ranges westwards beyond Germany. However, this observation should be qualified in that both species are uncommon even in their endemic ranges and their galls are inconspicuous (Ambrus 1974).

Quercus cerris is able to self-seed as far north as England and may eventually invade much of northern Europe, with major consequences for the cynipid fauna. At least 20 cynipid species attack *Q. cerris* only (Buhr 1965; Ambrus 1974). We know of range extensions only for four species; *Neuroterus macropterus* (Hartig) had reached Halle in eastern Germany by 1962, *A. grossulariae* has currently spread at least as far as Frankfurt, and *A. aestivalis* Giraud and *C. nitida* are also known from

Germany (Buhr 1965). As *Q. cerris* and other oaks become more wide-spread and less patchy, many of these species may become invaders.

In addition to the cases of *Q. cerris* feeders, there are many cynipids feeding on other oaks present in northern and western Europe which are restricted to southern Europe. Are these limited in their range by climate and natural enemies or by inability to disperse over the same geographic barriers which constrained *Q. cerris*? Only one southern gall wasp species which feeds on *Q. petraea* and *Q. robur* has arrived in Britain in recent years (the hedgehog gall wasp, *A. lucidus*; Hartig. Stone and Sunnucks 1992). We need detailed knowledge of the dispersal patterns of natural cynipid populations and of the distribution of their hosts in an environment dominated by human intervention before the biogeography of gall wasps in Europe can be understood.

Acknowledgements

The work at Silwood Park is funded by the Department of the Environment and the NERC. We thank Dr R.K. Wayne of the Institute of Zoology, Regent's Park for his support in all aspects of the genetic work and Susan Haines for technical assistance. We thank Dr M.J. Crawley, Dr R.S. Hails, and Professor J.H. Lawton for their support at Silwood Park.

This study would not have been possible without the help of many individuals throughout our study area. We thank everyone who has provided us with biogeographical information or gall samples and the following in particular. Dr C. Nelson, C.P. Kelly, Dr Mahmut Erolglu, Dr E. Altenhofer, Dr Werner Heitland, Dr Burgis, and Professor H. Pschorn-Walcher; the Directors and staff of the botanical gardens in Rostock, Wörlitz, Berlin, Frankfürt, Munich, and Stuttgart. The British Plant Gall Society, and, in particular, Howard Price, Arthur Chater, and Dr Anthony Biggs; The BRISC campaign. We thank Dr R.R. Askew and Dr M. Shaw for their taxonomic help.

References

Ambrus, B. (1974). Cynipida-gusbacsok-cecidia cyniparum. In *Fauna Hungariae*, Vol. 12, (Hymenopters 2), part 1/a (serial number 116). Academic Press, Budapest.

Askew, R.R. (1984). The biology of gall wasps. In *The biology of gall insects*, ed. T.N. Ananthakrishnan pp. 223–71), Oxford and IBH Publishing Co, New Delhi.

Beijerinck, M.W. (1897). Sur la cécidiogenese et la génération alternante chez le *Cynips calicis*. Observations sur la galle de l'*Andricus circulans*. *Archives Néerlandes des Sciences Exactes et naturelles*, **30,** 387.

Bennet, K.D. (1986). The rate of spread and population increase of forest trees during the postglacial. *Philosophical Transactions of the Royal Society of London, Series B*, **314,** 523–31.

Buhr, H. (1965). *Bestimmungstabellen der Gallen (Zoo und Phytocecidien) an Pflanzen Mittel und Noreeuropas*, Vol. 2, pp. 763–1572. V.E.B. Gustav Fischer Verlag, Jena, Germany.

Carter, R.N. and Prince, S.D. (1981). Epidemic models used to explain biogeographic distribution limits. *Nature*, **293,** 644–5.

Dingle, H. (1985). Migration and life histories. In *Migration: mechanisms and adaptive significance*, (ed. M.A. Rankin), pp. 27–42. University of Texas Press, Port Aransas.

Docters van Leuwen, W.M. (1959). Generatiewisseling en wisseling van waardplant bij vier Nederlandse galwespen. *De Levende Natuur*, **62,** 149–61.

Florence, L.Z., Johnson, P.C., and Coster, J.E. (1982). Behavioural and genetic diversity during dispersal: analysis of a polymorphic esterase locus in southern prine beetle, *Dendroctonus frontalis*. *Environmental Entomology*, **11,** 1014–18.

Gauss, R. von (1977). Zur Massenvermehrung der Knopperngallwespe *Andricus quercuscalicis Burgsd. im Jahre 1974 im Forstamt Stuttgart. Zeitschrift für angewandte Entomologie*, **82,** 277–84.

Graur, D. (1985). Gene diversity in Hymenoptera. *Evolution*, **39,** 190–9.

Hails, R.S. (1988). The ecology of *Andricus quercuscalicis* and its natural enemies. PhD thesis, University of London.

Hails, R.S. and Crawley, M.J. (1991). The population dynamics of an alien insect: *Andricus quercuscalicis* (Hymenoptera: Cynipidae). *Journal of Animal Ecology*, **60,** 545–62.

Harris, H. and Hopkinson, D.A. (1976). *Handbook of enzyme electrophoresis in human genetics*. Elsevier, New York.

Harrison, S. (1989). Long distance dispersal and colonization in the Bay Checkerspot butterfly, *Euphydryas editha bayensis*. *Ecology*, **70,** 1236–43.

Hartl, D.L. (1980). *Principles of population genetics*. Sinauer Associates, Sunderland, MA.

Hayhow, S.G. (1983). The knopper gall in Yorkshire. *The Sorby Record*, **21,** 79–81.

Hebert, P.D.N., Finston, T.L., and Foottit, R. (1991). Patterns of genetic diversity in the sumac gall aphid, *Melaphis rhois*. *Genome*, **34,** 747–62.

Huntley, B.H. and Birks, J.B. (1983). *An atlas of past and present pollen maps for Europe: 0–13,000 years ago*. Cambridge University Press, Cambridge.

Jalas, J. and Suominen, J. (1987). *Atlas Florae Europeae. Distribution of vascular plants in Europe*, Vol.2. Cambridge University Press, Cambridge.

Larew, H.G. (1987). Oak galls preserved by the eruption of Mount Vesuvius in A.D. 79, and their probable use. *Economic Botany*, **41,** 33–40.

Lawton, J.H. (1986). The effect of parasitoids on phytophagous insect communities. In *Insect parasitoids*, (ed. J.K. Waage and D. Greathead), pp. 265–89. Academic Press, London.

Maynard Smith, J. (1989). *Evolutionary Genetics*. Oxford University Press, Oxford.

McCauley, D.E. (1991). Genetic consequences of local population extinction and recolonization *Trends in Ecology and Evolution*, **6,** 5–8.

Moriya, S., Inoe, K., and Mabuchi, M. (1989). The use of *Torymus sinensis* to control Chestnut Gallwasp, *Dryocosmus kuriphilus*, in Japan. *Technical Bulletin of the Food and Fertiliser Technology Centre*, **118,** 1–12.

Notton, D.G. (1990). Parasitoids of the sexual and parthenogenetic generations of *Andricus quercuscalicis*. *Cecidology*, **4,** 15–17.

Pamilo, P., Rosengren, R., Vepsäläinen, K., Varvio-Aho, S.-L., and Pisarski, B. (1978). Population genetics of *Formica* ants. I. Patterns of enzyme gene variation. *Hereditas*, **89,** 233–48.

Payne, J.A. (1978). Oriental chestnut gall wasp: new nut pest in North America. *Proceedings of the American Chestnut Symposium*, 86–88. West Virginia University Press, Morgantown, Virginia.

Richardson, B.J., Baverstock, P.R., and Adams, M. (1986). *Allozyme electrophoresis: a handbook for animal systematics and population studies*. Academic Press, Sydney.

Ritchie, A.J. and Peters, T.M. (1981). The external morphology of *Diplolepis rosae* (Hymenoptera: Cynipidae: Cynipinae). *Annals of the Entomological Society of America*, **74,** 191–9.

Roberts, N. (1989). *The holocene: an environmental history*. Basil Blackwell Ltd, Oxford.

Sanderson, A.R. (1988). Cytological investigations of parthenogenesis in gall wasps (Cynipidae: Hymenoptera). *Genetica*, **77,** 189–216.

Stille, B. (1985*a*). Host plant specificity and allozyme variation in the parthenogenetic gall wasp *Diplolepis mayri* and its relatedness to *D. rosae*. *Entomologia Generalis*, **10,** 87–96.

Stille, B. (1985*b*). Population genetics of the parthenogenetic gall wasp *Diplolepis rosae* (Hymenoptera: Cynipidae). *Genetica*, **67,** 145–51.

Stille, B. and Dävring, L. (1980). Meiosis and reproductive strategy in the parthenogenetic gall wasp *Diplolepis rosae* (L.) (Hymenoptera: Cynipidae). *Hereditas*, **92,** 353–62.

Stone, G.N. and Sunnucks, P.J. (1992). The hedgehog gall *Andricus lucidus* (Hartig 1843) confirmed in Britain. *Cecidology*, **7,** 30–5.

Stone, G.N. and Sunnucks, P.J. (1993). Genetic consequences of an invasion through a patchy environment—the cynipid gall wasp *Andricus quercuscalicis* (Hymenoptera: Cynipidae). *Molecular Ecology*, **2,** 251–68.

Swofford, D.L. and Selander, R.B. (1981). Biosys 1: a FORTRAN program for the comprehensive analysis of data in population genetics and systematics. *Journal of Heredity*, **72,** 281–3.

Wright, S. (1978). *Evolution and the genetics of populations*, Vol. **4.** *Variability within and among natural populations*. University of Chicago Press, Chicago.

22. The communities associated with the galls of *Andricus quercuscalicis* (Hymenoptera: Cynipidae) an invading species in Britain: a geographical view

K. SCHÖNROGGE*, G.N. STONE*,
B. COCKRELL†, and M.J. CRAWLEY*

*Department of Biology and NERC Centre for Population Biology, Imperial College at Silwood Park, Ascot, Berkshire, UK † c/o Malcolm, Department of Biological Sciences, Western Michigan University, Kalamazoo, MI 49008, USA

Abstract

The agamic knopper gall of *Andricus quercuscalicis* in Britain provides an opportunity to investigate the guild associated with a cynipid gall in the process of its development. Species reared in association with the knopper gall include inquilines, which develop in the gall without having direct contact to the gall-causing lava, and parasitoids, both of the gall wasp and of the inquilines. Most of the members of the guild associated with *A. quercuscalicis* are polyphagous and bi- or multivoltine (that is, they attack more than one gall and have two or more generations per year) and, therefore, need alternative hosts, constituting ecological links between *A. quercuscalicis* and the local cynipid community in general. An extensive rearing programme of galls collected in Britain and on the European continent was carried out to examine qualitative and quantitative variations on different spatial scales.

Data on the patterns of parasitoid species diversity, abundance, and trophic relationships, in both the native range of *A. quercuscalicis* and the recently invaded regions, are presented. Comparisons are made between parasite-induced mortalitites experienced by the gall wasp in its native range and the exploitation of the invader by the native parasitoid fauna of Britain. Possible further development of the assemblages associated with the galls of *A. quercuscalicis* in Britain is discussed.

Plant Galls (ed. Michèle A. J. Williams), Systematics Association Special Volume No. 49, pp. 369–89. Clarendon Press, Oxford, 1994. © The Systematics Association, 1994.

Introduction

Species new to particular geographic areas can serve as convenient natural tests of ideas about the assembly of ecological communities (Diamond 1986). Whether an invading species will be successful or fail will depend on its life history parameters, its tolerance of abiotic factors, and its ability to out compete native species for resources or enemy free space (Cornell and Hawkins 1993).

It is generally agreed that herbivorous insect species are regulated by their natural enemies (Hairston *et al.* 1960; Askew 1961*a*). An invading species should therefore become a focus for the assembly of native species which can utilize it as a new resource. In the case of most herbivorous insects the natural enemies are mainly parasitoids. Moreover, the parasitoid complexes focused on endophytic hosts, such as gall wasps or leaf miners, tend to be dominated by generalists (Price and Pschorn-Walcher 1988; Pschorn-Walcher and Altenhofer 1989). The arrival of a new host species in a community of species interconnected by polyphagous parasitoids should therefore increase the complexity of the whole local system.

In many instances invading insects become a serious pest having escaped their natural enemies (Carl 1972; Payne 1978; Otake *et al.* 1982). As *A. quercuscalicis* does no economically important damage, its invasion has been allowed to proceed undisturbed and the predictions made about herbivorous insects as invaders can therefore be tested, using this species.

Communities are described as not only having a definite functional unity with characteristic trophic structures and patterns of energy flow, but also having a compositional unity, in that certain species will occur together from one year to the next (Anderson and Kikkawa 1986). We consider a cynipid gall as the focal point of a community, the members of which are the gall wasp itself, its parasitoids, inquilines feeding on the gall tissue, and their parasitoids. Askew (1961*a*) showed that the terms inquiline and parasitoid often do not describe the various possible trophic relationships very well. Inquilines may well act as predators by killing the gall wasp larva and parasitoids sometimes consume gall tissue before attacking the gall wasp, thus securing the protection of a fully developed gall (Askew 1960; Sellenschlo and Wall 1984). Parasitoid species can also be found attacking other parasitoid larvae of the same species (autoparasitism) or of other parasitoid species (hyperparasitism).

The species assemblages associated with cynipid galls are reported to be consistent over a wide geographical range (Schröder 1967; Askew and Shaw 1985). The galls themselves, on the other hand, are known to vary in their occurrence and density from tree to tree (Askew 1962; Hails and Crawley 1991). Whether parasitoid and inquilines respond to this inhomogeneity on small spatial scales would be of particular interest for biological control where density dependence is essential for generalists as control agents (Hassell 1985).

When the gall wasp *A. quercuscalicis* arrived in England, it found a suitable environment and no native competitors. During the 1970s this gall wasp, due to its high reproductive potential, spread through most of England and Wales and has become an abundant cynipid species in many places. *Andricus quercuscalicis* can therefore be considered a successful invader. Previous studies on the parasitoid complex associated with *A. quercuscalicis* have shown that parasitoid species are beginning to exploit this new host, and also that the population dynamics of *A. quercuscalicis* are regulated rather by the fruit crop of its host tree than by parasitoids (Hails *et al.* 1990; Hails and Crawley 1991; see also Hails, Chapter 23, this volume).

The life cycle of *Andricus quercuscalicis*

The life cycle of *A. quercuscalicis* involves two alternating generations every year. The galls of the parthenogenetic (agamic) generation, commonly known as the knopper gall, develops on the acorns of English oak (*Quercus robur*) from July until September. The galls fall to the ground before the end of September where they overwinter. The agamic females emerge in February and March and fly to Turkish oak (*Q. cerris*) trees where they oviposit into buds bearing catkins (male flowers). The galls on the male flowers contain males and females of the sexual generation. Galls of the sexual generation are very much smaller than the agamic galls (typically 1–2.5 mm in length). In May and June the sexual females oviposit in the female flowers of *Q. robur*, inducing the agamic galls. The life cycle is shown in Fig. 22.1. (Examples of other species with similar lifestyle are provided in Chapter 21).

Agamic females emerge from samples of knopper galls for up to 3 years after the galls fall. There is considerable geographic variation in the proportion of agamic females emerging each year (Fig. 22.2). The proportion of females emerging in the first year is clearly higher for samples from Britain than for the six samples from the continent ($F_{(1,34)} = 339.6$, $p < 0.005$). After 2 years, approximately 75 per cent of the agamics emerge throughout their range.

Methods

The results presented here are based on rearings of galls of *A. quercuscalicis*. Two different rearing methods allowed different questions about the community associated with the gall to be asked.

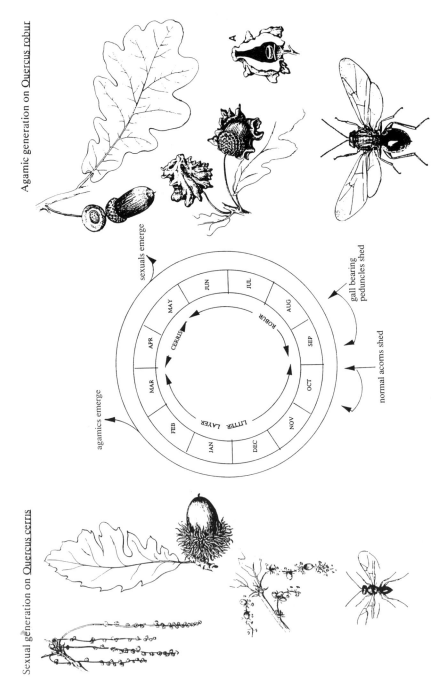

Sexual generation on Quercus cerris

Agamic generation on Quercus robur

sexuals emerge

agamics emerge

gall bearing
peduncles shed

normal acorns shed

MAY
JUN
JUL
AUG
SEP
OCT
NOV
DEC
JAN
FEB
MAR
APR

CERRIS

ROBUR

LITTER LAYER

Fig. 22.1. The life cycle of A. quercuscalicis. Cross-section through the agamic galls is shown at A.

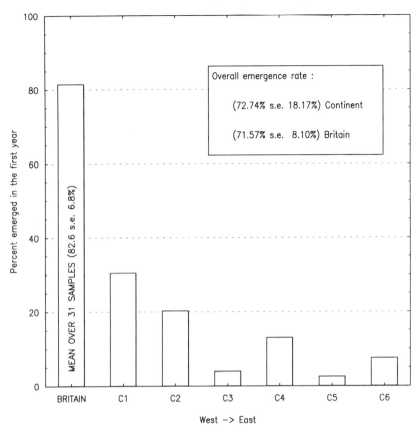

Fig. 22.2. Emergence patterns of the agamic females of *A. quercuscalicus*. C1–C6 = individual sites in continental Europe

The morphology of the knopper gall allows us effectively to separate the community of the other parts of the gall from factors affecting the gall former in the inner cell (Fig. 22.1, at A). A total of 5193 galls were cut open and the inner cell and the wall of each gall reared individually. In this way it is possible to separate those species feeding on the gall former from those developing in the gall wall. Moreover, it is possible to exclude 'visitors', like earwigs and lacewings, which live in the empty space around the inner cell of the galls, but have no apparent trophic relationship to any of the other inhabitants.

Mass rearings have been used to obtain a list of inhabitants which is as complete as possible. Patterns of emergence from these rearings also provide data about the emergence periods for the various species.

The collections have been carried out on different spatial scales. Twenty-four samples from within Silwood Park (National grid

Fig. 22.3. Collection sites of the agamic galls of *A. quercuscalicis* in Britain and on the European continent.

SU945690) represent the smallest scale. Twenty-seven samples come from sites throughout England and Wales (which represent the medium scale) and seven samples come from Germany, Austria, and Slovenia (Fig. 22.3) (representing the largest scale).

The community of the agamic gall of *A. quercuscalicis*

Because *A. quercuscalicis* has two generations which appear at different times of the year, and induces galls of different shapes on different trees, it is sensible to look at the communities associated with each gall separately.

Without doubt the longest species list of inquilines and parasitoids in cynipid galls was published by Fulmek (1968). Unfortunately it seems to be impossible to trace back the individual records and to clear the list of repetitions due to synonymous species names. The rearing methods used are not known. More recent species lists of inhabitants of knopper galls have been published by Collins *et al.* (1983) for the European continent and Hails *et al.* (1990) for Britain. These authors point out that there is very little overlap of species between Britain and continental Europe. This is remarkable, because most of the species which have been recorded from knopper galls on the continent are known to be present in other native galls in Britain.

Hails *et al.* (1990) list the parasitoid species in chronological order of detection and in this way give some indication of species which appeared very early and which ones appeared over consecutive years in the early stages of the invasion. The first parasitoid record was of *Torymus cyanea* Walker (Martin 1982), which is a typical parasitoid species for leaf galls of the genus *Cynips*, although *T. cyanea* has never since been reared from knopper galls. The list published by Hails *et al.* (1990) also includes two species of Gelidae (Ichneumonidea) and one species of the Diapriidae (Proctotrupoidea), groups which are very rarely recorded from galls.

More common species in earlier rearings are two Pteromalidae, *Mesopolobus amaenus* (Walker) and *Mesopolobus jucundus* (Walker), which were also reared over a number of consecutive years. Both species are also known from a number of different native cynipid galls (Askew 1961*a,b*). All these earlier studies recorded very low attack rates for most species. It therefore seems reasonable to interpret these attacks as rare events of 'accidental' oviposition in an unusual host.

Parasitoids and inquilines reared from knopper galls collected autumn 1990

Table 22.1 shows the species we reared from the agamic galls collected in autumn 1990. The parasitoid species we found to attack the knopper gall are members of five families (Pteromalidae, Torymidae, Ormyridae, Eupelmidae, and Eulophidae) of the superfamily Chalcidoidea and one belongs to the Gelidae (Ichneumonidea). All Chalcidoidea were previously recorded from *A. quercuscalicis* or other galls (Pfützenreiter and Weidner 1958; Askew 1961*a*; Fulmek 1968) and are therefore known cynipid parasitoids.

The record for *Gelis formicarius* is remarkable as it confirms the record by Hails *et al.* (1990) from 1986. So far only one ichneumonid (*Orthopelma mediator* (Thunberg)) is known in Europe which is a regular inhabitant of cynipid galls, attacking *Diplolepis rosae* (L.) (Askew 1984). The other two unusual parasitoids for cynipid galls in Hails *et al.* (1990), *Spilomicrus stigmaticalis* Westwood (Proctotrupoidea, Diapriidae) and *Mastrus castaneus* (Taschenberg) (Ichneumonidea, Ichneumonidae) were not repeated in our rearings.

Table 22.1 summarizes the results of the mass and separated rearings. The rearing method used divides the recorded species into two sur-

Table 22.1. Species reared from agamic galls of *A. quercuscalicis* in Britain and on the European continent. 'I': reared from the inner cell; 'O': from the outer wall. 'I + O': the species was reared from the inner cell as well as from the walls of the galls. ?: The species emerged only from mass rearings or single rearings and it is impossible to tell where they developed.

Continent Family	Continent Genus	Continent Species	Britain Genus	Britain Species	Inner cell (I) Outer wall (O)
Cynipidae	*Synergus*	*gallaepomiformis*	*Synergus*	*gallaepomiformis*	O
		umbraculus			O
		sp.			I
Pteromalidae	*Mesopolobus*	*jucundus*	*Mesopolobus*	*jucundus*	O
				amaenus	I (4) and O (17)
			Cecidostiba	*semifasciata*	—
	Cecidostiba	*adana*			O
Eurytomidae	*Eurytoma*	*brunniventris*	*Eurytoma*	*brunniventris*	O
	Sycophila	*biguttata*	*Sycophila*	*biguttata*	I (48) and O (3)
Torymidae	*Torymus*	*nitens*	*Torymus*	*nitens*	O
				dorsalis	O
	Megastigmus	*stigmatizans*	*Megastigmus*	*stigmatizans*	I
Eupelmidae	*Eupelmus*	*urozonus*	*Eupelmus*	*urozonus*	O
Eulophidae	*Aulogymnus*	*trilineatus*			I

prisingly clear groups. Only two species, *Sycophila biguttata* (Swederus) and *Mesopolobus amaenus*, emerged from both the inner cell and the outer wall, and even so they showed a preference for one place or the other. It may be significant that on all four occasions in which *M. amaenus* specimens emerged from inner cells, no inquilines emerged from the galls.

The parasitoid species reared from the wall of the gall probably attacked the larvae of inquilines. Compared to the complexity of trophic relationships in cynipid galls described by Askew (1961*a*), the first outline of the food web in knopper galls appears comparatively simple.

In contrast to earlier studies we found very much the same species in the galls from the continent and in British galls (Table 22.1). This is mainly due to rearing of four species new to the list of parasitoids of agamic *A. quercuscalicis* in Britain; *Ormyrus nitidulus* (Fabricius), and *Megastigmus stigmatizans* (Fabricius), reared from the inner cell and *Mesopolobus jucundus* and *Eurytoma brunniventris* Ratzeburg, reared from the outer wall. The 'expected' parasitoid species (that is the species which can be found in knopper galls on the continent) are beginning to exploit this gall in Britain too.

Only two species known from continental samples of knopper galls have not been reared from British samples: *Cecidostiba adana* Askew which has not been recorded from Britain (Kloet and Hincks 1978) and *Aulogymnus trilineatus* (Mayr) which is known in Britain from galls of *Andricus fecundator* (Hartig) (Askew 1960).

Parasitoid and inquiline infestation in knopper galls

The mortality inflicted on the gall wasp and on the inquilines is interesting for a number of reasons. The mortality of *A. quercuscalicis* has a direct impact on the population dynamics of the gall wasp and the parasitoids developing on *Synergus* larvae are a potential source of mortality for alternative hosts in other cynipid galls. The arrival of *A. quercuscalicis* might therefore change the death rates due to parasitism experienced by several components of the native gall fauna. All the results below are based on the rearing results and emergence during 1991. They are therefore not complete and not intended to be final.

The mortality caused by parasitism is deduced from the frequencies with which the parasitoids emerge from the galls. To estimate mortality correctly it is important to state that all species listed in Table 22.1 are known to be solitary parasitoids, that is, one parasitoid larva = one host larva (Askew 1961*a*).

The rearing results from Silwood Park, Britain and the continent are listed in Table 22.2. The category 'parasitism inner cell (per cent)' represents the mortality inflicted on the gall wasp according to the results of the separated rearings. The relative abundance of the *Synergus* species at each site is shown under 'Inquiline infestation'. As all the parasitoid species attacking the inquilines are solitary parasitoids it is assumed that each parasitoid specimen represents an inquiline host. The sum of emerged parasitoids and inquilines therefore gives the number of hosts for the parasitoids. Another assumption that follows is that the parasitoids are indifferent to the two inquiline species *Synergus gallaepomiformis* (Boyer de Fonscolombe) and *Synergus umbraculus* (Olivier).

Inquiline infestation in knopper galls

Comparison of the continental, Silwood Park, and British samples from outside Silwood Park, shows again that the galls from British sites outside Silwood have lower levels of inquiline attack (Fig. 22.4). Inquiline attack rates in galls from Silwood Park are similar to those at continental sites (Table 22.2). Although the median of 0.69 inquilines per gall was higher

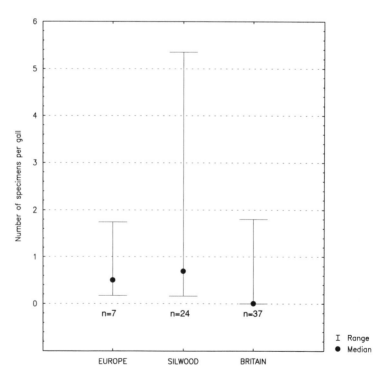

Fig. 22.4. Inquiline infestation in the agamic galls of *A. quercuscalicis* on the European continent, in Silwood Park, and in Britain (excluding Silwood Park).

Table 22.2. Attack rates and inquiline infestation in the agamic galls of *A. quercuscalicis* on the European continent in Silwood Park and in Britain (excluding Silwood Park).

	Europe				Silwood Park				Britain			
Site	Sample size	Parasitism inner cell (%)	Inquiline infestation (specimens gall)	Parasitism on Inquilines (%)	Sample size	Parasitism inner cell (%)	Inquiline infestation (specimens gall)	Parasitism on Inquilines (%)	Sample size	Parasitism inner cell (%)	Inquiline infestation (specimens gall)	Parasitism on Inquilines (%)
1	866(*)	5.6	0.27	26.3	362	1.2	5.4	19.1	178	3.9	0.8	56.9
2	920	4.2	0.5	17.7	395	1.3	1.8	21.2	556	0.2	1.5	16.1
3	842	3.6	0.6	6.9	858	1.2	0.4	3.5	463	–	1.8	17.2
4	384	2.3	0.4	4.5	489	1.4	1.0	5.6	376	5.1	0.3	5.8
5	251	1.2	0.2	16.0	396	1	1.9	24.6	412	0.2	1.2	1.2
6	457(*)	7.2	1.7	45.2	926	1.1	2.9	8.9	572	–	0.1	–
7	294	13.6	1.5	92.2	302	1	1.5	18.9	290	1	0.4	1
8					308	0.3	3.1	4.9	627	1	0.1	6.9
9					179	1.1	3.9	5	287	–	0.2	15.6
10					426	1.4	0.7	9.6	270	–	0.8	–
11					968	1.9	0.3	1.4	244	2.5	0.3	–
12					111	0.9	0.7	1.4	938	1.3	0.1	–
13					42	2.4	0.6	4.6	356	1.1	0.2	–
14					229	2.2	1.7	4	334	–	0.1	–
15					127	0.8	0.3	22.8	394	–	0.1	–
16					156	–	1.2	0.5	553	–	0.2	–
17					150	2.7	0.2	–	71	1.4	1	–
18					10	–	3.2	46.9	202	0.5	0.1	–
19					5	–	2.8	–	221	0.5	–	–
20					529	0.4	0.6	0.6	257	1.2	–	–
21					251	0.8	0.3	–	450	0.2	–	–
22					307	0.3	0.5	4	890	0.1	–	–
23					176	2.8	0.4	–	127	0.8	–	–
24					256	–	0.2	–				

* Samples gathered within the presumed native distribution of *A. quercuscalicis.*

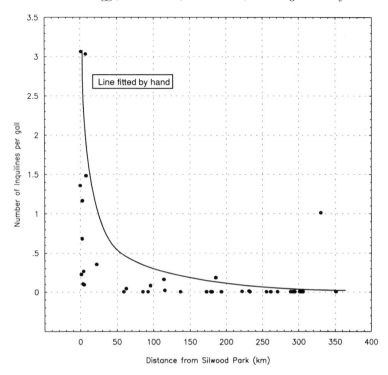

Fig. 22.5. Relative abundance of inquilines in knopper galls as a function of the distance to the area with the highest infestation (Silwood Park).

than for the continental samples, the wide range (5.35–0.16 specimens per gall) is even more remarkable and indicates considerable patchiness in inquiline attack rates (Fig. 22.4). The galls from 14 of the samples from the rest of Britain did not produce any inhabitants other than the gall wasp (Table 22.2). Generally this table displays the situation expected from the information available before this study started. Only a few species were reared from every individual sample and the abundances were relatively low.

Silwood Park is unusual by British standards because of the high levels of inquiline infestation which consequently have an impact on the species richness and abundance in knopper galls. It is interesting to ask whether there is a pattern in the distribution of high inquiline infestation. By plotting the relative abundance of the inquilines in the knopper galls against the distance from Silwood Park a clear relationship emerges (Fig. 22.5). The samples from Silwood Park have been collapsed to a single value, to keep them on a comparable spatial scale. Taking 0.5 inquilines per gall (the median of the continental samples, Figure 22.4) as a standard, the samples with high infestations are restricted to places 10–30 km away from Silwood Park.

While *S. gallaepomiformis* is dominant in all British samples, *S. umbraculus* is more abundant on the continent (Table 22.2). In the two samples gathered within the presumed native distribution of *A. quercuscalicis* (marked with * in Table 22.2), *S. umbraculus* appears to be more abundant than *S. gallaepomiformis*.

Parasitism on the inquilines in knopper galls

Of the parasitoid species attacking the inquilines in Britain, *M. jucundus* is the dominant species (Fig. 22.6). That *Eurytoma brunniventris* and *Eupelmus urozonus* Dalman occur only when *M. jucundus* is present suggests that they might be hyperparasitoids of the latter. Both species are known as generalists in other native cynipid guilds (Askew 1961a). *Torymus nitens*, *M. dorsalis*, and *C. semifascia* have been very rare but also occurred in a few samples from outside Silwood Park and should be therefore regarded as members of the guild.

The dominant parasitoid of inquilines on the continent is *C. adana*, which has the same position in the community that *M. jucundus* has in Britain. The low species richness of inquiline parasitoids from the

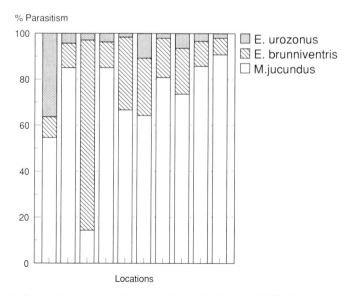

Fig. 22.6. Proportions of parasitism by *Mesopolobus jucundus* (MJ), *Eurytoma brunniventris* (EB), and *Eupelmus urozonus*, (EU) in British knopper galls.

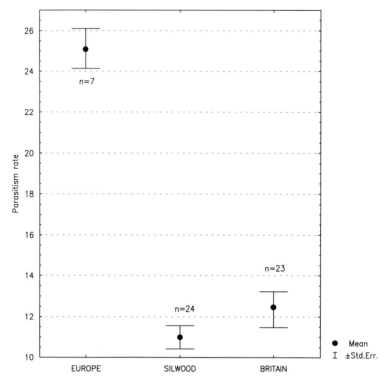

Fig. 22.7. Parasitism on the inquilines in knopper galls on the European continent, in Silwood Park, and in Britain (excluding Silwood Park).

continent may be due to the relatively lower sample size. The levels of parasitism inflicted on inquilines in continental samples is most impressive.

The 92.24 per cent parasitism on inquilines in the sample from Ludwigsburg (Germany) means that the proportion of the *S. gallaepomiformis* and *S. umbraculus* populations developing in knopper galls has been all but exterminated . Parasitoid attack on the inquilines is, as expected, highest on the continent (Fig. 22.7), even when the sample from Ludwigsburg (92.24 per cent inquiline mortality) was excluded as an outlier (ANOVA with binomial errors: $\chi^2 = 268.9$, df $= 2$ $p < 0.005$; only British samples which produced inquilines were included in the analysis).

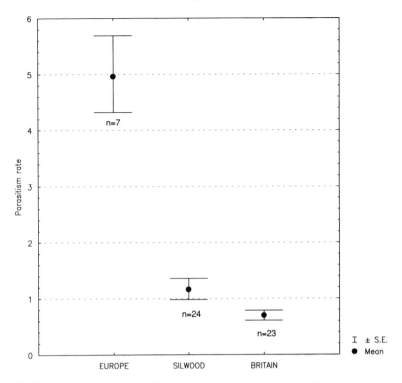

Fig. 22.8. Attack rate on the gall former in knopper galls on the European continent, in Silwood Park, and in Britain (excluding Silwood Park).

Parasitism in the larval chamber of the gall wasp

The mortality inflicted directly on the gall wasp is surprisingly low even in continental samples. The highest parasitism rate was found in galls from Ludwigsburg with a parasitism rate of 13.61 per cent. This was inflicted only by *M. stigmatizans*. In Britain, attack rates did not exceed 5.1 per cent and samples from 21 locations produced no evidence for parasitoids attacking the gall wasp larva.

Though the parasitism in galls from the continent is less than expected, it is less within Silwood Park and even less in other parts of Britain ($\chi^2 = 255$, df $= 2$, $p < 0.005$) (Fig. 22.8).

The community of the sexual gall of *A. quercuscalicis*

The only parasitoid recorded from the sexual galls on the continent is *Mesopolobus tibialis* (Pfützenreiter and Weidner 1958). In Britain three

Table 22.3.   Parasitoid species reared from sexual galls of *A. quercuscolicis* collected in Germany, Austria, Czechoslovakia, and Hungary

Family	Genus	Species	Reared from British galls (Y/N)
Pteromalidae	*Mesopolobus*	*tibialis*	Y
		fuscipes	Y
		xanthocerus	Y
	Cecidostiba	*adana*	N
Eulophidae	*Tetrastichus*	*sp1*	N
	Tetrastichus	*sp2*	N
	Aulogymnus	*gallarum*	N
Torymidae	*Torymus*	*nitens*	N

species of the Pteromalidae, (*Mesopolobus fuscipes* (Walker), *M. tibialis* (Westwood), and *M. xanthocerus* (Thomson) have been recorded regularly and a fourth species, *M. dubius* (Walker) was rare (Collins *et al.* 1983; Hails 1989).

The sexual galls on the catkins of *Q. cerris* were collected in spring 1992 at 12 locations throughout continental Europe. One thousand catkins were collected at each location. The rearing results are interpreted under the assumption that there is no mortality in the rearings or, if so, the mortality is proportionally the same for all species including the gall wasp. The dissection of 100 galls from a location in Hungary revealed only one adult insect, whether gall wasp or parasitoid, per gall (there are no inquilines in the sexual galls). Under these conditions the total number of galls and the number of galls per catkin can be calculated from the number of emerged adult cynipids and parasitoids.

Samples were collected from three German locations, Ludwigsburg, Stuttgart, and Munich, one from the Wiener Wald (Austria), two from Czechoslovakia, and six from Hungary. Table 22.3 lists all the species reared. In addition to the three *Mesopolobus* species known from Britain, four parasitoid species were reared. The records of *C. adana* and *Aulogymnus gallarum* L. represent new host records for these species. Two unidentified species of the subfamily Tetrastichinae were also reared from the sexual galls.

Figure 22.9 shows the overall parasitism rates of the sexual galls and the proportion each parasitoid species contributes to the total number of parasitoid adults found in the rearings, ranked by the location of the sample sites from west to east. The two samples from which no parasitoid emerged, one in Czechoslovakia and one in Hungary, also had the lowest galling rate (0.006 and 0.029 galls per catkin). Only two sites in Hungary showed a considerably higher parasitism rate than the 20–30 per cent parasitism recorded regularly over 8 years in Britain (Hails and Crawley

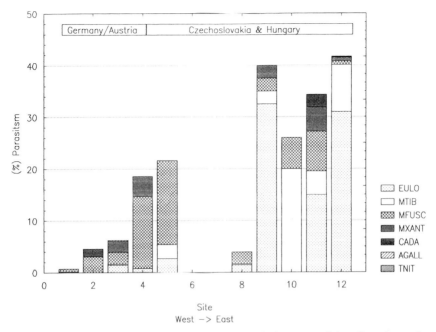

Fig. 22.9. Parasitism rates on the sexual galls of *A. quercuscalicis* collected on the continent. (EULO. Tetrastichinae sp.1 and sp.2; MTIB, *Mesopolobus tibialis*; MFUSC, *Mesopolobus fuscipes*; MXANT, *Mesopolobus xanthocerus*; CADA, *Cecidostiba adana*; AGALL, *Aulogymnus gallarum*; TNIT, *Torymus nitens.*)

1991). Parasitoid attack rates in the samples from Germany and Austria were surprisingly low by comparison.

Discussion

Apart from *Callirhytis glandium* the knopper gall is the only cynipid gall in Britain occurring on the acorns of *Q. robur*. *Andricus quercuscalicis* therefore has an unusual life history compared to the native galls in Britain. Polyphagous parasitoids are usually expected to be restricted in their host range by the host life history stage, taxonomy, or mode of exploitation of the host plant (Lawton 1985). Even though British parasitoid species known to attack the gall on the continent must have the morphology to do so in Britain, they might lack appropriate foraging behaviour to exploit the new gall. This might explain the number of species and low attack rate found in earlier studies of the community of the agamic gall (Collins *et al.* 1983; Hails *et al.* 1990). One result of this study is that more and new parasitoid species have entered the guild of the agamic gall of *A. quercuscalicis* in Britain since the last published rearing results from 1988 (Hails *et al.* 1990). The list of inhabitants compiled for knopper galls from Britain is becoming increasingly similar to that for continental Europe. The question to ask, therefore, is what are the characteristics of parasitoids which exploit the new gall successfully and/or what are the habitats where *A. quercuscalicis* is discovered and utilized earlier or later?

If one assumes that the ability of inquilines to exploit *A. quercuscalicis* is genetically determined, then the high rates of attack by inquilines in sites near Silwood Park may have a genetic basis. Windsor Park is a remnant of an ancient woodland and centre of diversity for a number of forest insects. Genetic diversity is often highest at a centre of distribution, as it is for *A. quercuscalicis* (see Sunnucks *et al.*, Chapter 21, this volume). It would therefore be more likely that *Synergus* specimens of genotypes that enable these individuals to exploit knopper galls, are present in these places.

An alternative hypothesis involves the accidental introduction of the inquiline species from the continent during intensive work on this species in Silwood Park. These hypotheses could possibly be distinguished using the molecular methods presented in Chapter 21 on the inquilines.

It is perhaps not surprising that parasitism rates in samples from the native range on the continent are higher than in those from Britain. More surprising are the low parasitism rates on the inner cell of the knopper galls even on the continent. The maximum of 5 per cent is very low and we have no evidence of density dependence, so it is unlikely that parasitoids attacking the agamic generation regulate the population dynamics of *A. quercuscalicis*. Parasitism on the sexual galls reaches substantially higher levels (43 per cent), but appears to be geographically very variable. In Britain parasitism rates in the sexual generation have been relatively constant over 8 years at approximately 25 per cent (Hails 1989). *Andricus quercuscalicis* in Britain is apparently only restricted by the acorn crop on *Q. robur* as parasitism appears not to be density-dependent, a precondition for generalists regulating a host population (Hails and Crawley 1991; Hassell 1985).

Variability in parasitism rates on the continent might be explained by a mechanism described by Washburn and Cornell (1981) for the gall wasp *Xanthoteras politum*. This wasp occurs in early succession stages in woodlands or after disturbances such as fire. The species disappears during the succession process and natural enemies sometimes drive host populations to extinction. Because the host trees of *A. quercuscalicis* are patchy in their distribution on the continent and the annual host change is obligate for the gall wasp, even high abundances in one year, do not guarantee the presence of the species in the following year. Personal observation suggests that in two locations where the galls were abundant in 1990 the galls were absent in 1991. The uncertainty of the occurrence in the sexual or agamic generation might present the natural enemies of those galls with a similar situation as described for *X. politum*. But if parasitoid attacks do reach very high levels on the continent, in either of the generations, we have yet to sample such populations.

Another source of mortality yet to be investigated is mortality during dispersal, either due to unsuccessful searching for a suitable host tree or to predation. This factor is expected to be important for a species with obligatory host change.

Dryocosmus kuriphilus on chestnut, a cynipid which is native to China and which was introduced to Japan around 1940, presents a similar case to the invasion of *A. quercuscalicis*. This species is regarded in Japan as one of the most serious pests on chestnut (Moriya *et al.* 1989). After 38 years, over the period 1978–1981, 10 parasitoid species native to Japan had discovered the new host. Mortalities inflicted by these species varied between years, seasons, and individual trees (0–74 per cent) (Otake *et al.* 1982). Interestingly, six out of the 10 parasitoid species are also known from Europe and three of them (*E. urozonus*, *E. brunniventris*, and *Ormyrus punctiger*) have been recorded to attack the agamic galls of *A. quercuscalicis*. This demonstrates the wide geographical ranges of some of the parasitoids. Even though the parasitism rates in *D. kuriphilus* could reach relatively high levels on particular trees in particular years the gall wasp population decreased only when a parasitoid species from China (*Torymus sinensis*) was released in Japan in 1983 (Moriya *et al.* 1989).

As regards possibilities for biological control, the agamic gall of *A. quercuscalicis* appears to be nearly invulnerable and is unlikely to be the stage in the life cycle where parasitism could have any stronger impact on the host population. Whether this is possible in the sexual generation remains unclear until more data can be gathered over a longer period of time.

Acknowledgements

This study was funded by the Department for the Environment. Personal acknowledgements are given in Chapter 21 by Paul J. Sunnucks, G.N. Stone, and K. Schönrogge.

References

Anderson, D.J. and Kikkawa, J.(1986). Development of concepts. In *Community ecology: pattern and process*, (ed. J. Kikkawa and D.J. Anderson), pp. 3–16. Blackwell, Melbourne.

Askew, R.R. (1960). The biology of the British species of the genus *Olynx* Förster with a note on seasonal colour forms in Chalcodoidea. *Proceedings of the Royal Entomological Society London*, **36,** 103–12.

Askew, R.R. (1961*a*). On the biology of the inhabitants of oak galls of Cynipidae in Britain. *Transactions of the Society for British Entomology*, **14,** 237–68.

Askew, R.R. (1961*b*). A study of the biology of the species of the genus *Mesopolobus* Westwood (Hym.: Pteromalidae) associated with cynipid galls on oak. *Transactions of the Royal Entomological Society London*, **113,** 155–73.

Askew, R.R. (1962). The distribution of galls of Neuroterus (Hym.: Cynipidae) on oak. *Journal for Animal Ecology*, **31,** 439–55.

Askew, R.R. (1984). The biology of gall wasps. In *The biology of galling insects*, (ed. T.N. Anathakrishnan), pp. 223–71. Oxford and I B H Publishing Co., New Dehli.

Askew, R.R. and Shaw, M.R. (1985). Parasitoid communities. In *Insect parasitoid* (ed. J. Waage and D. Greathead), pp. 225–64. Academic Press, London.

Carl, K.P. (1972). On the biology, ecology and population dynamics of *Caliroa cerasi* (L.). *Zeitschrift für angewandte Entomologie*, **71,** 58–83.

Collins, M., Crawley, M.J., and McGavin, G. (1983). Survivorship of the sexual and agamic generations of *Andricus quercuscalicis* on *Quercus cerris* and *Quercus robur*. *Ecological Entomology*, **8,** 133–8.

Cornell, H.V. and Hawkins, B.A. (1993). Accumulation of native parasitoid species on introduced herbivores: A comparison of 'Host-as-natives' and 'Hosts as invaders'. *American Naturalist*, **141,** 847–65.

Diamond, J. (1986). Overview: introductions, extinctions, exterminations and invasions. In *Community ecology*, (ed. J. Diamond, and T.J. Case,). Harper and Row, New York.

Fulmek L. (1968). Parasitinsekten der Insektengallen Europas. *Beiträge zur Entomologie*, **18,** 719–952.

Hails, R.S. (1989). Host size and sex allocation of parasitoids in a gall forming community. *Oecologia*, **81,** 28–32.

Hails, R.S. and Crawley, M.J. (1991). The population dynamics of an alien insect: *Andricus quercuscalicis*. *Journal of Animal Ecology*, **60,** 545–62.

Hails, R.S., Askew, R.R., and Notton, D.G. (1990). The parasitoids and inquilines of the agamic generation of *Andricus quercuscalicis* (Hym.: Cynipidae) in Britain. *Entomologist*, **109,** 165–72.

Hairston, N.G., Smith, F.E., and Slobodkin, L.B. (1960). Community structure, population control and competition. *American Naturalist*, **94,** 421–5.

Hassell, M.P. (1985). Parasitoids and population regulation. In *Insect parasitoids* (ed. J. Waage and D. Greathead), pp. 201–24. Academic Press, London.

Kloet, G.S. and Hincks, W.D. (1978). Hymenoptera. In *A check list of British insects*, Handbook for the identification of British Insects, Part 4, (ed. M.G. Fitton, M.W.R. de V. Graham, Z.R.J. Boucek, N.D.M. Fergusson, T. Huddleston, J. Quinlan, and O.W. Richards), pp. 1–159. Royal Entomological Society of London, Dorking.

Lawton, J.H. (1985). The effect of parasitoids on phytophagous insect communities. In *Insect parasitoids* (ed. J. Waage, and D. Greathead), pp. 265–87. Academic Press, London.

Martin, M.H. (1982). Notes on the biology of *Andricus quercuscalicis* (Burgsdorf), the inducer of the Knopper galls on the acorns of *Quercus robur* L. *Entomologist's Monthly Magazine*, **118,** 121–3.

Moriya, S., Inoue, K., and Mabuchi, M. (1989). The use of *Torymus sinensis* to control chestnut gall wasp, *Dryocosmus kuriphilus*, in Japan. *Technical Bulletin of the Fruit Tree Research Station*, **118,** 1–12.

Otake, A., Masakazu, S., and Seiichi, M. (1982). A study on parasitism of the chestnut gall wasp, *Dryocosmus kuriphilus* Yasumatsu by parasitoids indigenious to Japan. *Bulletin of the Fruit Tree Research Station*, **9,** 177–92.

Payne, J.A. (1978). Oriental chestnut gall wasp: new nut pest in north America. In *Proceedings of the American Chestnut Symposium*, (ed. W.L. Macdonald, F.C. Cech, J. Luchok, and C. Smith), pp. 86–8. West Virginia University Press. Morgantown, Virginia.

Pfützenreiter, F. and Weidner, H. (1958). Die Eichengallen im Naturschutzgebiet Favorite Park und ihre Bewohner. *Veröffentlichungen der Landesstelle für Naturschutz und Landschaftspflege Baden-Würtemberg*, **26,** 88–130.

Price, P.W. and Pschorn-Walcher, H. (1988). Are galling insects better protected than exposed feeders? A test using Tenthredinid sawflies. *Ecological Entomology*, **13** 195–205.

Pschorn-Walcher, H. and Altenhofer, E. (1989). The parasitoid community of leaf mining sawflies (Fenusini and Heterarthrini): a comparative study. *Zoologischer Anzeiger*, **222,** 37–56.

Schröder, D. (1967). *Diplolepis* (= *Rhodites*) *rosae* (L.) and a review of its parasite complex in Europe. *Commonwealth Institute of Biological Control Technical Bulletin*, **9,** 93–131.

Sellenschlo, U. and Wall, I. (1984). *Die Erzwespen Mitteleuropas*. Bauer, Keltern.

Washburn, J.O. and Cornell, H.V. (1981). Parasitoids, patches, and phenology: their possible role in the local extinction of a cynipid gall wasp population. *Ecology*, **62,** 1597–607.

23. The population dynamics of the gall wasp *Andricus quercuscalicis*

ROSEMARY S. HAILS

Imperial College at Silwood Park, Ascot, Berkshire, UK

Abstract

The gall wasp *Andricus quercuscalicis* invaded Britain from the continent in the early 1960s. The life cycle involves both alternation of generations and alternation of host plants. This study draws on 8 years of population data and considers the factors that limit and regulate the distribution and abundance of *A. quercuscalicis*. This gall wasp, now abundant, appears to be resource-limited in the agamic generation due to competition for acorns in the 'low years' of the acorn cycle. Factors influencing the survivorship of the sexual generation are considered in detail. Although the sexual generation does not appear to be resource-limited, there is evidence of density-dependent mortality during gall establishment. This arises partly as a consequence of the oviposition pattern of the agamic females, because the distribution of eggs is highly aggregated between and within trees. As a result of density-dependent mortality, the distribution of mature galls is less highly clumped. Although the agamic knopper gall suffers only low rates of parasitism from a few species, several native parasitoids attack the sexual gall. The total percentage of parasitism is remarkably consistent from year to year even though the contribution from any one species fluctuates considerably.

Introduction

1. The population dynamics of invasion

Many species of plants, insects, and other taxa have successfully invaded Britain, for example, garden escapes, some agricultural pests, and biological control agents imported to control those pests. Some of these invasions have been natural and some intentionally assisted by man. Data on the success and failure of establishment of biocontrol agents

Plant Galls (ed. Michèle A. J. Williams), Systematics Association Special Volume No. 49, pp. 391–403. Clarendon Press, Oxford, 1994. © The Systematics Association, 1994.

suggest that there is a large number of failures (Crawley 1989). Unfortunately, the very nature of the process means that we have few data on the failure of natural invasions. However, the cynipid gall wasp *Andricus quercuscalicis* (Burgsdorf 1783) provides a particularly fine example of a successful invader, which has been allowed to reach high population densities due to its lack of economic importance. This species not only allows the process of invasion to be studied, but also the potential impact upon a close-knit native cynipid community (Schönrogge *et al.* Chapter 22, this volume).

There has been an increasing awareness of the importance of spatial processes in the dynamics of populations (for example, Taylor 1988; Walde and Murdoch 1988). This is likely to be particularly true for populations which display patchy, aggregated distributions. Cynipid gall distributions are known to vary considerably in their occurrence from tree to tree, trees being characteristic in favouring particular gall species (Askew 1962; Hails and Crawley 1991). Trees are, therefore, acting as natural and distinctive patches for cynipids. Trees may also be broken down into finer natural units (for example, buds or shoots). Such a naturally patchy environment may well be important in structuring the dynamics of gall-forming species.

This chapter presents the dynamics of a successful invader and examines the spatial processes that limit and regulate its distribution and abundance.

2. Invasion of A. quercuscalicis into Britain

Andricus quercuscalicis was first recorded in Britain by Claridge (1962), although anecdotal records date back 10 years earlier than this (the late Ted Ellis, personal communication). One reason for its recent arrival in this country is that it is an obligate host alternator, attacking the Turkey oak, *Quercus cerris* L., as well as the English oak, *Q. robur* L. The Turkey oak was only introduced into Britain approximately 200 years ago as an ornamental tree. It has since escaped and become established over much of southern and midland Britain. This 'assisted' invasion of one of the host trees paved the way for the 'natural' invasion of the insect. The biogeographical patterns of invasion by *A. quercuscalicis* across Europe and into Britain are currently being investigated by Sunnucks *et al.* (Chapter 21, this volume).

Since its arrival in Britain, the wasp has spread to occupy those regions in which the Turkey oak, *Q. cerris*, is common (McGavin 1981) and it has reached high population densities. One generation attacks the acorns of the English oak, *Q. robur*, forming large and conspicuous galls. The proportion of acorns attacked can be very high and this caused considerable comment in the early 1980s when the wasp came to public attention (Crawley 1984). Our study started after the gall wasp had already become established in its current range and had reached high densities.

3. Life cycle of A. quercuscalicis

The life cycle of *A. quercuscalicis* is described in full by Schönrogge *et al.* (Chapter 22, this volume) and is summarized in Fig. 23.1. The sexual galls develop on the male buds during bud burst in April. These galls are much smaller and simpler in form than those of the agamic generation, one flask-shaped gall replacing two of the four anthers in a whorl. One catkin may carry up to 20 galls, but attack rates vary considerably both within and between trees (Hails and Crawley 1992). Each gall contains a sexual male or female. The males emerge first and battle for position over the female galls. The females are mated as soon as they break free from their galls and then disperse to *Q. robur*. The female flowers of *Q. robur* are freshly pollinated at this stage (late May) and the sexual females lay their eggs between what will become the acorn and its cup. The large knopper galls develop over the summer. There is one additional feature: a proportion of agamic females

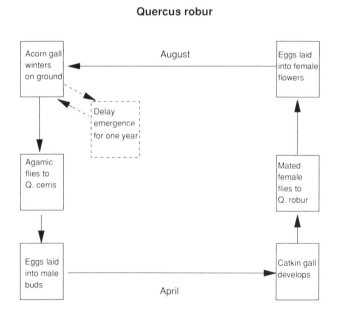

Fig. 23.1. The life cycle of *Andricus quercuscalicis*.

(approximately 30 per cent in Britain) delay emergence for 1 year, remaining inside the acorn gall in the litter layer. Recent data illustrate that there is geographic variation in this delayed emergence, with up to 80 per cent emerging after the first year (Schönrogge *et al.* Chapter 22, this volume).

The dynamics of a bivoltine, host-alternating insect

This system presents a challenging problem for the quantitative ecologist. Most theoretical models of insect populations deal with univoltine systems. This species is not only bivoltine, but also host-alternating. The first approach to modelling such a population is to attempt to predict the abundance of one generation from the abundance of the previous stage. A coupled set of non-linear difference equations is presented below, encapsulating the key features of the life cycle of this insect:

$$S_{t+1} = [\lambda_a f(A_t) p_e A_t + \lambda_a f(A_{t-1})(1 - p_e) \, d \, A_{t-1}] \times (1 - m_a) \quad (23.1)$$

Sexual generation in year $t+1$ | contribution from agamics in year t | contribution from agamics in year $t+1$ | migration mortality

$$A_t = (1 - p_p) \times \lambda_s g(S_t) S_t \times h(R_t) \times (1 - m_s) \quad (23.2)$$

agamic generation in year t | parasitism | other sexual generation mortalities | resource function | migration mortality

The first equation predicts the sexual generation in year $t+1$ from the abundance of the agamic generations of the previous 2 years (to simplify, the small proportion of agamics that delay their emergence for 2 years or more are ignored). The proportion of agamics emerging in the first year is p_e and, therefore, the proportion emerging in the second year is $1 - p_e$. The parameter λ_a and the function $f(A_t)$ refer to the intrinsic rate of increase and to any density-dependent mortalities for the agamic generation in the specified year. The additional overwintering mortality that is incurred by those agamics which delay emergence for 1 year is denoted by d. The final term involves m_a, the mortality of agamics suffered during migration as they disperse from *Q. robur* to *Q. cerris*.

The second equation predicts the agamic generation in year t from the abundance of the sexual generation earlier in the same year. As the sexual generation does not display delayed emergence, there is no time delay in this equation. The parameter λ_s and the function $g(S_t)$ refer to the intrinsic rate of increase and density-dependent mortalities in the sexual generation. Parasitism appears as separate, density-independent term. There is some evidence that the agamic generation is resource-limited and so a resource function, $h(R_t)$, has been included. This will be discussed in some detail later. Finally, m_s is the migration mortality for the sexuals as they migrate from *Q. cerris* to *Q. robur*.

Four key features of these equations will now be examined in greater detail.

1. Gall establishment in the sexual generation: a component of g(S$_t$)

The first mortality to be considered involves the establishment of the small catkin galls on *Q. cerris* in the spring. Dissection of the male flower buds of *Q. cerris* early in the year allows us to census the egg population as it is laid by the agamic females. Because this monitoring is necessarily destructive, we only obtain snapshots of the population through time. Egg densities increase over a period of a few weeks, reaching a maximum shortly before bud burst. At the time of peak egg densities, the distribution of eggs across buds is highly clumped, with most eggs being found in relatively few buds. The degree of clumping was quantified by fitting a negative binomial distribution to the frequency distribution of eggs per bud in each tree. This provides us with k, a parameter of this distribution, which describes the degree of clumping for the eggs in a given tree. A relatively high k would result from a distribution that is *less* clumped (that is, eggs would be more evenly distributed across buds), whereas a relatively low k would describe a highly clumped distribution. In this way, changes in the shape of the distribution can be monitored through time.

As bud burst occurs, the eggs hatch and the sexual galls become established. The distribution of galls can be described in the same way as the egg populations for each tree. If the rate of egg mortality had been the same in each bud, then k, the measure of aggregation (or clumpedness), would not change. In other words, random deaths would result in no change in k (Pielou 1977). However, if egg mortality were greater in buds with higher egg densities, aggregation would decrease (so that k would increase). A comparison of the egg and gall population distributions will, therefore, provide an insight into the pattern of mortalities that have occurred in the intervening period.

Collection of such data requires a sufficient range of egg densities to fit a negative binomial probability density function. Figure 23.2 illustrates the results for those trees and years in which this was possible. In seven out of nine cases, k increased when comparing egg and gall populations. This evidence suggests that the rate of gall establishment is dependent upon egg density within the bud.

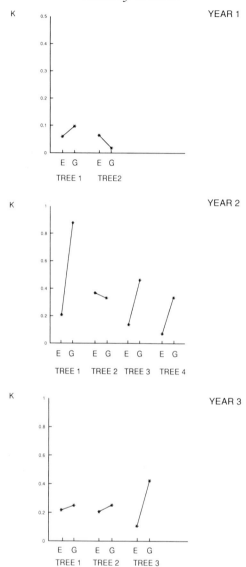

Fig. 23.2. Comparing k of the negative binomial distribution for egg (E) and gall (G) distributions in different trees and different years. In seven out of nine cases k increases between the egg and gall populations, suggesting density dependent mortality in the establishments of galls.

2. Parasitism in the sexual generation: p_p

The sexual generation is attacked by a number of polyphagous parasitoids, mostly in the genus *Mesopolobus* (Pteromalidae). The principal species are

Mesopolobus fuscipes (Walker), *M. xanthocerus* (Thomson) and *M. tibialis* (Westwood). These parasitoids attack the cynipid in the late larval/pupal stages. It has been shown that the gall distribution is aggregated, that is, there are many low density patches and relatively few high density patches.

Is parasitism rate affected by gall density? This question can be asked in many different ways. For example, the parasitoid may respond to gall density at one spatial scale but not another. Gall distributions and parasitism at the level of the catkin, bud, shoot, and twig should therefore be considered (Hails and Crawley 1992). Much parasitism is found to be spatially density-dependent, but parasitism by any given species does not exhibit a consistent relationship with density. An illustration of this is provided by *M. fuscipes*: in Fig. 23.3a, there is a positive relationship with density. Those galls in high density patches are more likely to be parasitized than galls in low density patches. Such patterns are obtained when parasitoids are attracted to or spend more time in high density patches. However, in a different tree but at the same spatial scale, the same species of parasitoid shows an inverse relationship between parasitism and gall density (Fig. 23.3b). Such patterns may be obtained when parasitoids arrive at a patch and lay a fixed number of eggs and become egg-limited in high density patches. The third pattern is density-independent parasitism, in which there was no relationship between parasitism and gall density.

In order that patterns of spatial density-dependence can have an impact upon the population dynamics of the gall wasp, they must translate into temporal density-dependence. However, the evidence we have from 8 years of data is that this is not the case. Total percentage parasitism is remarkably constant, fluctuating only between 20 and 30 per cent—although the contribution of individual species may vary considerably (Fig. 23.4). For this reason, parasitism was included as a density-independent term in the coupled equations.

3. Migration mortalities: m_a and m_s

Mortalities incurred during dispersal are notoriously difficult to measure and we have two such migration mortalities in these equations. Attempts have been made at mark–release–recapture studies, but in spite of large numbers of mark–releases (> 1000), no recaptures have been made. Although we cannot quantify these losses in absolute terms, it is possible to say something about their relative magnitude. Due to the ratio of *Q. robur* to *Q. cerris* in the study area (approximately $10:1$), it is likely that $m_a \gg m_s$.

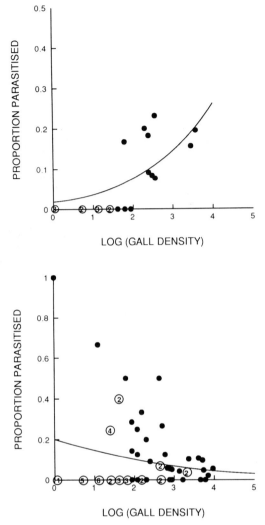

Fig. 23.3. The relationship between the proportion of galls parasitized by *M. fuscipes* and gall density in two trees within the same year. The analysis was conducted in logits and back transformed to produce the curved line. (a). $\text{Ln}(p/q) = 3.993 + 0.7345\,\text{ln (gall density)}$ and (b) $\text{Ln}(p/q) = -1.388 - 0.399\,\text{ln(gall density)}$, where p is the proportion parasitized and $q = 1 - p$. The first graph is an example of positive density dependence, whilst the second is an example of inverse density dependence. After Hails and Crawley (1992).

4. The resource function: $h(R_t)$

The resource function involves the agamic generation. Agamic gall density and acorn density have been recorded on 30 trees at Silwood

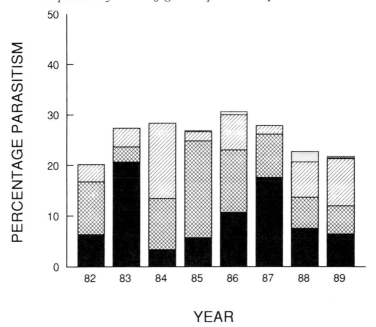

Fig. 23.4. Percentage parasitism of the sexual generation of *A. quercuscalicis* over an 8-year period. Four species of polyphagous parasitoids attack the sexual gall: *M. fuscipes* (■), *M. xanthocerus* (▨), *M. tibialis* (▧), *M. dubius* (▦). After Hails and Crawley (1992).

Park for more than a decade (Fig. 23.5). Acorn crops on *Q. robur* exhibit an approximately alternating yield (possibly a defence against seed predators). Gall densities appear to track these acorn fluctuations very closely. In acorn-poor years, the suggestion is that they are resource-limited. If gall density is correlated with acorn density, there is a strong linear relationship (Fig. 23.6). This provides strong circumstantial evidence for resource-limitation in acorn-poor years. However, it would be unwise to extrapolate this linear relationship to very high acorn densities, as there may well be a predator satiation effect in acorn-rich years; there is a distinct suggestion that there is a curvilinear relationship in Fig. 23.6. This hypothesis is supported by the fact that there is a negative relationship between percentage galling and acorn density ($r = -0.635$, $p < 0.05$). Individual trees differ greatly in their susceptibility to gall wasp attack, and these differences are consistent from year to year (Hails and Crawley 1991). Indeed, one tree produced a consistently high acorn crop, but was never galled. These differences were not correlated with other measurable factors, such as acorn density or location of nearest Turkey oak and we suspect that these differences in susceptibility result from genetic differences between the trees. This hypothesis has yet to be tested.

Rosemary S. Hails

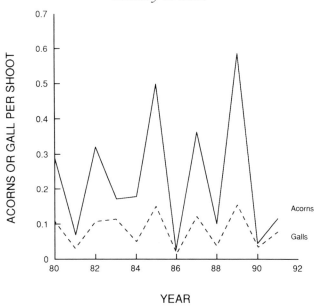

YEAR

Fig. 23.5. Fluctuations in acorn crop and gall density over a 12-year period. The solid line represents total resource density (that is acorns plus those acorns which were attacked) whilst the dotted line represents galls. Both acorns and galls were measured per shoot and the mean calculated over the same 30 trees each year at Silwood Park. After Hails and Crawley (1992).

Linking models with data: temporal trends in *A. quercuscalicis*

1. *Generation to generation density trends*

When attempting to model the dynamics of a bivoltine insect, there are a number of options. Two questions need to be answered: the first concerns the closeness with which the two generations are coupled and the second concerns the regulatory and/or limiting factors that are important in determining the dynamics of the two generations. The two generations may be quite closely coupled, in which case a knowledge of one will provide predictive power about the other. Alternatively, the two generations may be only loosely coupled, in which case it may be more practical to describe the dynamics of one generation without reference to the other. For our data from Silwood Park, the number of agamics in the autumn is not correlated with the number of sexuals earlier that year, nor is the number of sexuals related to agamic density the previous year, (Fig. 23.7). The two difference equations presented earlier appear to be very loosely coupled, so that a knowledge one genera-

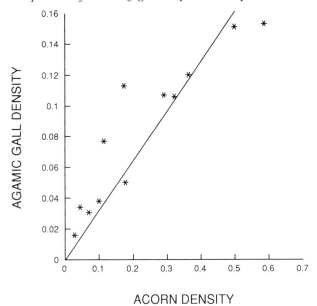

Fig. 23.6. The correlation between acorn density and gall density ($r = 0.728$, $p < 0.01$). Galls = 0.318 acorns with the line passing through the origin.

tion gives very little predictive power for the other even within the same year.

This lack of correlation between insect generations contrasts strikingly with the strong relationship between resource density and agamic gall density presented earlier (Fig. 23.6). Consequently, a much simpler and more direct way of describing the dynamics of this system may be to collapse the two equations into one, so bypassing the sexual generation. The agamic generation may then be described in terms of the resource function in the current and previous year as follows.

$$A_{t+1} = p_e \, h(R_t) + (1 - p_e) \, d \, h(R_{t-1}) \tag{23.3}$$

where A_{t+1} is the adult agamic density emerging in the spring of year $t + 1$ and all other parameters are as described previously.

2. Regulating factors in the sexual and agamic generations of A. quercuscalicis

Density-dependence in gall establishment was found in both the sexual and the agamic generations of this gall wasp. In the sexual generation it arose as a result of the oviposition pattern of the agamic females. Eggs were laid in a highly aggregated distribution, in spite of the fact that resources do not appear to be a limiting factor. In any one year, less than 10 per cent of available anthers of *Q. cerris* are destroyed by galling.

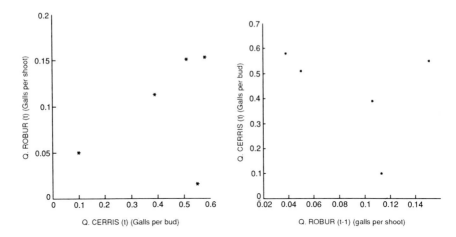

Fig. 23.7. (a) A recruitment plot, using galls per bud in the sexual generation to explain galls per shoot in the agamic generation the following autumn. (b) A recruitment plot using galls per shoot in the agamic generation the previous year to explain galls per bud in the sexual generation the following spring. Neither plot shows any significant correlation, illustrating that the two generations are very loosely coupled. After Hails and Crawley (1992).

The density-dependent mortality that results from these clumped egg distributions is weak. In contrast, in the agamic generation, there is strong density-dependent competition for acorns, illustrated by the directly proportional relationship between acorn and gall densities. Consequently, a knowledge of the resource function would allow prediction of agamic gall densities.

Given this strong regulating mortality in the agamic generation, the impact of any other density-dependent mortalities in the sexual generation would be difficult to predict and may well have no influence on the overall dynamics of the wasp.

References

Askew, R.R. (1962). The distribution of galls of *Neuroterus* (Hym.: Cynipidae) on oak. *Journal of Animal Ecology*, **31**, 439–55.
Claridge, M.F. (1962). *Andricus quercuscalicis* (Burgsdorf) in Britain (Hym.: Cynipidae). *Entomologist*, **95**, 60–1.
Crawley M.J. (1984). Big stories from little acorns grow. *Antenna (Bulletin of the Royal Entomological Society of London)*, **8**, 22–6.
Crawley, M.J. (1989). Plant life-history and the success of weed biological control

projects. In *Proceedings of the VIIth International Symposium on the Biological Control of Weeds*, (ed. E.S. Delfosse) pp. 17–26. Ist. Sper. Patol. Veg. (MAF).

Hails, R.S. and Crawley, M.J. (1991). The population dynamics of an alien insect: *Andricus quercuscalicis* (Hymenoptera: Cynipidae). *Journal of Animal Ecology*, **60**, 545–62.

Hails, R.S. and Crawley, M.J. (1992). Spatial density dependence in populations of a cynipid gall-former *Andricus quercuscalicis*. *Journal of Animal Ecology*, **61**, 567–83.

McGavin, G.C. (1981). *Andricus quercuscalicis* (Burgsdorf): an important new gall former. *Antenna*, **5** 19–20.

Pielou, E.C. (1977). *Mathematical ecology*. John Wiley, New York.

Taylor, A.D. (1988). Parasitoid competition and dynamics of host—parasitoid models. *American Naturalist*, **132**, 417–36.

Walde, S. and Murdoch, W. (1988). Spatial density dependence in parasitoids. *Annual Review of Entomology*, **33**, 441–66.

24. Mineral nutrition of galls induced by *Diplolepis spinosa* (Hymenoptera: Cynipidae) on wild and domestic roses in central Canada

GIUSEPPE BAGATTO and
JOSEPH D. SHORTHOUSE

Department of Biology, Laurentian University, Sudbury, Ontario, Canada

Abstract

The cynipid wasp *Diplolepis spinosa* induces large, multichambered galls on stems of the wild rose *Rosa blanda* throughout central Ontario, Canada and recently has become established on the domestic rose *Rosa rugosa*. The life history of *D. spinosa* and composition of the associated community of parasitoids will be described.

In this paper we compare various attributes of galls found on both hosts at two sites 260 km apart in central Ontario. Galls on domestic roses at both sites occur in dense clumps whereas those on wild roses tend to be widely distributed. Those on domestic roses are significantly larger and heavier and contain more *D. spinosa* than do galls on wild roses. Parasitoids are less dominant in galls on domestic roses than they are in galls of wild roses.

Concentrations of the mineral nutrients copper, zinc, iron, manganese, magnesium, and calcium within gall tissues were determined for both domestic and wild galls at the two sites. These minerals are known to be essential for plant growth and recent studies on other gall systems found that galls are physiological sinks for mineral nutrients as they are for other macronutrients and carbon assimilates. In the present study, we found that concentrations of the above mineral nutrients were not significantly different between galls on domestic roses at the two sites, nor were they different between galls from wild roses at the two sites. However, concentrations of these nutrients in galls collected from domestic roses were significantly less than those on wild roses. Correlations between the size, weight, and parasitism of galls on both domestic and wild roses, with the concentrations of mineral nutrients are discussed.

Plant Galls (ed. Michèle A. J. Williams), Systematics Association Special Volume No. 49, pp. 405–28. Clarendon Press, Oxford, 1994. © The Systematics Association, 1994.

Introduction

Galls induced by cynipid wasps are recognized as the most complex in the guild of gall-inducing insects. There are approximately 800 species of cynipids in North America and 170 in Europe (Dreger-Jauffret and Shorthouse 1992) and, like all other groups of gall inducers, their galls are distinguished from other growth and developmental anomalies by the existence of specific nutritional relationships between the gall inducer and the host plant. All benefits in the association appear to go to the cynipids because the host plant provides their larvae with both high quality food and a shelter.

Although much has been written about cynipids and the anatomy of their galls (see references in Meyer and Maresquelle 1983; Meyer 1987; Bronner 1992; Rohfritsch 1992), there are many basic aspects of cynipid biology that remain unknown. For example, the mechanism whereby cynipids solicit gall structures that are characteristic to the inducer and yet are markedly different from galls of closely related species is poorly understood. Furthermore, much remains to be learned about the process of gall induction, that most critical stage in gall biology when the insect gains control of growth processes of the attacked organ.

Another aspect of cynipid gall biology that has received little attention is gall physiology. This is mainly because physiological studies are best undertaken under controlled conditions and, unfortunately, the galls of most cynipids are difficult to culture in the laboratory. Furthermore, one needs detailed knowledge of the life cycle of the insect along with a basic understanding of gall attributes in order to undertake physiological studies and this information is available for very few species. Even so, it is known that cynipid galls, along with other gall inducers, act as physiological sinks, drawing assimilates and nutrients to the cells of the gall (see references in Abrahamson and Weis 1987).

Several researchers have measured carbon assimilates or the macronutrients such as nitrogen, phosphorus, and potassium attracted to the galls of various insects (Abrahamson and Weis 1987), but little research has been undertaken on the accumulation of essential micronutrients and macronutrients within gall tissues. However, we recently completed one such study on the chalcid gall of *Hemadas nubilipennis* Ashmead on lowbush blueberry (*Vaccinium angustifolium* Aiton) and even though we relied on field-collected galls rather than those cultured in the laboratory, we were able to show that the gall is a physiological sink for various mineral nutrients (Bagatto and Shorthouse 1991). This multi-chambered gall, about the size of a kidney bean, is anatomically similar (West and Shorthouse 1989) to cynipid galls and because several cynipid wasps of the genus *Diplolepis* and their galls are under investigation by the second author, we decided to undertake a similar study on the nutritional physiology of one of these galls, that of the stem-galler *Diplolepis spinosa* (Ashmead). This cynipid induces large multi-chambered galls on the wild rose *Rosa blanda* Ait. throughout central and northern Ontario, Canada and was suited for such a study because it is relatively common and its galls remain on the stems throughout the winter, allowing for large collections in the early spring prior to adult emergence. An added bonus for a nutritional study is that *D. spinosa* has become established on the domestic shrub rose *R. rugosa* Thunb. (Shorthouse 1988) providing the opportunity to compare the acquisition of mineral nutrients by galls found on hosts growing under different conditions. Furthermore, the presence of parasitoids in galls of both wild and domestic roses provided the opportunity to study the effects of inducer enemies on the role of galls as physiological sinks for micronutrients.

Thus, the purpose of this study is to compare various attributes of the gall of *D. spinosa* on both wild and domestic roses at two sites in central Ontario as they relate to the acquisition of mineral nutrients. We hypothesized that the identification of various mineral nutrients within the tissues of these galls could be used to study the role of galls as physiological sinks and that levels of mineral nutrients would be related to the size of galls, that is, larger galls would accumulate more mineral nutrients than smaller galls on both wild and domestic roses. We also hypothesized that parasitoids feeding on larvae of the inducers would interfere with the role of gall inducers in gall functioning and that galls with parasitoids would have less mineral nutrients than would non-parasitized galls.

Biology of cynipids and the form and function of their galls

Like all other gall-inducing insects, cynipids cause the formation of an atypical plant structure which provides the immatures with shelter and high quality food. This is done by the cynipids soliciting a wounding response by tissues of the host organ with the result that the affected cells become the centre of a morphogenic field that overrides normal developmental events (Rohfritsch and Shorthouse 1982). Although adult females deposit substances at oviposition and their freshly deposited eggs cause a lytic reaction and some cell division among adjoining cells, the stimulus for gall formation comes exclusively from the feeding larvae (Rohfritsch 1992). It is also known that cynipid larvae release fluids as they feed, however, it is thought that variation in the physical contact with cells by the larvae is responsible for the structurally distinct galls induced by each species (Magnus 1914; Meyer and Maresquelle 1983).

Cynipid galls are either single chambered or multichambered with each larva maturing in a separate chamber. Cynipid galls are also classified as prosoplasmas which means that they have characteristic external forms and tissues differentiated into well-defined zones, in contrast to the more primitive kataplasmas which are characterized by low-differentiated tissues and lack of constant external shape (Dreger-Jauffret and Shorthouse 1992). It also has been suggested that the wide variety of gall structures may be an integral part of the defensive strategy that benefits gall inducers by preventing generalist enemies from developing stereotyped search-and-attack behaviours (Cornell 1983).

There are two basic modes of reproduction among the cynipids. Some species exhibit heterogeny where alternating generations of the same species induce two distinct types of galls (Askew 1984), whereas other species are bisexual in which both sexes occur. The life cycles of sexual cynipids, such as those of the genus *Diplolepis*, are similar. All are univoltine with the adults emerging from the previous year's galls in the early spring. As with most gall insects, successfully soliciting a response from adjoining cells depends on the accurate timing and location of oviposition with each species having a narrow 'window of opportunity' in which eggs must be laid. Eggs are placed in contact with immature, undifferentiated tissues of the host because such cells are most susceptible to manipulation by the larvae. Once the larvae hatch, they feed on specialized cells lining their chambers throughout the summer and enter a prepupal stage in the autumn. They remain in the prepupal stage throughout the winter and then enter a pupal stage, which lasts approximately 1 week, once spring temperatures rise above 15°C. Adults chew their way to the outside and immediately begin searching for oviposition sites.

Developmental events for most cynipid galls follow three basic phases, referred to as initiation, growth, and maturation (Rohfritsch 1992). Initiation is the key phase in gall ontogeny, when cynipids stimulate the proliferation of parenchyma cells which quickly surround the feeding sites of each larva. Cells nearest the larvae become structurally and physiologically modifed forming the cytoplasmically dense cells referred to as a nutritive cells (Bronner 1992). These cells are rich in enzymes and soluble compounds such as glucose, sucrose, and amino acids (Bronner 1992). They line the inside surface of all chambers throughout gall growth and are not only the sole source of food for cynipid larvae, but also serve as the source of all other cells that form the gall. Cynipid larvae do not consume entire nutritive cells, but rather their mouthparts tear into the cells and they imbibe liquids by sucking movements. Of great importance in the feeding process is that as the nutritive cells are consumed, adjoining parenchyma cells, most of which contain starch granules, quickly develop similar cytological features and in turn are fed upon by the larvae. This process continues until the chambers of mature larvae are lined with collapsed cells.

Gall growth is the phase when biomass of the gall is vastly increased by division and enlargement of both gall parenchyma and the layer of nutritive cells. Vascular bundles appear within gall parenchyma soon after gall initiation and are joined to those of the host organ. All assimilates and nutrients pass from the host plant into the gall via these bundles and the movement of these substances is under the control of the larvae. Cynipids feed only minimally during the growth phase, but it is important to recognize that both the growth of the gall and physiological movement of assimilates and nutrients occurs only as long as the larvae are alive and active (Rohfritsch 1971). That is, within hours of the inducer being killed, growth and differentiation of gall tissues ceases and assimilates and nutrients are no longer directed into the gall.

The maturation phase occurs while the larvae are in the last instar. This is the main trophic phase of cynipids as they actively feed on the masses of nutritive tissues that have accumulated around them. A sheath of sclerenchyma cells often develops around the layers of nutritive cells which act as transfer tissue for water and solutes (Fourcroy and Braun 1967). The sclerenchyma sheath is synthesized only if the larvae are actively feeding; if the larvae are killed just before gall maturation, the sclerenchyma does not differentiate.

The growing tissues of cynipid galls function in similar ways to those of normal plant organs. That is, masses of cells proliferate and enlarge and are fed nutrients from the host plant through a network of vascular bundles. A major difference is that the development of galls is under the control of the insects within and not the host plant itself. However, the main functions of a gall to the cynipid are to shelter the larvae within from abiotic factors, provide a certain amount of protection from natural enemies, and provide a continuous supply of high quality food. Indeed, cynipid larvae exert a mobilizing effect on their host plants, drawing food materials to the gall. According to McCrea *et al.* (1985), galls can effect carbon flow through the host plant by either blocking the normal flow of resources and, thus, accumulate sugars, lipids, or other nutrients or they actively redirect resources from other parts of the host. Fourcroy and Braun (1967), for example, found that the leaf gall of *Aulax glechomae* L. reversed normal resource movement such that attacked leaves imported assimilates from other parts of the plant.

Galls of most cynipids are susceptible to attack by various species of minute wasps that feed on either larvae of the inducer or on tissues of the gall and are a major source of inducer mortality (Shorthouse 1973; Washburn and Cornell 1981; Askew 1984). Cynipid galls are apparent, predictable resources for natural enemies and inquilines, and the assemblage of these inhabitants associated with populations of galls induced by the same species of cynipid, result in microcommunities referred to as component communities (Claridge 1987). Such communities have been the subject of many ecological studies (Askew 1961, 1984; Shorthouse 1973; Jones 1983; Askew and Shaw 1986). Most parasitoids attacking cynipids deposit eggs early in gall development and interactions between the resulting larvae occur throughout the summer until the final composition of the community is established by autumn. It is this assemblage that re-establishes the community the following year. The rate of attack on the gall inducers is variable, ranging from relatively low levels to as high as 99 per cent (Askew 1975; Washburn and Cornell 1979, 1981). There is also considerable fluctuation from year to year and from site to site and it is clear that parasitoids and inquilines play an important role in the regulation of the gall-inducer population sizes and densities (Cornell 1983). Most parasitoids overwinter in their host galls and then exit by chewing their way free the following spring.

Several authors have suggested that gall inducers have been selected to modify plant tissues to maximize features of the gall that provide protection from enemies (Washburn and Cornell 1979; Jones 1983). For example, the increased mass of rapidly growing, multichambered galls may have at one time reduced the probability of parasitoids reaching the inducer larvae. Jones (1983) found that larger, multichambered galls have lower percentages of parasitism than small galls with fewer larvae and that the innermost larvae are less frequently attacked than larvae near the periphery . Jones (1983) also suggested that communal oviposition may have evolved as a means of increasing gall size. According to Abrahamson and Weis (1987), the evolutionary response to selection for increased gall size could occur through change in the gall inducer's ability to stimulate tissues of the host and to time their oviposition to the peak of plant reactivity.

Biology of *Diplolepis spinosa* and its gall

Diplolepis spinosa induces one of the most common galls in central and western Canada and northern United States and like all other species of *Diplolepis*, is restricted to the genus *Rosa*. There are approximately 30 species of *Diplolepis* in the Nearctic region and 10 in the Palaearctic region and the size and structure of their galls is characteristic of the species of gall inducer (Beutenmuller 1907; Shorthouse 1993). They range from lens-shaped structures on leaves to multichambered growths the size of lemons on roots. All species of *Diplolepis* are highly organ specific and the type of gall induced by each species is a result of the placement of eggs on either immature leaf or stem tissues. Thus, each species of *Diplolepis* is characterized as being either a leaf-, stem-, bud-, or root-gall inducer.

The taxonomy and diagnostic characters of *D. spinosa* are described in

Fig. 24.1. Habitus of *Rosa blanda* (wild roses) at the Timmins site. Note cane-shaped plants and galls (arrows).
Fig. 24.2. Habitus of *Rosa rugosa* (domestic roses) at the Timmins site. Note the more branched plants and the abundance of galls.

Shorthouse (1988, 1993). This species is found only on the wild rose *R. blanda* in eastern North America and *R. woodsii* Lindl. throughout the prairie provinces of Canada, but is not found on the more common *R. acicularis* in either eastern or western Canada. In central Ontario *R. blanda* grows in clearings in forests, in pastures, and along roadways and railroad lines. The stems of *R. blanda* are usually cane-like (Fig. 24.1) and they commonly grow in clusters, but individual specimens dispersed across large expanses of land are also found. *Rosa blanda* averages between 0.5 and 1 m in height, with plants growing in clusters usually taller. Galls of *D. spinosa* are usually found widely dispersed in habitats of *R. blanda* with no more than one or two galls per plant (Fig. 24.1). However, on occasion, clusters of up to five galls per plant are found. The occurrence of galls on single plants growing hundreds of metres from other galled plants illustrates the vagility of *D. spinosa*.

The second host of *D. spinosa*, the domestic shrub rose *R. rugosa*, is commonly grown in gardens and is a much sturdier (Fig. 24.2), more vigorous-growing, branched, leafy, and floriferous shrub than is *R. blanda*. *Rosa rugosa* usually grows 1–2 m high and is extremely hardy; its ability to tolerate cold winter temperatures makes it one of the most popular shrub roses in Ontario (Shorthouse 1988). It is grown also in rural areas, such as around farm buildings and here too is frequently a host for *D. spinosa*. Extremely dense populations of *D. spinosa* galls (up to 30 galls per m^2 are commonly found on *R. rugosa* in both urban and rural areas (Fig. 24.2).

Galls of *D. spinosa* on both wild (Fig. 24.3) and domestic (Fig. 24.4) roses are spherical or irregularly rounded and clothed with stout, sharp spines, although sometimes the surface is smooth. Galls on wild roses average 2.3 cm in diameter (range 1.5–4.0 cm) and usually are sparsely distributed in rose patches. Those on *R. rugosa* average 3.0 cm in diameter (range 1.8–5.0 cm) and commonly are found in dense clusters.

The life cycle of *D. spinosa* is similar to that of other species in the genus. Eggs are laid from mid-May to late June at the base of leaflets within leaf buds averaging 0.5 cm in length. The galls grow rapidly, incorporating all tissues of the bud and become firmly attached either to stems of mature plants or to 1-year-old sucker shoots. Immature galls are yellowish-green, soft, and usually clothed with dense, slender spines. When mature, they are reddish-brown or dull purple and so hard that pressure with a sharp knife is necessary to split them open. Larval chambers of galls on both wild and domestic roses are elongate and dispersed about the gall, but often are clustered near the centre (Fig. 24.5). A random sample of 44 galls on wild roses from central Ontario had a mean of 16.5 ± 4.1 (± SE, range 3–35) larval chambers per gall, whereas a sample of 95 galls from *R. rugosa* from Sudbury, Ontario had a mean of 32.5 ± 1.5 (± SE, range 9–73) larvae per gall.

Galls of *D. spinosa* are susceptible to attack by five species of parasitoids throughout the range of the galls in Ontario and all species attacking wild galls are also found in domestic galls. The most common species are *Eurytoma spongiosa* (Bugbee), *Orthopelma occidentale* Ashmead, either *Torymus solitarius* (Osten Sacken), *T. bedeguaris* (L.) *T. flavacoxa* Osten Sacken), or *T. chrysochlorus* (Osten Sacken); less common inhabitants are

Fig. 24.3. Typical gall of *Diplolepis spinosa* on *Rosa blanda* at the Timmins wild site.
Fig. 24.4. Typical cluster of galls of *Diplolepis spinosa* on *Rosa rugosa* at the Timmins domestic site.
Fig. 24.5. Dissected gall of *Diplolepis spinosa* collected in the early spring showing arrangement of inducer larvae and pupae in their chambers.
Fig. 24.6. Dissected gall of *Diplolepis spinosa* collected in the early spring showing larvae and pupae of *Eurytoma spongiosa* in inducer chambers.

Tetrastichus sp. and *Habrocytus* sp. In all cases, the parasitoids emerge from the previous years galls within 2–4 weeks of the gallers and oviposit within immature galls in the early growth phase that are less than 1 cm in diameter. Eggs are deposited either on the inside surface of the chambers or on the larvae of *D. spinosa* and, in all cases, the larvae of the *D. spinosa* are no more than one-quarter full-grown when the parasitoids hatch and begin to feed. Because larvae of the parasitoids *E. spongiosa*, *O. occidentale*, and the *Torymus* mature to a size almost equal to that of *D. spinosa* and they consume only one *D. spinosa* larva, they must feed on the immature larvae of the inducer without killing it. That is, these three species of parasitoids remove fluids from the host *D. spinosa* throughout most of the gall's growth and maturation phase while the *D. spinosa* larva continues to feed and mature. The *D. spinosa* larva is only killed and consumed once it is fully grown, such that by the end of season, chambers with larvae of *D. spinosa* attacked by parasitoids contain only the larva of the parasitoid (Fig. 24.6). Several larvae of *Tetrastichus* sp. attack each *D. spinosa* larva, but it is not known when they attack or kill their hosts. Mature larvae of *Habrocytus* sp. are substantially smaller than mature larvae of *D. spinosa* and although maturing larvae of *Habrocytus* are frequently found feeding on maturing larvae of *D. spinosa*, it is not known when the inducer is killed.

Study sites

Galls from both wild and domestic roses used in the study were collected at two sites in central Ontario in May 1991, prior to the emergence of any adults. One site, referred to in this chapter as the Chelmsford site, is located approximately 20 km NW of Sudbury. *Rosa blanda* here was found scattered in an abandoned pasture of approximately 5 ha, a habitat typical for this species. Galls were found widely dispersed over the roses at this site. Galls on the domestic rose, *R. rugosa*, were collected on the same day from four roses growing in a typical urban site (between a lawn and a roadway in the front yard of a residence) approximately 5 km east of the wild site; the four roses at this site were 4 years old at the time the galls were initiated in the spring of 1990.

The second site is referred to as the Timmins site. Galls on *R. blanda* here were from a large patch growing along a railroad line 1 km south of South Porcupine, which is approximately 5 km east of Timmins. Galls from approximately 10 domestic roses came from plants in typical urban settings (private gardens) in both South Porcupine and Timmins.

Materials and Methods

Galls at all four sites were removed from their host plants by snipping the galls near their point of attachment. A total of 45 galls were collected at the Chelmsford wild site, 281 at the Chelmsford domestic site, 300 at the Timmins wild site, and 74 at the Timmins domestic site. At the two wild sites, galls were collected by walking haphazardly throughout the roses and removing every gall observed. At the two urban sites, all galls were removed from the attacked roses. The galls were returned to the laboratory and each was numbered, measured, and weighed. The maximum diameter of each gall was taken with a digital calliper, and each gall was weighed (wet weight) with a Mettler balance. Each gall was then placed into an individual whirl-pak® plastic bag and stored at room temperature. Adult wasps were allowed to exit their galls and die in the bags before they were removed, identified and counted. No further emergents were obtained by the middle of June and at this time 20 galls from each collection were chosen such that a range of sizes from smallest to largest were included, however, no attempt was made to choose galls based on the number of inducers or parasitoids. Each gall was dried and weighed (dry weight) and then ashed in a muffle furnace at 500°C and the remaining ash weighed. Organic and mineral weights were defined as follows: mineral weight = ash weight and organic weight = dry weight − ash weight.

Each of the 20 ashed galls was then analysed for the mineral nutrients Cu, Ni, Zn, Mn, Mg, and Ca by the methods described in Bagatto and Shorthouse (1991). Briefly, samples were digested in aqua regia and levels of mineral nutrients determined by flame atomic absorption spectrophotometry using a Perkin–Elmer atomic absorption spectrophotometer model 303. Blank samples and analysis of standard reference material were performed concurrently. Descriptive statistics were calculated for each variable and one-way ANOVAs were computed to determine site (Chelmsford and Timmins) and type (wild and domestic) differences. In addition, a factor analysis was computed for all variables to isolate and identify causal factors underlying the relationships among variables. The factor analysis used a VARIMAX rotation of the factor matrix.

Results

Domestic galls from both urban sites were larger and heavier than galls on roses at the two wild sites. Indeed, galls on domestic roses were approximately 25 per cent larger than wild galls (Fig. 24.7a) and twice as heavy (Fig. 24.7b). Dry weights of domestic galls were also significantly heavier than wild galls (Fig. 24.7c). Of interest, galls on the domestic roses at Chelmsford were larger and heavier (both wet and dry weight) than were galls on the domestic roses at Timmins, whereas the size and weight of wild galls were the same at both sites.

Galls from domestic roses at the two sites also contained a significantly larger number of inhabitants (chambers) per gall than did the wild galls at both sites (Fig. 24.8a), however, the numbers of inhabitants in both wild and domestic galls were greater at the Chelmsford site than they were at both Timmins sites. This same relationship occurred for the

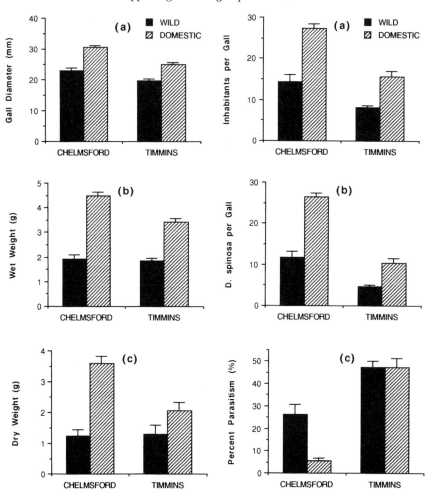

Fig. 24.7. (a) Mean diameter, (b) wet weight, (c) dry weight per gall for wild and domestic galls from the Chelmsford and Timmins sites.
Fig. 24.8. (a) Mean number of inhabitants per gall, (b) mean number of *Diplolepis spinosa* per gall, and (c) per cent parasitism per gall for wild and domestic galls from the Chelmsford and Timmins sites.

number of *D. spinosa* per gall (Fig. 24.8b), that is, there were substantially more *D. spinosa* larvae in galls of domestic roses at Chelmsford than in wild galls at the Chelmsford site. Approximately 25 of the 27 gall inhabitants in domestic galls at the Chelmsford site were gall inducers compared to 10 of the 15 gall inhabitants in domestic galls from the Timmins site (Fig. 24.8a and b).

Parasitoids were far more dominant in the wild galls at Chelmsford

Table 24.1. Factor analysis of morphometric and insect inhabitant variables of galls collected from wild and domestic roses at the Chelmsford and Timmins sites ($n = 700$ galls)

Chelmsford				Timmins			
Wild		Domestic		Wild		Domestic	
F1[a]	F2	F1	F2	F1	F2	F1	F2
X1[b]		X1		X1		X1	X1
X2		X2		X2		X2	X2
X3	X3	X3		X3	X3	X3	
X4		X4		X4		X4	
	X5		X5		X5		X5

[a] Hypothesized, unmeasured, and underlying variables which are presumed to be the sources of the observed variables; often divided into unique and common factors.
[b] Variables used to build the factor matrix: X1, gall diameter; X2, wet weight; X3, number of insect inhabitants per gall; X4, number of gall inducers per gall; X5, number of parasitoids per gall.

than they were in the domestic galls at Chelmsford (Fig. 24.8c). That is, a mean of 25 per cent of all inhabitants per gall were parasitoids in the population of wild galls from Chelmsford, whereas only 5 per cent of the inhabitants per gall were parasitoids in the Chelmsford domestic galls. Parasitoids represented approximately half of all gall inhabitants in both the wild and domestic galls at the Timmins sites (Fig. 24.8c).

Factor analysis of gall diameter, wet weight, number of inhabitants per gall, number of *D. spinosa* per gall, and the number of parasitoids per gall for the 700 wild and domestic galls collected at Chelmsford and Timmins illustrates that two underlying factors, F1 and F2, are responsible for the covariation among the observed variables (Table 24.1). Most of the variance is explained by F1 which grouped gall diameter, wet weight, total number of inhabitants, and the total number of inducers for wild and domestic galls at the Chelmsford and Timmins sites. That is, the size and weight of galls at all four sites is related to both the total number of inhabitants and the number of inducers (Table 24.1). The number of parasitoids per gall was grouped in a separate factor (F2) meaning that the size and weight of galls are not correlated with the presence of parasitoids.

Wild galls accumulated significantly lower concentrations of copper ($7 \mu g \ g^{-1}$) than domestic galls ($12 \mu g \ g^{-1}$) at both the Chelmsford and Timmins sites and site differences were not significant (Fig. 26.9a). Concentrations of zinc in wild and domestic galls showed contrasting patterns of accumulations at the two sites (Fig. 24.9b), that is, levels of zinc are similar for wild galls ($22 \mu g \ g^{-1}$) at the two sites whereas domestic galls from Timmins contained a significantly greater amount of zinc ($30 \mu g$

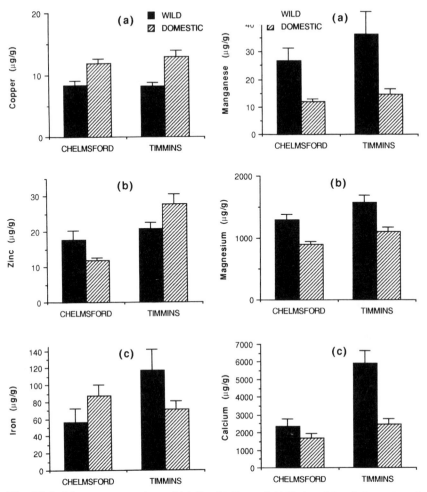

Fig. 24.9. Mean concentrations of (a) Cu, (b) Zn, and (c) Fe in wild and domestic galls from the Chelmsford and Timmins sites.

Fig. 24.10. Mean concentrations of (a) Mn, (b) Mg, and (c) Ca in wild and domestic galls from the Chelmsford and Timmins sites.

g^{-1}) than domestic galls from Chelmsford (11 μg g^{-1}). Concentrations of iron were not significantly different among wild and domestic galls with 80 μg g^{-1} from both wild and domestic galls from Chelmsford and the domestic galls from Timmins (Fig. 24.9c); however, the concentration of iron in the wild galls from Timmins was 120 μg g^{-1}.

The patterns for manganese, magnesium, and calcium accumulations were similar for both wild and domestic galls at the two sites (Fig. 24.10a, b, and c) with concentrations of these elements significantly greater in

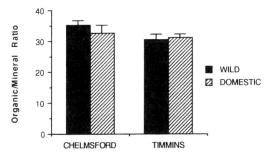

Fig. 24.11. Mean organic to mineral ratios for wild and domestic galls from the Chelmsford and Timmins sites.

wild galls at the two sites. In addition, calcium displayed a site difference being significantly greater in wild galls $(6000 \mu g \ g^{-1})$ at Timmins compared with wild galls from Chelmsford $(2500 \mu g \ g^{-1})$. Ratios of organic to mineral mass (organic weight/ash weight) between wild and domestic galls at the two sites were not significantly different (Fig. 24.11), with the ratios averaging 31.

Factor analysis of gall diameter, weight and dry weight, number of inhabitants per gall, number of inducers per gall, number of parsitoids per gall, copper, zinc, manganese, iron, magnesium, calcium, O/M ratios, ash weight, and organic weight for the 80 wild and domestic galls at the four sites resulted in four underlying factors (F1, F2, F3 and F4) being responsible for the covariation among the observed variables (Table 24.2). Most of the variance could be explained by F1 which grouped gall diameter, wet weight, number of inhabitants, number of gall inducers, ash weight, and organic weight for wild and domestic galls at both sites. The number of parasitoids per gall was grouped with concentrations of the mineral nutrients in separate factors (F2, F3, and F4) and was, in general, not related to the morphometric characteristics of the galls.

Discussion

When galls of *D. spinosa* are found on the domestic shrub rose *R. rugosa*, they appear in much denser clusters than they do on wild hosts. This is especially the case within 2 or 3 years of *R. rugosa* being planted and the gall wasps becoming established. The reason for the abundance is likely related to both the physiological condition and architectural characteristics of *R. rugosa* compared to *R. blanda*. We suspect that the robust rate of growth and the abundance of oviposition sites on the much bushier *R. rugosa* roses compared to the wild roses is responsible for the

Table 24.2. Factor analysis of morphometric, insect inhabitant, and mineral nutrient variables of galls collected from wild and domestic roses at the Chelmsford and Timmins sites ($n = 80$ galls)

Chelmsford Wild				Chelmsford Domestic				Timmins Wild				Timmins Domestic			
F1[a]	F2	F3	F4	F1	F2	F3	F4	F1	F2	F3	F4	F1	F2	F3	F4
X1[b]				X1				X1				X1			
X2				X2				X2				X2			
X3				X3			X3	X3				X3			
X4	X4				X4			X4				X4			
X5						X5		X5				X5			X5
	X6					X6		X6	X6				X6		X6
	X7	X7		X7			X7			X7			X7		
			X8	X8									X8		
	X9	X9		X9			X9	X9					X9		
	X10			X10						X10	X10				
		X11		X11				X11							
		X12						X12	X12					X12	
X13		X13						X13	X13					X13	X13
X14								X14	X14	X14		X14	X14		
X15						X15		X15				X15			

[a] Hypothesized, unmeasured, and underlying variables which are presumed to be the sources of the observed variables; often divided into unique and common factors.

[b] Variables used to build the factor matrix: X1, gall diameter; X2, wet weight; X3, dry weight; X4, number of insect inhabitants per gall; X5, number of gall inducers per gall; X6, number of parasitoids per gall; X7, copper ($\mu g\,g^{-1}$); X8, zinc ($\mu g\,g^{-1}$); X9, manganese ($\mu g\,g^{-1}$); X10, iron ($\mu g\,g^{-1}$); X11, magnesium ($\mu g\,g^{-1}$); X12, calcium ($\mu g\,g^{-1}$); X13, organic/mineral ratio; X14, ash weight; X15, organic weight.

increased abundance. We also suspect that recently established *R. rugosa* have more buds suitable for oviposition than do wild roses. Furthermore, the ability of adults of *D. spinosa* to locate individual, isolated plants and beds of *R. rugosa* so quickly after their being established is amazing, especially when other species of *Diplolepis* are recorded as being hesitant and poor fliers (Kinsey 1920). Vast numbers of adults are undoubtedly leaving galled wild roses each spring and are either flying or being transported by the wind such that they can locate domestic roses in urban areas. Once established on *R. rugosa*, it is evident that the populations of both *D. spinosa* and their galls increase rapidly.

The substantially larger (Fig. 24.7a) and heavier (Fig. 24.7b and c) galls on domestic roses, compared to those on wild roses is likely to be due to the more vigorous growing characteristics of *R. rugosa*, that is, galls on vigorously growing roses would have more resources to draw upon than galls on wild roses that are much slower growing. Furthermore, it is likely that domestic roses are watered and fertilized by gardeners and the resulting vigour would enhance gall growth. However, Hartley and Lawton (1992), studying two cynipids that induce single-chambered galls on the leaves of oak, found that increasing the level of nitrogen to host trees did little to enhance survivorship of the inducers. Perhaps further studies will reveal differences in the nutritional requirements of single-chambered vs multi-chambered galls. Obviously much remains to be learned about the variation in galls growing at different sites and under different conditions, however, for our gall system one would assume that galls will be less abundant on the more stressed wild roses than on the larger and less stressed domestic roses. Waring (1986) came to a similar conclusion in studies of other galls and suggested that well-watered plants are better resources for some gall insects.

A rapidly growing population of *D. spinosa* would also increase the chances for multiple ovipositions on the same buds (Kelleher 1988) and galls of increased size with increased *D. spinosa* per gall (Fig. 24.8b) would result. Also, we suspect that *D. spinosa* would become established on new urban roses for several years before the parasitoids locate them and build up their populations. This could explain why the Chelmsford galls have more *D. spinosa* (Fig. 24.8b) and fewer parasitoids (Fig. 24.8c) per gall than do the Timmins domestic galls that were on much older plants. It also is possible that the Timmins wild site is older than the Chelmsford wild site and, if so, would explain the differences in chambers occupied by *D. spinosa* or parasitoids (Fig. 24.8b and c). We suggest that the percentage of chambers occupied by parasitoids and the diversity of the component community increases with time for galls on both wild and domestic roses.

Several authors have discussed the possibility of galls providing protection against parasitoids (for example, Weis *et al.* 1985) and suggested a correlation with diversity of gall types (for example, Askew 1961; Cornell 1983). This is the so-called 'enemy hypothesis' of Price *et al.* (1986) and there are some galls where the degree of enemy attack does decline with increasing gall size (Jones 1983; Stille 1984; Weis *et al.* 1985), that is, if the larvae of the inducer escape detection when the galls are small and within reach of parasitoids, they are protected once imbedded inside thick-walled maturing galls. Our results, in contrast, indicate that for the gall of *D. spinosa*, an increase in gall size and the associated number of larval chambers per gall, does not reduce susceptibility to attack by parasitoids (Table 24.1). However, Kelleher (1988) found some correlation between larger gall size and a reduction in parasitism for an isolated population of *D. spinosa* galls on domestic roses. We also showed that parasitoids do not influence the size of *D. spinosa* galls, that is, galls with all chambers inhabited by parasitoids are just as large as galls with all chambers inhabited by inducers (Table 24.1). This is in contrast to other gall systems where parasitoids cause galls to remain small (Askew

1975; Washburn and Cornell 1979). However, parasitoids of the *H. nubilipennis* gall act in a similar manner to those in the *D. spinosa* gall and do not influence gall size (Shorthouse *et al.* 1990).

All species of parasitoids associated with galls of *D. spinosa* oviposit when the galls are immature, except for *Torymus* which can oviposit into fully-grown galls (Kelleher 1988). Parasitoids in galls of other species have also been shown to oviposit in mature galls (Askew 1961; Washburn and Cornell 1979). Eggs of all *D. spinosa* parasitoids, with the possible exception of *Torymus*, hatch when the inducer is approximately half grown (Shorthouse, unpublished data) and obviously if the inducer was killed at this stage, the gall would stop growing. Furthermore, it is important for the inducer host to reach approximately its mature size before it is killed and completely consumed because the larvae of full-grown parasitoids are about the same size as full-grown *D. spinosa* larvae. We suggest that the attacked inducer larvae must remain alive and continue stimulating gall cells while the immature parasitoid feeds. Parasitoids that feed in this manner and allow the host to continue feeding and grow beyond the stage attacked are referred to as 'koinobionts' by Askew and Shaw (1986).

The gall of *D. spinosa*, especially with its occurrence on both wild and domestic roses, proved ideal for studying mineral nutrients associated with a cynipid gall. In a previous study (Bagatto *et al.* 1991) we were able to show that galls of *D. spinosa* differentially accumulate mineral nutrients compared to non-galled tissues. For example, levels of copper and iron were 10 times greater in tissues of the gall compared to non-galled stem tissues allowing us to conclude that these galls are physiological sinks for mineral nutrients. In this 1991 study, relationships between mineral nutrients and the size of galls, along with community composition, were not examined, however, in this current study, gall size and the number of inhabitants associated with both wild and domestic galls at the two sites were considered allowing us to further resolve the relationships between these parameters and mineral nutritional status of the galls.

The present study also has shown that galls of *D. spinosa* on domestic roses are significantly larger than wild galls and that the size of galls is influenced by the number of inducers present. We had hypothesized that larger galls would be stronger physiological sinks for mineral nutrients than would smaller galls and this was the case; organic and ash content was greater in large galls than in small galls at all sites (Table 24.2), that is, on a weight basis, as the number of inducers per gall increases, there is an increase in the amount of organic matter (for example, carbon, nitrogen, sulphur, etc.) and ash (for example, copper, zinc, magnesium, etc.) and positive correlations among these variables are highly significant. Of interest, the organic to mineral (ash) ratios were not significantly different at all four sites (Fig. 24.11). These ratios would suggest that the total ash content, which is comprised mainly of the macronutrients calcium, magnesium, and potassium and the micronutrients copper, zinc, iron, and manganese, expressed on a gram dry weight basis, is the same for wild and domestic galls at both sites. Thus, the ability of the gall to act as a physiological sink for mineral nutrients is the same for all mineral nutrients regardless of size of the gall or site; there is no relative increase in mineral nutrients as a function of gall size.

However, the conclusion that similar amounts of mineral nutrients are accumulated regardless of gall size is somewhat misleading, because analysis of total mineral content does not allow one to assess differences in individual nutrients that make up the total mineral pool of the gall. Indeed several factors such as the physiology of the host or the inducers and interactions of the minerals may act to reduce or enhance the accumulation of mineral nutrients within galls. For example, when we compared the levels of individual mineral nutrients on a gram dry weight basis for the wild and domestic galls at the two sites, variations were apparent (Figs 24.9 and 24.10). Concentrations of copper are less in the wild galls compared to domestic galls, whereas concentrations of manganese, magnesium, and calcium are greater in wild galls than in domestic galls. Furthermore, there was no consistent pattern in the levels of zinc and iron for wild and domestic galls at the two sites. At first glance, this would suggest a difference in the physiological role of the nutrients in domestic and wild galls; however, our understanding of the physiological role of mineral nutrients in gall systems is poor. Mineral nutrients in most plant systems are essential cofactors in energetic and metabolic pathways, protein synthesis, hormonal control, and membrane stability (Marschner 1986) and, presumably, many of the physiological roles of mineral nutrients in the tissues of galls are the same as those in normal plants. However, it was not our intention here to elucidate these roles.

We also hypothesized that the presence of parasitoids would influence the accumulation of mineral nutrients, that is, galls with parasitoids would have less mineral nutrients than galls inhabited by inducers. Of interest, our results indicate the opposite, that is, there is a positive correlation between galls containing parasitoids and the levels of some mineral nutrients. Although the factors which are derived from a correlation matrix of all the variables outlined in Table 24.2 illustrate that the primary factor (F1) influencing gall size, the organic and mineral content, and the total nutrient pool, is controlled by the gall inducer, the relative composition of the mineral pool is influenced by the number of parasitoids within the gall. Thus, the secondary and tertiary factors (F2 and F3), which relate levels of various mineral nutrients to the number of parasitoids suggest that the unexplained variance associated with the mineral nutritional status of the gall involves the influence of parasitoids.

That is, positive correlations are found between levels of all mineral nutrients and the number of parasitoids (Table 24.2), however, there are differences between wild and domestic galls at the two sites. For example, copper, manganese, and iron concentrations in the galls of wild and domestic roses are greater in galls that are highly parasitized. Similarly, parasitized wild and domestic galls from Timmins contained greater ash and mineral content. Unfortunately, we were not able to determine how and to what degree the mineral nutrients are influenced by parasitoids and we suspect that our sample size of 20 galls from each site is not sufficiently large given the number of variables considered in this study. Further investigations on the influence of parasitoids on mineral nutrient acquisition in this gall and others, should separately consider samples of galls within a specific size range and containing similar numbers of inducers and parasitoids.

However, even with the size of our samples and the fact that we did not choose galls with known inhabitants (for example, all inducers or all parasitoids) for mineral nutrient analysis, our findings strongly suggest that parasitoids rearrange, alter or convert mineral nutrients within their host insect and the tissues of the galls they inhabit. Such concentrations or rearrangements are not likely for the nutritional needs of the parsitoids but instead are an incidental affect of parasitoid behaviour. Vinson and Iwantsch (1980) suggested that insect hosts represent the 'container' which dictates the nutritional resources available and provides all the nutrients necessary for the growth and development of a parasitoid. In the gall of *D. spinosa*, it appears that the parasitoids indirectly influence the nutritional status of the gall by altering the feeding behaviour of their hosts. We suspect that because parasitoids feeding on immature *D. spinosa* cannot kill their hosts until they are sufficiently mature and provide the biomass necessary for the parasitoid to reach a similar size, the parasitoids stimulate their hosts to feed on and cause the proliferation of more gall cells than normal.

In a study of another gall, that of the chalcid *H. nubilipennis*, Bagatto (1992) showed that higher concentrations of various mineral nutrients were found within the tissues of the parasitoids than within tissues of the inducers and concluded that the parasitoids feed on the inducers through-out their larval development and do not kill them until the inducer has nearly matured. We suspect that a similar feeding strategy occurs within the parasitized galls of *D. spinosa*, that is, parasitoids stimulate *D. spinosa* to increase their food intake to support the growing larval parasitoid on their exterior. One effect of the inducer increasing the power of the gall's physiological sink when parasitized over that of a non-parasitized inducer, is that an additional flux of mineral nutrients would be attracted to the gall tissues. At the time we selected galls for mineral analysis, we did not suspect that parasitized galls might have more mineral nutrients than non-parasitized galls and, unfortunately, we did not keep the adults of both the inducers and parasitoids for separate analysis. However, if such an analysis is performed and our conclusions are correct, higher levels of mineral nutrients will be found in the tissues of the parasitoids than in non-parasitized inducers, just as occurred with the larvae of inducers and parasitoids associated with blueberry galls (Bagatto 1992).

The mechanisms whereby gall inducers influence the movement of nutrients, as well as other assimilates, from their host plants into the tissues of their galls, remains a poorly known aspect of gall biology (Hartley and Lawton 1992), but an important one worthy of future research as we strive to understand the intimate relationships between these specialized herbivores and their hosts. Galls are in nearly all respects a plant organ and given the vast amount of physiological research devoted to the mineral nutrition of plants, many of the same techniques can be employed to understand fundamental aspects of gall physiology. The results we have reported here suggest that future studies of the accumulation of mineral nutrients within gall tissues from initiation to maturity, as well as accumulation within larvae of the inducers from soon after hatching to the end of the summer season, would be particularly revealing. Furthermore, our suggestion that parasitoids stimulate the accumulation of more nutrients within the tissues of parasitized inducers and their galls than do non-parasitized hosts in galls containing only inducers, has important implications for the overall understanding of interactions between the three trophic levels (Price *et al.* 1980). It might also mean a new approach in the use of gall inducers in the biological control of weeds if it can be shown similarly that parasitized gall inducers inflict more host damage by stimulating an increased nutrient drain than do non-parasitized inducers in non-parasitized galls.

Acknowledgements

This study was supported by research grants from the Natural Sciences and Engineering Research Council of Canada and Laurentian University Research Fund awarded to the second author. We thank Louise Paquette and Ninha Maia for their help with rearing gall inhabitants and analysing tissues for mineral nutrients, Michael Malette for his advice on factor analysis, and Mary Roche for printing the photographs. We also thank W.G. Abrahamson and O. Rohfritsch for their comments on the manuscript.

References

Abrahamson, W.G. and Weis, A.E. (1987). Nutritional ecology of arthropod gall makers. In *Nutritional ecology of insects, mites, spiders, and related invertebrates*, (ed. F. Slansky Jr and J.G. Rodriguez). pp. 235–58. John Wiley & Sons, New York.

Askew, R.R. (1961). On the biology of the inhabitants of oak galls of Cynipidae (Hymenoptera) in Britain. *Transactions of the Society for British Entomology*, **14**, 237–68.

Askew, R.R. (1975). The organization of chalcid-dominated parasitoid communities centered upon endophytic hosts. In *Evolutionary strategies of parasitic insects and mites*, (ed. P.W. Price), pp. 130–53. Plenum, New York

Askew, R.R. (1984). The biology of gall wasps. In *Biology of gall insects*, (ed. T.N. Ananthakrishnan), pp. 223–71. Oxford & IBH Publishing Co., New Delhi.

Askew, R.R. and Shaw, M.R. (1986). Parasitoid communities: their size, structure and development. In *Insect parasitoids*, (ed. J. Waage and D. Greathead), pp. 225–64. Academic Press, London.

Bagatto, G. (1992). Concentrations of minerals (Cu, Ni, Zn, Fe, Mn, Mg, and Ca) in tissues of lowbush blueberry (*Vaccinium angustifolium*) and an insect gall induced by *Hemadas nubilipennis* (Hymenoptera: Pteromalidae) and its inhabitants in metal-contaminated and non-contaminated sites. Unpublished Ph D thesis. Queen's University, Kingston, Ontario.

Bagatto, G. and Shorthouse, J.D. (1991). Accumulation of copper and nickel in plant tissues and an insect gall of lowbush blueberry, *Vaccinium angustifolium*, near an ore smelter at Sudbury, Ontario, Canada. *Canadian Journal of Botany*, **69**, 1483–90.

Bagatto, G., Zmijowskyj, T.J., and Shorthouse, J.D. (1991). Galls induced by *Diplolepis spinosa* influence distribution of mineral nutrients in the shrub rose. *HortScience*, **26**, 1283–4.

Beutenmuller, W. (1907). The North American species of *Rhodites* and their galls. *Bulletin of the American Museum of Natural History*, **23**, 629–51.

Bronner, R. (1992). The role of nutritive cells in the nutrition of cynipids and cecidomyiids. In *Biology of insect-induced galls*, (ed. J.D. Shorthouse and O. Rohfritsch), pp. 118–40. Oxford University Press, New York.

Claridge, M.F. (1987). Insect assemblages—diversity, organization and evolution. In *Organization of communities: past and present*, (ed. J.H.R. Gee and P.S. Giller), pp. 141–62. Blackwell, Oxford.

Cornell, H.V. (1983). The secondary chemistry and complex morphology of galls formed by the Cynipinae (Hymenoptera): why and how? *American Midland Naturalist*, **110**, 225–32.

Dreger-Jauffret, F. and Shorthouse, J.D. (1992). Diversity of gall-inducing insects and their galls. In *Biology of insect-induced galls*, (ed. J.D. Shorthouse and O. Rohfritsch), pp. 8–33. Oxford University Press, New York.

Fourcroy, M. and Braun, C. (1967). Observations sur la galle de *l'Aulax glechomae* L. sur *Glechoma hederacea* L. II. Histologie et rôle physiologique de la coque sclérifiée. *Marcellia*, **34**, 3–30.

Hartley, S.E. and Lawton, J.H. (1992). Host-plant manipulation by gall-insects: a test of the nutrition hypothesis. *Journal of Animal Ecology*, **61,** 113–19.

Jones, D. (1983). The influence of host density and gall shape on the survivorship of *Diastrophus kincaidii* Gill. (Hymenoptera: Cynipidae). *Canadian Journal of Zoology*, **61,** 2138–42.

Kelleher, M.J. (1988). Influence of parasitoids on a population of *Diplolepis spinosa* (Ashmead) (hymenoptera: Cynipidae) found on *Rosa rugosa* Thunb. (Rosaceae) in Sudbury, Ontario. Unpublished M Sc. thesis Laurentian University, Sudbury, Ontario.

Kinsey, A.C. (1920). Life histories of American Cynipidae. *Bulletin of the American Museum of Natural History*, **42,** 319–57.

Magnus, W. (1914). *Die Entstehung der Pflanzengallen verursacht durch Hymenopteren.* G. Fischer, Jena.

Marschner, H. (1986). *Mineral nutrition in higher plants.* Academic Press, New York.

McCrea, K.D., Abrahamson, W.G., and Weis, A.E. (1985). Goldenrod ball gall effects on *Solidago altissima*: C translocation and growth. *Ecology*, **66,** 1902–7.

Meyer, J. (1987). *Plant galls and gall inducers.* Borntraeger, Berlin.

Meyer, J. and Maresquelle, H.J. (1983). *Anatomie des Galles.* Borntraeger, Berlin.

Price, P.W., Boutin, C.E., Gross, P., McPheron, B.A., Thompson, J.N., and Weis, A.E. (1980). Interactions among three trophic levels: influence of plants on interactions between insect herbivores and natural enemies. *Annual Review of Ecology and Systematics*, **11,** 41–65.

Price, P.W., Waring, G.L., and Fernandes, G.W. (1986). Hypotheses on the adaptive nature of galls. *Proceedings of the Entomological Society of Washington*, **88,** 361–3.

Rohfritsch, O. (1971). Développement cécidien et rôle due parasite dans quelques galles d'arthropodes. *Marcellia*, **37,** 233–339.

Rohfritsch, O. (1992). Patterns in gall development. In *Biology of insect-induced galls*, (ed. J.D. Shorthouse and O. Rohfritsch) pp. 60–86. Oxford University Press, New York.

Rohfritsch, O. and Shorthouse, J.D. (1982). Insect galls. In *The molecular biology of plant tumors*, (ed. G. Kahl and J.S. Schell), pp. 131–52. Academic Press, New York.

Shorthouse, J.D. (1973). The insect community associated with rose galls of *Diplolepis polita* (Cynipidae, Hymenoptera). *Quaestiones Entomologicae*, **9,** 55–98.

Shorthouse, J.D. (1988). Occurrence of two gall wasps of the genus *Diplolepis* (Hymenoptera: Cynipidae) on the domestic shrub rose, *Rosa rugosa* Thunb. (Rosaceae). *Canadian Entomologist* **120,** 727–37.

Shorthouse, J.D. (1993). Adaptions of gall wasps of the genus *Diplolepis* (Hymenoptera: Cynipidae) and the role of gall anatomy in cynipid systematics. *Memoirs of the Entomological Society of Canada*, **165,** 139–63.

Shorthouse, J.D., MacKay, I.F., and Zmijowskyj, T.J. (1990). Role of parasitoids associated with galls induced by *Hemadas nubilipennis* (Hymenoptera: Pteromalidae) on lowbush blueberry. *Environmental Entomology*, **19,** 911–15.

Stille, B. (1984). The effect of host plant and parasitoids on the reproductive success of the parthogenetic gall wasp *Diplolepis rosae* (Hymenoptera: Cynipidae). *Oecologia*, **63,** 364–9.

Vinson, S.B. and Iwantsch, G.F. (1980). Host regulation by insect parasitoids. *Quarterly Review of Biology*, **55,** 143–5.

Waring, G.L. (1986). Galls in harsh environments. *Proceedings of the Entomological Society of Washington*, **88,** 376–80.

Washburn, J.O. and Cornell, H.V. (1979). Chalcid parasitoid attack on a gall wasp population *Acraspis hirta* (Hymenoptera: Cynipidae) on *Quercus prinus* (Fagaceae). *Canadian Entomologist*, **111,** 391–400.

Washburn, J.O. and Cornell, H.V. (1981). Parasitoids, patches and phenology; their possible role in the local extinction of a cynipid gall wasp population. *Ecology*, **62,** 1597–1607.

Weis, A.E., Abrahamson, W.G., and McCrea, K.D., (1985). Host gall size and oviposition success by the parasitoid *Eurytoma gigantea*. *Ecological Entomology*, **10,** 341–8.

West, A. and Shorthouse, J.D. (1989). Initiation and development of the stem gall induced by *Hemadas nubilipennis* (Hymenoptera: Pteromalidae) on lowbush blueberry, *Vaccinium angustifolium* (Ericaceae). *Canadian Journal of Botany*, **67,** 2187–98.

25. Biochemical modification of the phenotype in cynipid galls: cell membrane lipids

MARGRET H. BAYER

Fox Chase Cancer Center, Institute for Cancer Research, Philadelphia, PA 19111, USA

Abstract

Histological modifications of the normal growth of higher plants coincide with changes at the biochemical level, affecting cell differentiation and the cellular transport system. Since membranes play a vital part in cell morphogenesis and transformation, the biochemical composition of cell membranes from cynipid galls and from normal host tissues was studied. The fatty acid composition of phosphoglycerides as well as the activity of lipolytic enzymes were analysed in the vernal oak galls of the bisexual generation of *Andricus palustris*. Phosphoglycerides, glycosylacylglycerols, and fatty acids were identified by thin-layer and gas chromatography. Phospholipids of membranes from cynipid galls contained four times as much palmitate $(16:0)$ and seven and three times as much linoleate $(18:2)$ and linolenate $(18:3)$, respectively, than normal tissues. *In vitro* labelling of developing leaf galls with $[^{14}C]$-acetate and $[^3H]$-oleate $(18:3)$ was used to study the label distribution in monogalactosyldiacylglyceride (MGDG), digalactosyldiacylglyceride (DGDG), and phosphoglycerides. The incorporation of both labels was higher into leaf than into gall galactolipids, whereas the incorporation of oleate was slightly increased in the phospholipids of cynipid galls. In general, acyl chain unsaturation of the individual endogenous phosphoglycerides was higher in gall tissues. Immature cecidia tissues incorporated larger amounts of $[^{14}C]$acetate and $[^3H]$oleate into polar lipids than did the mature galls. The conclusion drawn from these data is that the distribution and composition of membrane glycerolipids and phosphoglycerides is affected by the interaction of the gall insect with the host plant tissue. Such modifications in the structure of membrane lipids may be responsible, in part, for the regulation of some of the metabolic activities initiated in the host by the gallicolous insect.

Plant Galls (ed. Michèle A. J. Williams), Systematics Association Special Volume No. 49, pp. 429–46. Clarendon Press, Oxford, 1994. © The Systematics Association, 1994.

Introduction

Cecidogenesis (gall formation) provides us with an exceptionally vivid example of the extraordinary abilities of the plant for modifications of the normal phenotype and for the distinct morphogenetic alterations, induced by the gall insect.

Plant galls, elicited by insects to supply the developing larvae with food and shelter, have often been referred to as hyperplasias, overgrowths, or as abnormal growths. In earlier papers, however, cecidogenesis was occasionally likened to plant tumourigenesis (Jones 1935). In all these abnormal growths the morphogenetic changes occur concomitantly with physiological and biochemical modifications at the cellular level.

In this chapter, we will focus on one of the many changes that are inherently associated with cecidogenesis, namely, the modification of plant cell membranes during growth and development of the cynipid gall. Prior to a discussion of some of these alterations, however, I will focus on the general concept of cell transformation in plants and on some of the features which differ between galls and plant tumours.

Cecidogenesis and tumourigenesis: a comparison

The normal growth of higher plants and animals is precisely regulated by distinct and characteristic developmental processes. In spite of their finite nature, however, these growth processes constitute a rather flexible pattern and changes that occur during growth and development may lead to atypical, abnormal, or pathological structures. In particular, in plants, regulatory processes can easily be modified by environmental factors (such as temperature, radiation, humidity, and nutritional conditions), by pathogenic infections, or by changes in the genetic constitution of the organism. Alterations in the normal phenotype may be very slight or may appear very modest or they can lead to significant manifestations of new morphological structures. Typical changes from a normal to an abnormal development in plants may be the result of, for example, bacterial and viral infections, pathogenic fungi, or animals (for example, nematodes and arthropods). The structures take on a characteristic shape and often exhibit well-defined cell differentiation and tissue organization. Nowhere are such changes more pronounced than in the highly organized plant galls, elicited by insects and their larvae. These zoocecidia represent new and distinct features, arising at an otherwise normally growing plant or plant organ through the interaction between gall insect and the host tissue. These galls should not be mistaken for genuine tumour formations, however. While the most rapidly growing plant tumours are generally unorganized and form amorphous cell

Table 25.1. Cecidogenesis and plant tumour formation

	Insect galls	Crown gall disease	Genetic plant tumours
Induction	Chemical (injury)	*Agrobacterium tumefaciens*; Ti plasmid DNA	Imbalance of parental genes in hybrids
Proliferation	Dependent on continuous stimulation	Autonomous	Autonomous
Cell stage during infection (*in vivo*)	Growing phase (embryonic to maturity)	Growing phase to maturity	Not applicable
Tissue affected	Meristem, parenchyma	Intercellular space	Meristem (cambium?)
Tissue growth	Differentiation, organization, specialization	Mostly amorphous; teratomas with organ differentiation	Mostly amorphous, some teratomas
Histology of growth pattern	Cell types arranged to form a functional whole	Mostly unorganized growth; some organization of clustered cell types	Mostly unorganized, some teratomas
Location	All organs, especially buds and leaves	All organs, especially at wounds (shoots, roots)	Stressed growth areas
Growth after grafting to host tissue	Not done	Excellent	Excellent
In vitro growth on basic medium	Some cell types divide; loss of organ specialization	Excellent	Excellent

Table 25.2. Classification of cecidogenesis/tumour formation

	Zoocecidia	Crown gall	Genetic tumours
Alteration	Phenotype	Genotype	Genotype
Growth progression	Self-limiting	Non-limiting	Non-limiting
Phenotypic plasticity	± Unlimited	Limited	Limited
Type of growth	Neoplasmatic	Neoplastic	Neoplastic

complexes (or, in some cases, teratomas), zoocecidia exhibit a distinct morphological organization while their growth is restricted and finite. Moreover, gall development is dependent upon the continued stimulation by either the pathogen or by specific pathogen-derived elicitors (Boysen-Jenson 1948; Garrigues 1956; Hori and Miles 1977; Rohfritsch 1980). These changes in growth pattern represent beautiful examples of dependent differentiation which have been classified by Braun (1959) as 'self-limiting overgrowths'. It has also been suggested that an insect gall with its 'determinate' growth, own polarity, and symmetry almost resembles a plant organ (Bloch 1965). A list of several features that distinguish the highly developed insect galls (cynipid galls) from crown galls (Braun 1978; Pengelly 1989) and plant tumours (Kostoff 1930) is given in Tables 25.1 and 25.2. Significant similarities can be found that are characteristic for at least two of the three listed abnormal growths. Therefore, the term 'neoplasmatic' has recently been proposed to distinguish these growths from the non-self-limiting neoplastic cell transformations (Bayer 1991).

Biochemical modification of the phenotype

Although numerous researchers have contributed to our understanding of the differentiation and development of insect galls (more recent reviews by Rohfritsch and Shorthouse 1982; Meyer and Maresquelle 1983), the gall-inducing signal molecules involved in insect–plant interactions remain obscure. We know that the growth hormones auxin and cytokinin are intimately involved in gall development and differentiation (Kaldewey 1965; Mills 1969; Engelbrecht 1971; van Staden and Davey 1978) and that several secondary metabolites accumulate in insects galls (Bronner 1977; Ishak *et al.* 1972; Cornell 1983; Nishizawa and Yamagishi 1983). However, little is known about biochemical changes that may regulate these events during gall formation. Modifications in the physiological properties of cells, especially those regarding the cellular membrane system, appear to be substantial, since they are involved in phytohormone binding, the transport of solutes, and in the activation by ions of a large segment of metabolism concerned with cell growth and division (Wood and Braun 1965). Furthermore, galls are important sinks for host carbon. They accumulate significant amounts of carbohydrates, lipids, and proteins to feed, protect, and house the developing larvae.

1. General consideration

Alterations in the biochemical make-up of fully transformed plant cells, such as crown gall disease (produced by *Agrobacterium tumefaciens*) and the genetic tumours on plant hybrids (Kostoff's genetic tumours) have been described (Smith 1972; Braun 1978; Bayer 1982; Kado 1984; Nester *et al.* 1984; Pengelly 1989). However, biochemical analyses of insect galls have been carried out much less frequently. In the work of Bronner (1977; Bronner *et al.* 1989) and Rey *et al.* (1980), chemical components in cynipid and cecidomyid galls were studied in histological sections. The authors detected cytological changes in developing gall tissues and found in the nutritive cells, lining the larval cavity, elevated levels of carbohydrates and lipids, in particular, triacyl- and diacylglycerides. Antibody labelling demonstrated high phosphatase activities in these cells. Ultrastructure studies by Rey and Moreau (1983) revealed numerous paracrystalline protein inclusions in the nutritive layer of *Diplolepis rosae* (L.) galls (on *Rosa canina* L.). These intracytoplasmic inclusions consisted of staggered, closely parallel filaments. The physiological significance of these protein inclusions, however, remained unresolved. In fact, these studies confirm, what is obvious from morphological observation: galls, serving the nutritional requirements of developing larvae, contain highly modified cells to satisfy this need. The data, however, do not address questions regarding possible mechanisms for these alterations or a regulatory role for the unique physiological relationships between galls and the supporting host tissue (leaves).

Galls act as sinks for nutrients and assimilates from the normal, gall-bearing leaf. The histoid complex cynipid galls (which are essentially chlorophyll-free), were used by Kirst and Rapp (1974) in experiments using ^{14}C-labelled CO_2 and labelled assimilates. The labelled products accumulated in the gall. Evidence suggests (Williams 1992) that source leaves can regulate the metabolism of sink leaves. Neither the messages(s) nor the mechanism by which this source–sink co-regulation occurs is as yet understood. In our studies on the transport of phytophormones in genetically transformed plants (Kostoff's genetic tumours; Kostoff 1930) we found that the auxin transport in hybrid tissue was distinctly different from that in the non-tumorous parents (for review of data, see Bayer, (1982). Such changes in cell to cell transport reside, in part, in modications of cell membranes (Martiny-Baron and Scherer 1989; Nickel et al. 1991) and have been implicated in the response to the phytohormone auxin

(Scherer and Andre 1989). This growth hormone has been implicated repeatedly in the induction and maintenance of the cecidogenic response during gall development (Tandon and Arya 1979, 1980).

Subsequently, we asked whether such cell membrane alterations may be testable in cynipid galls. We used for our studies the vernal, mono-thalamous succulent oak galls of the bisexual generation of *Andricus palustris* O.-S. (Bayer 1983, 1987, 1991). As had been shown by Losel (1978), cell membranes of stressed plant cells and of tumour tissues undergo modifications in the composition of their phospholipids. We studied the lipid component of cell membranes in galls and in the normal host leaf (mesophyll tissue). Such phospholipid changes were observed, among others, in leaves under water stress (Chetal *et al.* 1980) and as a result of crown gall transformation (Cockerham and Lundeen 1979). Recently, the roles of phospholipases and phospholipids in signal trans-duction in higher cells have been confirmed (Ferguson and Hanley 1991).

2. Membrane lipids

(a) Lipids, phospholipids, and phospholipase   Phospholipids and lipolytic cyd-rolase activities are components of plant cell membranes. They are responsible, in part, for transmembrane trafficking of ions, solutes, and cell metabolites (Moore 1982; Bishop 1983; Dennis 1989). We determined phospholipids in extracts containing the plasma membrane, the tonoplast, and the membrane systems of cell organelles, that is, chloroplasts, mitochondria, and endoplasmic reticulum (Bayer 1983, 1987, 1991). Freshly harvested immature galls contained, on average, 2.1 times more water than leaf laminae and mature galls. Therefore, data are expressed as per cent of the total lipids extracted from these tissues (see Tables 25.3 and 25.4). Phospholipids, extracted either with chloroform–methanol or with hot isopropanol (Folch *et al.* 1957) were collected after centrifugation and the free fatty acids (FFA) and lipids were extracted in the chloroform phase (Kates 1986). Lipids were separated by two-dimensional thin-layer chromatography (TLC) as described (Rouser *et al.* 1976; Bayer 1987). The total lipid content, including neutral as well as polar lipids, was determined by the phosphorus assay of Fiske–Subba Row (Bartlett 1959). As shown in Table 25.3, the total extractable lipid content was considerably lower in the mature oak galls than in the supporting leaves. The relatively low content of total lipids and of phosphoglycerides detected by us in cynipid galls was the result of the more mature developmental stage of these cecidia (the nutritive layer was partly atrophied) as well as of their low content of total membrane material (few chloroplasts and, generally, larger cell size).

The highest amount of lipid was detected in immature galls, that

Table 25.3.   Lipid and phospholipid content of gall and leaf tissues

Tissue	Total lipids (mg g FW^{-1})	Phosphorus content of total lipids	
		(μg mg lipid^{-1})	(μmol mg lipid^{-1})
Galls (mature)	2.4 ± 0.6	1.76	0.05
Galls (immature)	40.2 ± 3.8	4.31	0.16
Young leaves (gall-bearing)	38.5 ± 3.4	3.30	0.10

is, 40.2 mg (g FW)$^{-1}$ (FW, fresh weight) tissue, whereas mature galls (harbouring the pupa stages of the cynipid) contained only 2.4 mg (g FW)$^{-1}$ (Table 25.3). As had been shown by Rey *et al.* (1980) large amounts of storage lipids (di- and triglycerides) are present in the nutrient cell layer of developing oak galls. The low content of lipids detected by us in the mature cecidia coincides with the degeneration of the nutritive cell layer in these galls. A relatively high concentration of lipids and phospholipids could be extracted from leaves. However, this relates to the increased number of plastids in leaf mesophyll cells. (There are 10–15 times more plastids in leaf cells than in cells of young and mature galls.) Individual phosphoglycerides, identified by two-dimensional TLC and phosphorus assay, are listed in Table 25.4. The amount of phosphatidylethanolamine (PE) was approximately four times as high in normal tissues than in galls. In leaves, the increase in phosphatidylglycerol (PG, the major phospholipid of chloroplasts (Moore 1982) was due to the large amount of these organelles in mesophyll cells. Similarly, the amount of the mitochondrial phospholipid diphosphatidylglycerol (or cardiolipin, CL) and one of the major membrane phospholipids, phosphatidylinositol (PI), were higher in the leaf than in galls. Phos-

Table 25.4.   Phospholipids in galls and leaves and ratio of their distribution

	Gall (μg P mg lipid^{-1})	Leaf	G:L ratio
PC	0.52	0.51	1:1
PG	0.30	0.85	1:2.8
PE	0.17	0.70	1:4.1
CL	0.34	0.60	1:1.7
PI	0.17	0.30	1:1.7
PS	0.26	0.35	1:1.3
Total P Phospholipid	1.76	3.30	1:1.8

Table 25.5. Phospholipase activity in tissue extracts of 1 g gall or leaf tissue each

Enzyme source	Total substrate hydrolysed (units)	Free fatty acids (oleic and palmitic acids) (%)	Mono-and diglycerides (%)
Gall	10.2 ± 2.1	5.2 ± 1.4	2.8 ± 0.6
Leaf	2.7 ± 0.4	1.4 ± 0.5	1.1 ± 0.4
Control:ddH$_2$O	1.6 ± 0.5		

phatidylserine (PS), a minor, but ubiquitously distributed phospholipid, was found in small amounts in both tissue types.

Lipid acyl hydrolase (phospholipase) activity was measured by the amount of ^{14}C from palmitate- (16:0) or oleate-labelled substrates by TLC (Bayer *et al.* 1982). Gall tissues yielded higher lipolytic activities than leaves (Table 25.5). In order to determine the ratio of fatty acids (16:0, 18:1), hydrolysed by plant phospholipase, bacterial substrates were prepared with either palmitic or oleic acid as the labelled acyl side chain of the phospholipid moiety (Bayer 1983). Enzymes in extracts of gall and leaf tissues released twice as much oleic than palmitic acid from the substrate. We concluded from these data that the amount and composition of lipids and phospholipids varied between normal leaves and plant galls (Bayer 1983).

Such differences in phospholipid composition may be comparable with variations in the phosphoglyceride content detected in stressed and pathogenic plant tissues (Losel 1978; Chetal *et al.* 1980). The increase in lipolytic hydrolase activity detected in developing cynipid galls was reminiscent of earlier observations on the defective regulation of fatty acid synthesis and on the deranged control of enzyme biosynthesis in neoplastic cells (Wallach 1968). It has to be added, however, that the phospholipid degradation can, at least in part, be interpreted by the activity of phospholipase C or D followed by lipase activity.

(b) Fatty acid side chains of phosphoglycerides and acylglycolipids The quantitative differences in phospholipid species detected in cynipid galls and leaf tissues did not allow a definition of their acyl side chain composition. Therefore, we studied the fatty acids associated with the phosphoglycerides in cynipid galls and in normal tissues. We had anticipated that cell membranes of gall tissues might contain higher concentrations of unsaturated acyl side chains, in accordance with the idea that some neoplastic and stressed plant tissues contain elevated levels of unsaturated fatty acids (Losel 1978; Clarkson *et al.* 1980; Bishop 1983).

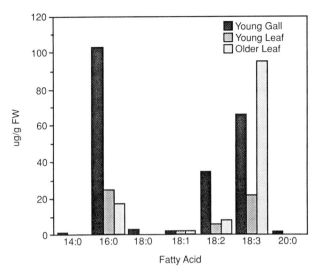

Fig. 25.1. Acyl-chain composition of endogenous phospholipids.

Fatty acids of the endogenous phospholipid pool were determined as well as the uptake of labelled oleic (18:1) and palmitic (16:0) acid into phospholipids of cecidia and leaf tissues. We also measured the uptake of oleate into the phospholipids of fully transformed plant tissues, that is *Agrobacterium tumefaciens*-induced crown gall tumour.

Cell membrane phospholipids, separated by TLC, were transesterified by 0.5 N methanolic base reaction and the fatty acid methyl esters identified by gas chromatography (Bayer 1987, 1991). As shown in Fig. 25.1, phospholipids derived from galls contained the largest amount of palmitate $(103.68 \mu g \ (g FW)^{-1})$ and linoleate $(34.27 \mu g \ (g FW)^{-1})$ when compared with those from leaves. A small, but significant, amount of stearic acid $(1.8 \mu g \ (g FW)^{-1})$ could be detected exclusively in the phospholipid portions from galls. Trace amounts (less than 1 ug) of myristate and arachidate were present only in gall extracts; a surprisingly large amount of linolenic acid $(65.87 \mu g \ (g FW)^{-1})$ was detected in these extracts. The high concentration of linolenic acid found in older leaves $(94.67 \mu g \ (g FW)^{-1})$ was based on the large number of chloroplasts present in these tissues. In contrast, the cynipid galls studied here were relatively chloroplast-free.

Similar procedures were used to compare the levels and acyl chain compositions of the major galactolipids, monogalactosyldiacylglycerol (MGDG) and digalactosyldiacylglycerol (DGDG). As can be seen in Table 25.6, fatty acids recovered from galactolipids were palmitic (16:0), stearic (18:0), oleic (18:1), linoleic (18:2), and linolenic acid (18:3). The fatty acid profiles of MGDG from leaves and galls were similar,

Table 25.6. Galactolipid fatty acids in oak leaves and cynipid galls

Fatty acids	Leaf	Gall
MGDG		
16:0	4.4 ± 1.2	–
18:0	–	–
18:1	–	–
18:2	10.2 ± 2.5	13.5 ± 5.5
18:3	84.5 ± 10.2	86.0 ± 9.8
Total μg FA (g FW)$^{-1}$	0.17	0.10
Double bond index	2.73	2.85
DGDG		
16:0	–	16.2 ± 3.5
18:0	–	–
18:1	–	12.6 ± 3.7
18:2	23.9 ± 7.7	43.1 ± 6.6
18:3	76.0 ± 11.8	27.9 ± 6.3
Total μg FA (g FW)$^{-1}$	1.16	0.48
Double bond index	2.75	1.82

Values are expressed as the weight per cent of total fatty acids in the individual galactolipid.

except that 16:0 was detectable only in leaf samples. The predominant fatty acid was 18:3 in both tissue types. The double bond index for MGDG in both tissues was the highest, that is 2.73 for leaf and 2.85 for gall tissue. The fatty acid composition of DGDG showed striking differences: whereas in leaf tissues only 18:2 and 18:3 could be detected, DGDG from galls also included 16:0 and 18:1. The relative amounts of 18:2 and 18:3 were reversed when compared with the fatty acid composition in leaves. This distribution reduced the double bond index to 1.82 in the galls.

For a more direct comparison between the individual glycerolipids and their acyl chains, fatty acids were expressed as the weight per cent of total fatty acid per individual phospholipid (Table 25.7). The largest amounts of fatty acids were associated with PE, PC (phosphatidylcholine) and DPG (diphosphatidylglycerol or cardiolipin, CL) (total μg FA (g FW)$^{-1}$). In PE, palmitate (16:0) accounted for the highest percentage in both tissues. Although only a very low level of trans-16:1 could be detected in these two lipids, the amount present in PC from gall could only be described as a trace ($<$ 1 per cent). The linoleate level in the normal tissue was approximately twice that in galls and an unidentified long-chain fatty acid of PE (not identical with arachidic acid, 20:0) was detected only in the cecidia. The double bond indices for fatty acids in PC were 1.03 for leaves vs 1.58 in galls and reflected the increase in 18:3 in the neoplasmatic tissues. The composition of fatty acids in DPG

Table 25.7. Phospholipid fatty acids in oak leaves and cynipid galls

Fatty acids	Leaf	Gall
PE		
16:0	49.5 ± 6.1	53.8 ± 7.0
16:1	3.1 ± 1.2	5.0 ± 1.9
18:0	23.2 ± 4.1	10.5 ± 2.3
18:1	–	2.2 ± 0.5
18:2	22.5 ± 6.6	9.3 ± 2.3
18:3	–	–
Unidentified long chain FA	–	16.1 ± 2.1
Total μg FA (g FW)$^{-1}$	1.14	4.62
Double bond index	0.48	0.25
PC		
16:0	39.6 ± 7.3	29.0 ± 6.5
16:1	2.3 ± 1.0	tr
18:0	9.3 ± 3.3	11.9 ± 4.4
18:1	7.4 ± 2.5	3.9 ± 1.5
18:2	15.9 ± 4.0	10.1 ± 2.5
18:3	20.7 ± 3.9	44.9 ± 7.4
Total μg FA (g FW)$^{-1}$	1.25	0.77
Double bond index	1.03	1.58
DPG (CL)		
16:0	53.1 ± 5.5	–
16:1	–	–
18:0	19.2 ± 2.8	20.5 ± 4.0
18:1	6.5 ± 2.0	8.2 ± 3.0
18:2	4.8 ± 1.8	–
18:3	16.1 ± 2.8	71.2 ± 8.0
Total μg FA (g FW)$^{-1}$	1.16	0.60
Double bond index	0.64	2.21
PG		
16:0	45.4 ± 8.4	41.0 ± 8.0
16:1	tr	–
18:0	34.5 ± 3.0	–
18:1	–	12.8 ± 2.5
18.2	6.0 ± 2.0	45.1 ± 4.2
18:3	12.2 ± 2.3	–
Total μg FA (g FW)$^{-1}$	0.13	0.34
Double bond index	0.48	1.03
PI		
16:0	35.2 ± 5.5	17.3 ± 4.4
16:1	–	–
18:0	20.5 ± 5.0	–
18:1	–	65.3 ± 7.9
18.2	30.8 ± 5.2	17.3 ± 4.4
18:3	13.2 ± 3.5	–
Total μg FA (g FW)$^{-1}$	0.05	0.10
Double bond index	1.01	0.99

Values are expressed as the weight per cent of total fatty acids in the individual phospholipid.
Abbreviations: FA, fatty acid; PC, phosphatidylcholine; PE, phosphatidylethanolamine; PG, phosphatidylglycerol; phosphatidylinositol; PS, phosphatidylserine; CL, cardiolipin; TLC, thin-layer chromatography; FW, fresh weight; 14:0, myristic acid; 16:0, palmitic acid; 18:0, stearic acid; 18:1, oleic acid; 18:2, linoleic acid; 18:3, linolenic acid; 20:0, arachidic acid; ddH$_2$O, double distilled H$_2$O.

in both tissue types was markedly different. No 16:0 or 18:2 were associated in galls, raising the double bond index to 2.21 in these tissues. Differences in the fatty acid composition of the minor phospholipids PG and PI were also detected, the most significant feature being the absence of 18:3 in PG and the high level of 18:1 in the PI of gall tissues. The total amount of fatty acid (μg FA (g FW)$^{-1}$) in these two phospholipids was very low.

Radioactive labelling of cynipid galls

1. Labelling with ^{14}C-acetate

Palmitic and oleic acid synthesis from ^{14}C-acetate occurs in leaf tissues under aerobic conditions (Stumpf and James 1963). We labelled fresh tissue samples of young, developing cynipid galls (still containing small feeding larvae) and of the corresponding young leaf laminae. The tissues were cut up and incubated for up to 2 h according to the procedure described by Slack *et al.* (1978) using 0.14 mM [2-^{14}C] acetate (25 μC), specific activity 59 mC mmol^{-1} (Amersham) in 0.35 M mannitol. Extracts were partitioned in chloroform–methanol and the lipid residues in the chloroform phase were analysed by TLC.

2. Labelling with ^{3}H-oleic acid

Plant galls, normal leaves, and crown gall tumours of *Lycopersicum esculentum* L. were cut into 1 mm slices and incubated on a shaker for 5 h at room temperature in 0.35 M mannitol, containing 100 μCi[^{3}H] oleic acid, specific activity 35 Ci m mol^{-1} (New England Nuclear).

In vivo experiments were carried out using young developing leaves (4–5 cm long) to which the immature growing galls were still attached. The freshly cut petioles were immersed into a watery solution of 100 μCi^{3}H oleate under daylight illumination. Special care was taken so that the solvent touched neither the leaf blade nor the gall. After 5 h, the label was chased with unlabelled oleate (30 μM) for 15 h, washed several times in distilled H$_2$O, and the tissues were cut separately and weighed.

3. Uptake of ^{14}C-acetate and ^{3}H-oleate into phospholipid moieties

^{14}C-Acetate uptake was highest in the phosphoglyceride portions of the immature galls and of crown gall tumours, that is, 16.3 and 17.6 μmol g tissue^{-1}, respectively (Fig. 25.2). Tissues of mature galls, in contrast, take

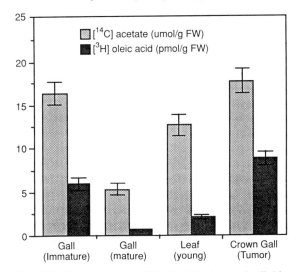

Fig. 25.2. Incorporation of labelled FA into polar lipids.

up only 5.2μmol acetate g tissue^{-1}. ^{3}H-Oleate levels were highest in developing immature galls and in crown gall tumours, 6 and 8.7 pmol g tissue^{-1}, respectively; considerably less ^{3}H-label was recovered from leaf (2 pmol g^{-1}) and mature gall tissues (0.5 pmol g^{-1}). The degree of fatty acid incorporation into individual phospholipids varied between gall, leaf, and crown gall tissues. The highest incorporation occurred into the two major plant phospholipids PC and PE.

4. Incorporation of ^{3}H-oleate into phospholipids of galls attached to leaves

We studied the incorporation of ^{3}H-oleic acid into whole leaves and into their attached galls, in an attempt to test the 'source–sink' relationship between leaves and galls. Oleate incorporation into gall phospholipids was decreased. The total amount of oleate in the phospholipid fraction of whole galls and leaves resembled those which were detected in incubated tissue slices. As determined by TLC and argentation chromatography, 85 per cent of the label was associated with oleic acid, indicative of position 2 for the oleoyl residue on the molecules. Since the young galls contain only small amounts of chlorophyll and a low photosynthetic capacity, they are attracting sinks for assimilates from the leaf blade (Kirst and Rapp 1974). The alteration in the phospholipid content, the composition of the acyl side chains, and an increased lipolytic activity in the cynipid galls, may reflect their involvement in the source–sink relationship between leaf and gall.

Conclusion

The molecular basis for the relationship between structure and function
of plant cell membranes has remained rather speculative. However,
several proposals have suggested a regulatory role for the fluid bilayer of
membrane lipids (see reviews by Mazliak and Kader 1980; Raison
1980; Trewavas 1986). Our results indicate that the phospholipid and
galactolipid portion of normal and of insect-transformed tissues differ not
only in the composition of the polar head groups, but also in the
composition of their acyl side chains. In this respect our data are in
agreement with earlier observations on an increased liquidity of plant
cellular membranes in stressed plant tissues (Losel 1978; Chetal *et al.*
1980).

Our results show that the degree of fatty acid saturation of the major
and minor lipids in neoplasmatic tissues differs from the composition
found in normal cells. In conclusion, the real uniformity that exists in
the phospholipid and fatty acid content of all cell membranes within the
same species (except for chloroplast membranes) can be altered by the
interaction of the gall insect with the host tissue. Distinct membrane
modifications occur in tissues transformed by gall insects. Such modifications of membrane lipid structure may be responsible, in part, for the
regulation of the many metabolic activities initiated in host tissues by the
developing gall insects. It remains to be tested which other lipid classes
(sterol and sterol derivatives) and enzymes or enzyme complexes are
involved in lipid regulation and which role they are playing in the
modification of essential components of biological membranes in plant
neoplasmatic tissues.

Acknowledgement

Data from previous publications of the author are reprinted here with
permission from the appropriate journal editors. I thank Ms Carole
Eisele for the typing of the manuscript. Part of the work was supported
by grant DCB-85–03684 from the National Science Foundation, grants
A1–10414–12, CA-06927, and PR-05539 from the National Institutes
of Health, and by an appropriation from the Commonwealth of Pennsyl-
vania.

References

Bartlett, G.R. (1959). Phosphorus assay in column chromatography. *Journal of
Biological Chemistry*, **234**, 466–8.

Bayer, M.H. (1982). Genetic tumors: physiological aspects of tumour formation in interspecies hybrids. In *Molecular biology of plant rumours*, (ed. G. Kahl and J.S. Schell), pp. 33–67. Academic Press, New York.

Bayer, M.H. (1983). Phospholipids and lipid acyl hydrolase (phospholipase) in leaf galls [Hymenoptera: Cynipidae of black oak (*Quercus robur* L.)]. *Plant Physiology*, **73,** 179–81.

Bayer, M.H. (1987). Fatty acid composition of glycerophospholipids in cynipid galls and crown gall tumors. *Plant Physiology and Biochemistry*, **25**(2), 155–61.

Bayer, M.H. (1991). Fatty acid composition of galactolipids and phospholipids in neoplasmatic plant tissues (cecidia) and normal leaf tissue. *Physiologia Plantarum*, **81,** 313–18.

Bayer, M.H., Costello, G., and Bayer, M.E. (1982). Isolation and partial characterization of membrane vesicles carrying markers of the membrane adhesion sites. *Journal of Bacteriology*, **149,** 758–67.

Bishop, D.G. (1983). Functional role of plant membrane lipids. In *Biosynthesis and function of plant lipids*, Proceedings 6th American Symposium Botany, (ed. N. Thomson, N. Mudd, and M. Gibbs), pp. 81–103, Waverly Press, Baltimore.

Bloch, R. (1965). Abnormal development in plants: a survery. *Handbuch Pflanzenphysiologie*, **15,** 156–83.

Boysen-Jenson, J.P. (1948). Formation of galls by *Mikiola fagi*. *Physiologia Plantarum*, **1,** 95–108.

Braun, A.C. (1959). Growth is affected. In *Plant pathology, an advanced treatise*, (ed. J.G. Horsfall and A.E. Dimond), pp. 189–248. Academic Press, New York.

Braun, A.C. (1978). Plant tumors. *Biochemica et Biophysica Acta*, **516,** 167–91.

Braun, A.C. and Wood, H.N. (1962). On the activation of certain essential biosynthetic systems in cells of *Vinca rosa* L. *Proceedings of the National Academy of Sciences of the United States of America*, **48,** 1776–82.

Bronner, R. (1977). Contribution à l'étude histochimique des tissus nourriciers des zoocécidies. *Marcèllia*, **40,** 1–134.

Bronner, R., Westphal, E., and Dreger, F. (1989). Chitosan, a component of the compatible interaction between Solanum dulcamara L. and the gall mite *Eriophyes cladophthirus* Nal. *Physiological and Molecular Plant Pathology*, **34,** 117–30.

Chetal, S., Wagle, D.S. and Nianawatee, H.S. (1980). Phospholipid changes in wheat and barley leaves under water stress. *Phytochemistry*, **19,** 1393–5.

Clarkson, D.T., Hall, K.C., and Roberts, J.K.M. (1980). Phospholipid composition and fatty acid desaturation in the roots of rye during acclimatization of low temperature. *Planta*, **149,** 464–71.

Cockerham, L.E. and Lundeen, C.V. (1979). Transformation induced alteration of cellular membranes in crown-gall tumour-cells. *Plant Physiology*, **64,** 543–5.

Cornell, H.V. (1983). The secondary chemistry and complex morphology of galls formed by the Cynipinae (Hymenoptera): why and how? *The American Midland Naturalist*, **110**(2), 225–34.

Dennis, D.T. (1989). Fatty acid biosynthesis in plastids. In *physiology, biochemistry and genetics of non-green plastids*, (ed. C.D. Boyer, J.C. Shannon and R.C.

Hardison), pp. 120–9. American Soc. Plant Physiologists. Rockville, Maryland.

Engelbrecht, L. (1971). Cytokinin activity in larval infected leaves. *Biochemie und Physiologie der Pflanzen*, **162**, 9–27.

Ferguson, J.E. and Hanley, M.R. (1991). The role of phospholipases and phospholipd-derived signals in cell activation. *Current Opinion in Cell Biology*, **3**, 206–12.

Folch, J., Lees, M., and Sloane-Stanley, G.H. (1957). A simple method for the isolation and purification of total lipids from animal tissues. *Journal of Biological Chemistry*, **226**, 497–509.

Garrigues, R. (1954). De l'existence d'un gradient chimique, agent d'action cécidogène. Huitième congrès international de Botanique, Paris. *Comptes Rendus des Sèances*, pp. 222–6.

Hori, K. and Miles, P. (1977). Multiple plant growth promoting factors in the salivary glands of plant bug. *Marcellia*, **39**, 399–400.

Ishak, M.S., el-Sissi, H.I. Nawar, M.A., and el-Sherbieny, A.E. (1972). Tannins and polyphenolics of the galls of tamarix aphylla. *Planta Medica*, **21**, 246–53.

Jones, D.F. (1935) The similarity between fasciations in plants and tumors in animals and their genetic basis. *Science*, **81**, 75–6.

Kado, C.I. (1984). Phytohormone-mediated tumorigenesis by plant pathogenic bacteria. In *Plant gene research. Genes involved in microbe–plant interactions*, (ed. D.P.S. Verma and T. Hohn), pp. 311–6. Springer Verlag, New York.

Kaldewey, H. (1965). Wachstumsregulatoren aus Pflanzengallen und Larven der Gallenbewohner. *Sonderabdruck aus den Berichten der Deutschen Botanischen Gesellschaft*, **2**, 73–84.

Kates, M. (1986). Techniques of lipidology. Isolation, analysis and identification of lipids. In *Laboratory techniques in biochemistry and molecular biology*, (ed. R.H. Burdon and P.H. van Krippenberg), pp. 186–324. Elsevier Science Publishers, Amsterdam.

Kirst, G.O. and Rapp, H. (1974). Zur Physiologie der Galle von Mikiola fagi Hts. auf Blättern von Fagus silvatica L. 2. Transport C markierter Assimilate aus dem befallenen Blatt und aus Nachbarblättern in die Galle. *Biochemie und Physiologie der Pflanzen*, **165**, 445–55.

Kostoff, D. (1930). Tumors and other malformations on certain Nicotiana hybrids. *Zentralblatt Bakterielle Parasitenkunde II*, **81**, 258–60.

Losel, D.M. (1978). Lipid metabolism of leaves of Poa pratensis during infection by Puccinia poarum. *New Phytologist*, **80**, 167–74.

Martiny-Baron, G. and Scherer, G.F. (1989). Phospholipid-stimulated protein kinase in plants. *Journal of Biological Chemistry*, **264**(30), 18052–9.

Mazliak, P. and Kader, J.C. (1980). Phospholipid exchange system. In *The biochemistry of plants. IV lipids: structure and function*, (ed. P.K. Stumpf and E.E. Conn), pp. 283–300. Academic Press.

Meyer, J. and Maresquelle, H.J. (1983). *Anatomie des galles*. Gebr. Borntraeger, Berlin.

Mills, R.R. (1969). Effect of plant and insect hormones on the formation of the goldenrod gall. *National Cancer Institute Monographs*, **31**, 487–91.

Moore Jr, T.S. (1982). Phospholipid biosynthesis. *Annual Review of Plant Physiology*, **33,** 235–96.

Nester, E.W., Gordon, M.P., Amasino, R.M., and Yanofsky, M.F. (1984). Crown gall: a molecular and physiological analysis. *Annual Review of Plant Physiology*, **35,** 387–413.

Nickel, R., Schutte, M., Hecker, D., and Scherer, G.F.E. (1991). The phospholipid platelet-activating factor stimulates proton extrusion in cultured soybean cells and protein phosphorylation and ATPase activity in plasma membranes. *Journal of Plant Physiology*, **139**(2), 205–11.

Nishizawa, M. and Yamagishi, T. (1983). Tannins and related compounds. Part 9. Isolation and characterization of polygalloylglucoses from Turkish galls (*Quercus infectoria*). *Journal of the Chemical Society Perkin Transactions.*, **I,** 961–5.

Pengelly, W.L. (1900). Neoplastic progression in plants. In *Comparative aspects of tumor development*, (ed. H.E. Kaiser), pp. 15–23. Kluwer Academic Publishers, Dordrecht.

Raison, J.K. (1980). Membrane lipids: structure and function. In *The biochemistry of plants IV lipids: structure and function*, (ed. P.K. Stumpf and E.E. Conn), pp. 57–80. Academic Press, New York.

Rey, L. and Moreau A. (1983). Sur les inclusions paracristallines des celules nourricières de la galle provoquèe par *Diplolepis rosae* L. sur *Rosa canina* L. *Journal of Ultrastructure Research*, **84,** 94–102.

Rey, L., Dubacq, J.P., and Tremolieres, A. (1980). Lipids of oak galls. *Phytochemistry*, **19,** 2569–70.

Rohfritsch, O. (1980). Relations hôte-parasite au début de la cécidogenèse. Colloque international de Cécidologie et de Morphogenèse pathologique, Strasbourg. *Bulletin de la Socièté Botanique Française*, **127**(1), 199–207.

Rohfritsch, O. and Shorthouse, J.D. (1982). Insect galls. In *Molecular biology of plant tumors*, (ed. G. Kahl and J.S. Schell), pp. 131–52. Academic Press, New York.

Rouser, G., Kritchev, G., and Yamamato, A. (1976). Column chromatographic and associated procedures for separation and determination of phosphatides and glycolipids. In *Lipid chromatographic analysis*, (ed. G.V. Marinetti), pp. 713–76. Marcel Dekker, New York.

Scherer, G.F. and Andre, B. (1989). A rapid response to a plant hormone: auxin stimulates phospholipase A2 *in vivo* and *in vitro*. *Biochemical and Biophysical Research Communications*, **163**(1), 111–17.

Slack, C.R., Roughan, P.G. and Balasingham, N. (1978). Labeling of glycerolipids in the cotyledons of developing oilseeds by [1-^{14}C]-acetate and [2-^{3}H]-glycerol. *Biochemical Journal*, **170,** 421–33.

Smith, H.H. (1972). Plant genetic tumors. *Progress in Experimental Tumor Research*, **15,** (ed. A.C. Braun), 138–64. S. Karger, Basel.

Stumpf, P.K. and James, A.T. (1963). The biosynthesis of long-chain fatty acids by lettuce chloroplast preparations. *Biochimica et Biophysica Acta*, **70,** 2–32.

Tandon, P. and Arya, H.C. (1979). Effect of growth regulators on carbohydrate metabolism of *Zizyphus jujuba* gall and normal stem tissues in culture. *Biochemie und Physiologie der Pflanzen*, **174,** 772–9.

Tandon, P. and Arya, H.C. (1980). Auxin autotrophy and hyperauxinity of

Eriophyes induced *Zizyphus* stem galls in culture. *Biochemie und Physiologie der Pflanzen*, **175**, 537–41.

Trewavas, A. (1986). Resource allocation under poor growth conditions. A major role for growth substances in developmental plasticity. *Symposium of the Society of Experimental Biology*, **40**, 31–76.

van Staden, J. and Davey, J.E. (1978). Endogenous cytokinins in the larvae and galls of Erythrina latissima leaves. *Botanical Gazette*, **139**, 36–41.

Wallach, D.F.H. (1968). Cellular membranes and tumor behavior: a new hypothesis. *Proceedings of the National Academy of Sciences of the United States of America*, **61**, 868–74.

Williams J.H.H. (1992). Sucrose: a novel plant growth regulator; P2. 34. *Journal of Experimental Botany*, Abstracts Plant and Related Cell Topics.

Wood, H.N. and Braun, A.C. (1965). Studies on the net uptake of solutes by normal and crown-gall tumor cells. *Proceedings of the National Academy of Sciences of the United States of America*, **54** 1532–8.

26. The fossil record of leaves with galls

ANDREW C. SCOTT, JONATHAN STEPHENSON,
and MARGARET E. COLLINSON

*Department of Geology, Royal Holloway University of London, Egham,
Surrey, UK*

Abstract

A few scattered examples of galls on plant fossils have been reported from the late
Palaeozoic and early Mesozoic. A more extensive record exists in the Cretaceous,
largely coincident with the first radiation of angiosperms and the earliest fossil
occurrences of several insect groups known today to induce gall formation. In the
Tertiary a higher diversity of gall types are encountered. This review incorporates
new data on morphological diversity in galls from several large assemblages of
Cretaceous and early Tertiary angiosperm leaves. These fossil galls have also been
compared with possible modern analogues. The implications of the new evidence
are briefly considered in the context of the evolution of this important plant–
arthropod interaction.

Introduction

Most discussions on the evolution of plant–animal interactions have not
extensively considered the fossil record, mainly because of a lack of
published data (Southwood 1973, 1985; Strong *et al.* 1984). Recent
interest in this field has highlighted the need to examine the fossil record
for the traces of plant–animal interactions and to report and illustrate
examples (Chaloner *et al.* 1991*a,b*).

Galls or 'cecidia' may be defined as 'all manifestations of growth,
whether positive or negative, and of abnormal differentiation induced on
a plant by an animal or plant parasite' (Meyer 1987). This excludes all
effects of ordinary plant feeding including the production of a callus and
hypertrophy of the surrounding cells. It also excludes all insect mines
and 'domatia' colonized by mites (Arachnida: Acari) and ants (Insecta:
Hymenoptera). Galls are the result of a physiological reaction induced

Plant Galls (ed. Michèle A. J. Williams), Systematics Association Special Volume No. 49,
pp. 447–70. Clarendon Press, Oxford, 1994. © The Systematics Association, 1994.

in the host plant tissues immediately surrounding the invasive parasite. According to Küster (1911) galls can be divided into two main morphological types: 'organoid' galls and 'histioid' galls, the latter being further separable into kataplasmatic galls and prosoplasmatic galls. Galls may also be classified according to the number of larvae or larval cells with three types: unilocular galls, bilocular galls, and plurilocular galls (Connold 1909). Dreger-Jauffret and Shorthouse (1992) provided descriptions and illustrations of basic gall types (based on mode of formation and initial position): filz, pit, blister, pouch, roll, fold, covering, mark, bud, and rosette galls. In fossil material the recognition of some of the above gall types is difficult, depending on the state of preservation. Usually the leaf and gall are compressed and the organic material has often decayed leaving only an impression fossil. Therefore, in this paper, we subdivide fossil galls into categories based on a number of gross morphological characteristics (see below).

Galls on fossil leaves

The recognition of galls on fossil leaves is based only on gross morphology and position. The fossils have subsequently been compared with modern material induced by arthropods. Descriptions of the fossils and comments on possible modern analogues are given in the Appendix. Illustrations of the leaves and galls are provided in Figs 26.1 and 26.2 with details of individual galls in Fig. 26.3. Table 26.1 indicates the known fossil history of living gall-inducing arthropods. Details of the fossil insect record may be found in Carpenter and Burnham (1985), Jarzembowski (1989), Ross and Jarzembowski (1993), and Wooton (1988).

All of the leaves that we describe and figure are impression fossils (lacking cuticles). These may reveal either the upper or lower leaf surface or some intermediate, mixed, fracture plane. In most cases only one specimen was collected from the two pieces (part and counterpart) resulting from the original split of the rock to reveal the fossil. In addition, venation is generally poorly preserved (first and second order only) on most of the specimens studied. The combination of these factors has resulted in lack of determination of the leaves to particular angiosperm groups. Details of the galls were difficult to distinguish. Where preservation allowed, the following gross morphological criteria were used to describe and categorize the galls.

1. *Size.* Diameter: small < 3 mm, medium 3–8 mm, large > 8 mm. Area: very small up to 1 mm^2, small 1–3 mm^2, medium 3–15 mm^2, large 15–35 mm^2, massive 35 mm^2 and larger.

2. *Shape.* Cone gall, spot gall, pouch gall, ball gall, irregular/cerebroid

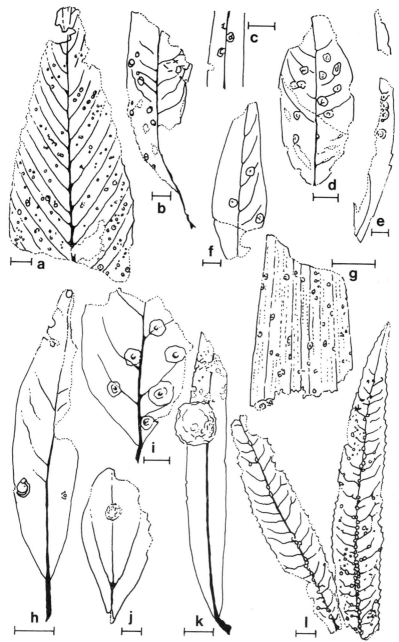

Fig. 26.1. Leaves showing gall types (for details see Appendix) (Scale bar = 1 cm.) (a) Type 1, PP5351; (b) type 2, v50936b; (c) type 3, PP8664; (d) type 4, PP13553; (e) type 8, F1861; (f) type 6, UP348; (g) type 11, PP8932; (h) type 5, v49728; (i) type 7, PP13471; (j) type 9, PP9196; (k) type 9, PP14154; (l) type 10, v47982. (a), (b), (g), (h) and (l) are Tertiary (middle Eocene);; (c), (f), (i)–(k) are Cretaceous ((c), (d) (j) Maastrichtian; (e), (f), (k) Cenomanian, (i) Turonian).

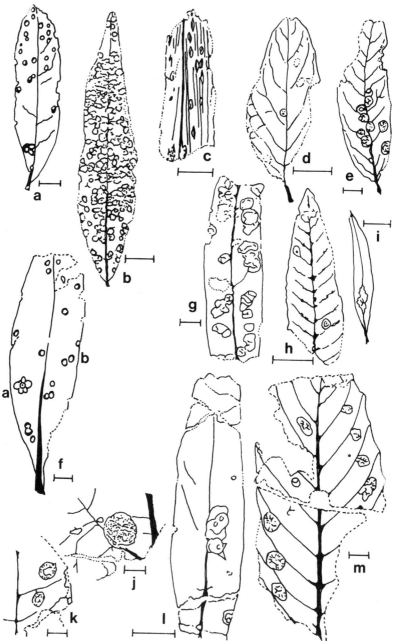

Fig. 26.2. Leaves showing gall types (for details see Appendix) (Scale bar = 1 cm.) (a) Type 12, v50320; (b) type 14, PP12481; (c) type 13, PP15860; (d) type 15, v50154; (e) type 16, v49947; (f) types 20(a) and 21(b), v52584; (g) type 17, PP7704; (h) type 18, v48051; (i) type 19, v40285; (j) type 25, v48524; (k) type 24, v49272; (l) type 22(a) and type 1(b), v49689; (m) type 23, PP5647/PP5653. All are Tertiary. (k) is late Palaeocene, others are middle Eocene.

gall, hollow gall, pit gall, pustule/pock gall, covering gall, bifoliate gall.

3. *Position on leaf.* Margin, on veins (a, primary vein; b, secondary veins; c, both), between veins, apical, basal, dispersed throughout leaf lamina, on petiole.

4. *Nature of gall wall.* Smooth, rough, hairy, with wings/ flanges, with nectaries.

5. *Nature of exit pore.* Absent, single (a, central; b, peripheral), more than one (a, central; b, peripheral; c, both).

6. *Comparison with recent specimens.* Suggested identifications for gall inducers rests entirely on comparison with modern examples induced by arthropods.

General reviews on galls in the fossil record may be found in Larew (1986, 1992) and Boucot (1990). Larew (1992) provides a table of published records of fossil galls in which there are only 26 pre-Pliocene records. All except three of these were on leaves, the one example on cones (Miocene *Sequoia*) being of special significance in actually containing the gall inducer, a cecidomyiid (Larew 1992). In some cases authors implied a high degree of certainty in the identification of gall inducers on leaves but we have found this impossible. Twenty-five gall types are described in this paper (based upon criteria listed in the previous section). We recognize that more than one of these types may have resulted from the activities of the same gall-inducing organism. Equally, the same gall type may have resulted from different gall inducers. The nature of the fossil material and the lack of a comparable classification of modern galls prevents a more precise assessment.

The fossil record

1. Palaeozoic

The earliest galls and the only recorded examples from the Palaeozoic, appear on *Odontoperis* leaves from the Permian (Potonié 1893; Conway-Morris 1981). We have not seen any convincing Carboniferous gall material (Chaloner *et al.* 1991*b*; Scott *et al.* 1992). None of the major modern gall-inducing insects have a Palaeozoic fossil record (Table 26.1).

2. Mesozoic: pre-Cretaceous

There are no well-documented examples of galls on leaves from Mesozoic, pre-Cretaceous, material (Larew 1992; Scott *et al.* 1992). Alvin *et al.*

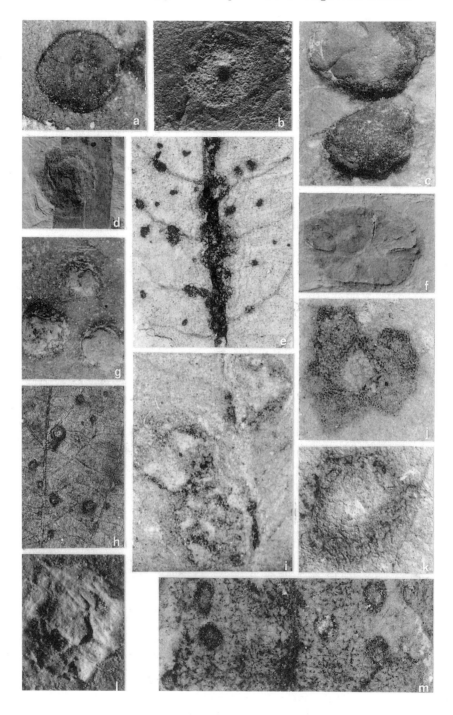

(1967) noted the presence of insect galls on an *Anomozamites* leaf from the middle Jurassic (Bajocian). Only two groups (both Coleoptera) of the major modern gall-inducing insects have a pre-Cretaceous fossil record (Table 26.1).

3. Mesozoic: Cretaceous

(a) Literature   The occurrence of galls increases significantly in the mid-Cretaceous, largely coincident with the first radiation of angiosperms and the first occurrences of important gall-inducing insects (Scott *et al.* 1992; Table 26.1). Brues (1946) mentioned that galls had been observed on Cretaceous leaves but provided no supporting evidence. Lesquereux (1892) and later Berry (1923) figured a gall from the Dakota Formation (late Albian–middle Cenomanian), North America, which Berry described as 'resembling an oak-leaf gall' inferring that it had been induced by a member of the Cynipidae (Hymenoptera). Berry (1934) described a specimen named *Vitis dakotana* Berry from the lower Lance Formation (Upper Cretaceous), South Dakota, North America, with numerous small galls which he inferred were induced by *Phylloxera* (Hemiptera: Homoptera: Phylloxeridae). Hickey and Doyle (1977) noted (p. 41) and illustrated (p. 52, Figs 49–51) galls on 'Sassafras' leaves from the Albian of the Potomac Group (Sub Zone IIB), at Stump Neck, Maryland. They suggested (Hickey and Doyle 1977, p. 41) that these might be the oldest known insect galls. Larew (1992) noted and figured additional examples of these galls on the same leaf type (now considered to be a primitive member of the Hamamelididae or Rosidae) and noted 'In size and imprint structure the galls resemble relatively complex cynipid spangle galls on extant oaks [e.g., small, circular impressions with a hard central capsule]'. These galls are 4–5 mm in diameter according to the magnifications on the illustrations. Larew (1992) noted a total of 49 leaves with galls from the Hickey and Doyle collections in the US National Museum. However, he gave no information on how many leaves had been examined, how many leaf types bore galls, the range of

---

Fig. 26.3.   Details of gall types shown in Figs 26.1 and 26.2. (For details see Appendix). (a) Type 18, v48051, × 8 (see Fig. 26.2(h)); (b) type 6, UP348, × 3.7 (see Fig. 26.1(f)); (c) type 17, PP7704, × 6 (see Fig. 26.2(g)); (d) type 9, PP14154, × 1.5 (see Fig. 26.1(k));(e) type 10, v47982, × 4 (see Fig. 26.1(l)); (f) type 23, PP5647/5653, × 5.5 (see Fig. 26.2(m)); (g) type 2, pp12107, × 6; (h) type 1, v53724, × 1; (i) type 16, v49947, × 7 (see Fig. 26.2(e)); (j) type 20, v59878, × 8; (k) type 24, v49272, × 6.5 (see Fig. 26.2(k)); (l) type 4, PP13553, × 14 (see Fig. 26.1(d)); (m) type 12, v50320, × 6 (see Fig. 26.2(a)). (b), (d), (h) and (l) and Cretaceous, (l) Maastrichtian others Cenomanian; (a), (c), (e)–(g), (i)–(k), (m) are Tertiary, (i) and (k) late Palaeocene, others middle Eocene.

gall diversity, or the range of stratigraphic levels involved (the Potomac Group ranges from the Barremian to the Cenomanian). Scott *et al.* (1992) illustrated, but did not describe, a number of galls from the Upper Cretaceous, mostly from North America. A combination of non-recognition and non-collection (Condon and Whalen 1983) probably accounts for the relatively poor record of galls in the literature on Cretaceous floras.

(b) New material   We present here illustrations (Figs 26.1 and 26.3) and descriptions (Appendix) of a range of new material which shows that galls are much more common on Cretaceous leaves than was previously recognized. The majority of the new specimens, representing seven gall types (Table 26.2), are from the Dakota Formation (late Albian–middle Cenomanian, see Crane and Dilcher (1984); here tabulated as Cenomanian) of North America and from the European Cenomanian (Kvaček, personal communication). Comparisons with modern material suggest that these Cretaceous galls were produced in response to gall mites (Acari: Eriophyidae), aphids or psyllids (Hemiptera: Homoptera: Aphididae or Psyllidae), gall midges (Diptera: Cecidomyiidae), and gall wasps (Hymenoptera: Cynipidae). Maastrichtian material yielded a similar diversity of galls with several types in common. Turonian material (of intermediate age) yielded only four gall types (Table 26.2) with five specimens. Four of these were probably induced by gall mites (Acari: Eriophyidae) and the other by gall midges (Diptera: Cecidomyiidae). Differences in gall diversity and specimen number during the Cretaceous age are partially accounted for by the variation in the numbers of specimens available for study (Table 26.2).

4. Tertiary

(a) Literature   We are not aware of any galls reported from the Palaeocene although there are a number of Eocene records from North America. Collins (1925) observed evidence of fossil galls and Berry (1916, 1923, 1930) described and figured examples of 'cone galls' and a petiolar gall, which were said to have been induced by Diptera, from the Claiborne Formation at Holly Springs. Berry (1924) illustrated a leaf spot fungus on leaves named *Combretum petraflumenses* Berry from the Fayette Sandstone, Texas. Brooks (1955) described more examples resembling modern 'cone' galls and also simple 'pouch' galls in leaf material from the Eocene Claiborne Formation at Puryear (Dilcher 1973). Barthel and Rüffle (1976) illustrated leaves of *Symplocos hallensis* Barthel, Kvaček & Rüffle from the Eocene of Geiseltal, Germany, with possible gall sites (plate 42,

Table 26.1.   First appearance in the fossil record of major groups of recent gall-inducing arthropods

Acari	Sil.		
Family: Tarsonemidae			
Eriophyidae			
Insecta			
Thysanoptera		Coleoptera	
Family:Phlaeothripidae	Eoc.	Family: Cerambycidae	L.Cret
		Buprestidae	M.Jur.
Hemiptera: Heteroptera		Curculionidae	L.Jur?
Family: Tingidae	L.Cret	Scolytidae	L.Cret.
Hemiptera: Homoptera		Lepidoptera	
Family: Cercopidae	U.Cret.	Family: Heliozelidae	
Membracidae	Olig.	Tortricidae	Eoc.
Psyllidae	L.Cret.?	Gelechiidae	
Aphididae	L.Cret.	Pterophoridae	Olig.
Thelaxidae	Eoc.	Orneodidae	
Pemphigidae	Eoc.	(=Alucitidae)	
Adelgidae	U.Cret.?	Cosmopterygidae	
		Coleophoridae	
Phylloxeridae	Eoc.	Aegeriidae	
Margarodidae	Eoc.	(=Sesiidae)	
Asterolecaniidae		Lycaenidae	Eoc.
Diaspididae	Eoc.		
Eriococcidae	Eoc.		
(= Brachyscelidae)			
		Diptera	
		Family: Cecidomyiidae	L.Cret.
Hymenoptera		Tephritidae	Mio.
Family: Tenthredinidae	L.Cret.	Lonchaeidae	
Cynipidae	U.Cret.	Chloropidae	U.Cret.
Eurytomidae	Eoc.	Agromyzidae	Mio.
Agaonidae	Olig.	Platypezidae	L.Cret.

Recent gall inducers derived from Meyer (1987) and Shorthouse and Rohfritsch (1992); their fossil records derived from Ross and Jarzembowski (1993).
Key: Plio, Pliocene; Mio, Miocene; Olig, Oligocene; Eoc, Eocene; Cret, Cretaceous; Jur, Jurassic; Sil, Silurian; L, Lower; M, Middle; U, Upper

Fig. 7). Stephenson and Scott (1992) illustrated, but did not describe, galls on angiosperm leaves from the Eocene of Bournemouth, UK.

    Scudder (1886) mentioned cynipid galls from the Oligocene of Florissant, North America and a gall was figured by Cockerell (1908) and Brues (1910). However, Kinsey (1919) dismissed the cynipid origin of these galls. Hoffman (1932) described a Miocene leaf of *Quercus cognatus* Knowlton which had 25 gall-like impressions. He compared these to modern cynipid and cecidomyiid galls. Lewis (1985) described galls on a possible oak leaf from the Miocene of Clarkia, North America, which he considered were induced by Cynipidae. Larew (1992) tabulated six additional examples of published galls on leaves from the Oligocene and Miocene of which the record from the Upper Oligocene of Hungary (Ambrus and Hably 1979) is the most significant being recently published and based on a known leaf type (*Daphnogene*, Lauraceae). This gall was said to have been induced by *Eriophes* (gall mites).

Berger (1949) illustrated galls on Pliocene *Quercus* leaves which he believed were induced by *Neuroterus* (Hymenoptera: Cynipidae). A variety of cecidia from the Pliocene have been described by Straus (1977) and Givulescu (1984) working on material from Willershausen, Germany and Chiuzbaia, Romania, respectively. Larew (1992) tabulated 27 distinct records and noted 34 impressions of galled leaves from the work of Straus (1977). Berry (1923) mentioned the variety of cecidomyiid galls in *Taxodium* leaves from the Pleistocene of Maryland, North America. Quaternary galls include those figured by Pentecost (1985) on *Alnus glutinosa* (L.) Gaertn. from a post-glacial tufa in Yorkshire, UK, which were probably induced by the gall mite *Phytopus laevis* (Acari: Eriophyidae). Larew (1992) tabulated one additional Pliocene record, one additional Pleistocene record, and five Holocene records of galls on leaves.

(b) New material The vast majority of new specimens illustrated (Figs 26.1, 26.2 and 26.3) and described (Appendix) here are from the Middle Eocene either from the Claiborne Formation of central North America (Dilcher 1973; Potter and Dilcher 1981) or the Bournemouth and Bracklesham Groups (early Middle Eocene) of southern England (see Collinson (1990*a*) and Collinson and Hooker (1987) for discussion of floras and stratigraphy). Very few Palaeocene or Early Eocene specimens have been observed. This may partly reflect extinctions associated with Cretaceous/Tertiary boundary events and it is noteworthy that only three of the Cretaceous gall types were encountered in the Eocene. However, the apparent absence of Palaeocene and Early Eocene records may also reflect the reduced amount of material available for study (Table 26.2).

Evidence from the comparison with modern material suggests that galls induced by gall mites (Acari: Eriophyidae) were the most abundant although dipteran (cecidomyiid) and possible hemipteran (cynipid) galls were also common. Galls of a type most commonly associated today with the Homopteran family Psyllidae (psyllids) were also present in material from the late Palaeocene and Middle Eocene. Gall diversity increased considerably in the Tertiary with 18 gall types in the Middle Eocene (Table 26.2).

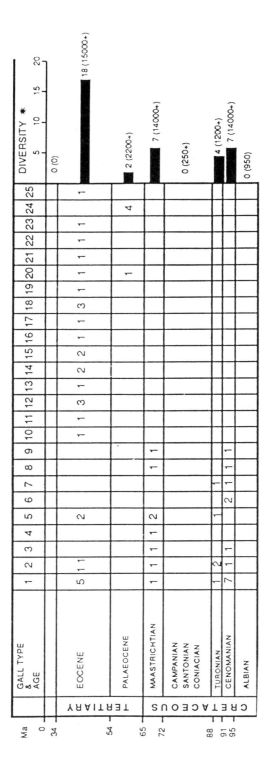

Table 26.2. The number of examples of each gall type and their stratigraphic distribution. Time scale from Hartland *et al.* (1900). *The numbers following the diversity histograms represent the total number of gall types found with, in brackets, the numbers of leaf fossils examined by Stephenson (1991). For further details of materials studied and gall types see Appendix 1, Figs. 26.1, 26.2, and 26.3.

Period	Ma	GALL TYPE & AGE	1	2	3	4	5	6	7	8	9	10	11	12	13	14	15	16	17	18	19	20	21	22	23	24	25	DIVERSITY*
TERTIARY	0 / 34																											0 (0)
TERTIARY	54	EOCENE	5	1			2					1	1	3	1	2	2	1	1	3	1	1	1	1	1		1	18 (15000+)
TERTIARY		PALAEOCENE																				1				4		2 (2200+)
CRETACEOUS	65 / 72	MAASTRICHTIAN	1	1	1	1	2			1	1																	7 (14000+)
CRETACEOUS	88	CAMPANIAN SANTONIAN CONIACIAN																										0 (250+)
CRETACEOUS	91	TURONIAN	1	2				1	1																			4 (1200+)
CRETACEOUS	95	CENOMANIAN	7	1	1			2	1	1	1																	7 (14000+)
CRETACEOUS		ALBIAN																										0 (950)

DIVERSITY* scale: 5   10   15   20

Discussion on new material

An attempt to identify the causal organism for the fossil galls documented here has been made by comparison with recent arthropod-induced galls. These comparisons must be viewed with caution as the criteria used for gall descriptions are at best rather crude, gall preservation is poor, and there is little published on modern leaf galls, especially those which do not have economic importance. Nevertheless a 'best match' has been recognized for the majority of the fossil galls. Possible erroneous comparisons with modern species may be partly overcome if the comparisons are considered at familial or ordinal level. Galls are induced by a variety of other organisms including viruses, bacteria, and fungi ('phytocecidia', Dreger-Jauffret and Shorthouse 1992) as well as protozoans and nematodes. Most nematode galls are on roots (Dreger-Jauffret and Shorthouse 1992) as are bacterial galls of the crown gall type (see Davey *et al.*, Chapter 2, this volume). However, it is certainly possible that the smaller leaf galls we have described here could have been induced by fungi or other gallers and not by arthropods. The larger galls, the more complex galls, and those for which a very similar modern analogue has been located, are probably insect induced.

We have examined several large assemblages of angiosperm leaves (three of which contained more than 14 000 specimens) from the Late Cretaceous and early Tertiary (Palaeogene). Although Larew (1992) noted 49 leaf impressions with galls in a Late Cretaceous collection he did not state how many leaves had been examined or any details of stratigraphy or gall diversity. Our data therefore seem to offer the only example from which preliminary patterns of evolution in gall diversity might be tentatively inferred.

A variety of gall types were already present by the mid-Cretaceous (Cenomanian). These were probably induced by gall mites, gall midges, aphids, and the gall wasps which together represent four of the seven modern cecidogeneous arthropod groups. Gall types 1 and 2, small to medium cone galls, probably induced by gall mites, are widespread from the Cenomanian age onwards. Gall type 5, a larger and irregular probable gall mite cecidia, is known from the Turonian and Maastrichtian of North America and is also found in the Middle Eocene of England. The other gall types recorded in the Cretaceous were not recorded in Tertiary assemblages.

Angiosperms diversified and came to dominate fossil leaf floras over a relatively short period of geological time (some 30 million years between the Barremian and the Albian) (Lidgard and Crane 1990). The diversity of gall types present in the Cenomanian leaves suggests either rapid coevolution with angiosperms or a previous history of galling other host plants with a switch to angiosperm hosts after angiosperms rose to dominance. Resolution of these conflicting hypotheses requires detailed study of pre-Cenomanian leaf assemblages.

An increased diversity of gall types (18 vs a maximum of seven in any Cretaceous sample) is present by the Middle Eocene (Table 26.2). In addition, a higher proportion of galled leaves are encountered within Eocene assemblages (40 specimens on 15 000 + Eocene leaves vs 14 specimens on 14 000 + Cenomanian leaves and eight specimens on 14 000 + Maastrichtian leaves, Table 26.2). This increased diversity and abundance follows the second major radiation of flowering plants which is reflected in early Tertiary floras and vegetation (Collinson 1990*b*). Understanding of the influence of Cretaceous–Tertiary boundary events in shaping future galler–host interactions must await study of an increased data set from Palaeoene and Early Eocene strata.

Leaf galls likely to have been induced by Coleoptera (weevils), Lepidoptera (mainly microlepidopteran moths), or Thysanoptera (thrips) have not been encountered even in the Eocene material. An apparent absence of fossil galls induced by these insect groups was also noted by Larew (1992). In the case of Lepidoptera and Thysanoptera our results may indicate a post-Palaeogene origin of the gall-inducing habit in these relatively advanced insect groups. The relatively primitive insect group Coleoptera rarely produce leaf galls today but attack other parts of their plant hosts such as roots and stems (Meyer 1987; Shorthouse and Rohfritsch 1992). Soft stems would have provided an easily accessed galling site for early prosoplasmatic gall inducers (Shorthouse 1986) and it is possible that stem galls induced by Coleoptera were among the earliest examples of plant galls. We are not aware of any fossil evidence pertinent to this hypothesis.

Early flowering plants included Magnoliidae and Hamamelididae in the Aptian–Albian plus Rosidae in the Late Cretaceous, all of which are now documented by floral (and other) remains (Crepet *et al.* 1991). In contrast Asteridae and Caryophyllidae and to a lesser extent Dilleniidae do not occur in any abundance until the Tertiary (Crepet *et al.* 1991). Patterns of systematic distribution of modern host plants of modern gall inducers (Roskam 1992, for example, Table 3.1) therefore suggest that many gall inducer–host relationships are very unlikely to have developed until the Tertiary (see also other papers in this volume). Some of these are probably post-Palaeogene (not earlier than 34 million years ago) as suggested in the previous paragraph.

We concur with the suggestion made by Larew (1992) that one useful approach in future research on fossil galls would be to target selected plant taxa. We suggest that suitable taxa should have a good fossil record and should be hosts to abundant, preferably distinctive, gall inducers at the present day. One example would be the Salicaceae with an excellent fossil record in the northern hemisphere Tertiary (Collinson 1992) and on which almost all of the modern sawfly galls (induced by *Pontania* and *Euura*) are found (Roskam 1992). Another example would be the Fagaceae, with a comparable fossil record (Crepet 1989; Kvaček and Walther 1989*a*, *b*; Mai 1989) and on which almost all modern cynipini (oak gall wasps) are found (Dreger-Jauffret and Shorthouse 1992; Roskam 1992). Stone (personal communication) is proposing to examine this latter example. The diverse fossil record of Fagaceae should enable recognition of any early occurrences of the oak gall wasp galls outside the genus *Quercus* thus constraining the evolutionary development of this major insect–plant interaction.

Acknowledgements

We thank Ed Jarzembowski for assistance in compiling Table 26.1, Joseph Shorthouse for the loan of his copy of Shorthouse and Rohfritsch, Michèle Williams for allowing us, at a late stage, to present a poster and include this paper in this volume, Kevin D'Souza for photographic assistance, and Sandra Muir for drafting. The Royal Society 1983 University Research Fellowship to M.E.C. and the NERC research studentship to J.S. are gratefully acknowledged. J.S. would like to thank P.R. Crane (Field Museum of Natural History, Chicago), D.L. Dilcher (Indiana University, now at University of Gainesville, Florida), J. Kvaček (National Museum, Prague, Czechoslovakia), Z. Kvaček (Charles University, Prague, Czechoslovakia), and C.H. Shute (Natural History Museum, London) for access to material in their care.

References

Alvin, K.L., Barnard, P.D.W., Harris, T.M., Hughes, N.F., Wagner, R.H., and Wesley, A. (1967). Gymnospermophyta. In *The fossil record*, (ed. W.B. Harland *et al.*), pp. 247–68. Geology Society of London, London.

Ambrus, B. and Hably, L. (1979). *Eriophyes daphnogene* sp. n., a fossil from the Upper Oligocene in Hungary. *Annales Historico-naturales Musei Nationalis Hungarici*, **71,** 55–6.

Barthel, M. and Rüffle, L. (1976). Ein Massenvorkommen von Symplocaceenblättern als Beispiel einer Variationsstatistik. *Abhandlungen des Zentralen Geologischen Instituts Paläontologische Abhandlungen*, **26,** 291–305.

Berger, W. (1949). Lebensspuren schmarotzender Insekten an Jungtertiaren Laubblättern. *Sitzungsberichte der Osterreichischen Akademie der Wissenschaften Mathematisch-Naturwissenschaftliche Klasse*, **1,** 789–92.

Berry, E.W. (1916). The Lower Eocene floras of southeastern North America. *United States Geological Survey Professional Paper*, **91,** 1–481.

Berry, E.W. (1923). Pathological conditions among fossil plants. In *Paleopathology:*

An introduction to the study of ancient evidences of disease, (ed. R.L. Moodie), pp. 99–109. University of Illinois Press, Urbana, Illinois.

Berry, E.W. (1924). The Middle and Upper Eocene floras of southeastern North America. *United States Geological Survey Professional Paper*, **92**, 1–201.

Berry, E.W. (1930). Revision of the Lower Eocene Wilcox flora of southeastern United States. *United States Geological Survey Professional Paper*, **193E**, 83–199.

Berry, E.W. (1934). A Lower Lance florule from Harding County, South Dakota. *United States Geological Survey Professional Paper*, **185F**, 127–33.

Boucot, A.J. (ed.) (1990). *Evolutionary paleobiology of behaviour and co-evolution*. Elsevier, Amsterdam.

Brooks, H.K. (1955). Healed wounds and galls on fossil leaves from the Wilcox deposits (Eocene) of Western Tennessee. *Psyche*, **62**, 1–9.

Brues, C.T. (1910). The parasitic Hymenoptera of the Tertiary of Florissant, Colorado. *Bulletin of the Museum of Comparative Zoology*, **54**, 1–125.

Brues, C.T. (1946). *Insect dietry*. Harvard University Press, Cambridge, MA.

Carpenter, F.M. and Burnham, L. (1985). The geological record of insects. *Annual Review of Earth and Planetary Sciences*, **13**, 297–314.

Chaloner, W.G., Harper, J.L., and Lawton, J. (ed.) (1991*a*). The evolutionary interactions of animals and plants. *Philosophical Transactions of the Royal Society of London*, **333B**, 177–305.

Chaloner, W.G., Scott, A.C., and Stephenson, J. (1991*b*). Fossil evidence for plant–arthropod interaction in the Palaeozoic and Mesozoic. *Philosophical Transactions of the Royal Society of London*, **333B**, 177–86.

Cockerell, T.D.A. (1908). Fossil insects from Florrisant, Colorado. *Bulletin of the American Museum of Natural History*, **24**, 59–69.

Collins, R.L. (1925). A Lower Eocene termite from Tennessee. *American Journal of Science*, ser. 5, **9**, 406–10.

Collinson, M.E. (1990*a*). Vegetational change during the Eocene in the coastal wetlands of southern England. In *Paleofloristic and paleoclimatic changes in the Cretaceous and Tertiary*, (ed. E. Knobloch and Z. Kvaček), pp. 135–39. Geological Survey, Prague.

Collinson, M.E. (1990*b*). Plant evolution and ecology during the early Cainozoic diversification. *Advances in Botanical Research*, **17**, 1–98.

Collinson, M.E. (1992). The early fossil history of Salicaceae: a brief review. *Proceedings of the Royal Society of Edinburgh*, **98B**, 155–67.

Collinson, M.E. and Hooker, J.J. (1987). Vegetational and mammalian faunal changes in the Tertiary of southern England. In *The origin of angiosperms and their biological consequences*, (ed. E.M. Friis, W.G. Chaloner, and P.R. Crane), pp. 259–304. Cambridge University Press, Cambridge.

Condon, M., and Whalen, M.D. (1983). A plea for collection and preservation of herbivor and pathogen damaged plant materials. *Taxon*, **32**, 105–7.

Connold, E.T. (1909). *Plant galls of Great Britain*. Adlard and Son, London.

Conway-Morris, S. (1981). Parasites and the fossil record. *Parasitology*, **82**, 489–509.

Crane, P.R. and Dilcher, D.L. (1984). *Lesqueria*: an early angiosperm fruiting axis from the mid Cretaceous. *Annals of the Missouri Botanical Garden*, **71**, 384–402.

Crepet, W.L. (1989). History and implications of the early North American fossil record of Fagaceae. In *Evolution, systematics, and fossil history of the Hamamelidae*, (ed. P.R. Crane and S. Blackmore), Systematics Association Special Vol. No. 40B, pp. 45–66. Clarendon Press, Oxford.

Crepet, W.L., Friis, E.M., and Nixon, K.C. (1991). Fossil evidence for the evolution of biotic pollination. *Philosophical Transactions of the Royal Society of London*, **333B**, 187–95.

Dilcher, D.L. (1973). A revision of the Eocene flora of southeastern North America. *Palaeobotanist*, **20**, 7–18.

Dreger-Jauffret, F. and Shorthouse, J.D. (1992). Diversity of gall-inducing insects and their galls. In *Biology of insect-induced galls*, (ed. J.D. Shorthouse and O. Rohfritsch), pp. 8–33. Oxford University Press, Oxford.

Givulescu, R. (1984). Pathological elements on fossil leaves from Chiuzbaia (galls, mines and other insect traces). *Dari de Seama ale Sedintelor*, **69** (1981), 123–33.

Harland, W.B., Armstrong, R.L., Cox, A.V., Craig, L.E., Smith, A.G., and Smith, D.G. (1990). *A geologic time scale 1989*. Cambridge University Press, Cambridge.

Hickey, L.J. and Doyle, J.A. (1977). Early Cretaceous fossil evidence for angiosperm evolution. *The Botanical Review*, **43**, 3–104.

Hoffman, A.D. (1932). Miocene insect gall impressions. *Botanical Gazette*, **93**, 341–2.

Jarzembowski, E.A. (1989). A century plus of fossil insects. *Proceedings of the Geologists Association*, **100**, 433–49.

Kinsey, A.C. (1919). Fossil Cynipidae. *Psyche*, **26**, 44–9.

Küster, E. (1911). *Die Gallen der Pflanzen*. Leipzig, S. Hirzel.

Kvaček, Z. and Walther, H. (1989*a*). Revision der mitteleuropäischen tertiären Fagaceen nach blattepidermalen Charakteristiken III. Teil *Dryophyllum* Debey ex Saporta und *Eotrigonobalanus* Walther & Kvaček gen, nov. *Feddes Repertorium*, **100**, 575–601.

Kvaček, Z. and Walther, H. (1989*b*). Paleobotanical studies in *Fagaceae* of the European Tertiary. *Plant Systematics and Evolution*, **162**, 213–29.

Larew, H.G. (1986). The fossil gall record, a brief summary. *Proceedings of the Entomological Society of Washington*, **88**, 385–8.

Larew, H.G. (1992). Fossil galls. In *Biology of insect-induced galls*, (ed. J.D. Shorthouse and O. Rohfritsch), pp. 51–9. Oxford University Press, Oxford.

Lesquereux, L. (1892). The flora of the Dakota Group. *United States Geological Survey Monograph*, **17**, 1–400.

Lewis, S.E. (1985). Miocene insects from the Clarkia deposits of Northern Idaho. In *Late Cenozoic history of the Pacific Northwest*, (ed. C.J. Smiley), pp. 245–64. American Association for the Advancement of Science, San Francisco.

Lidgard, S. and Crane, P.R. (1990). Angiosperm diversification and Cretaceous floristic trends: a comparison of palynofloras and leaf macrofloras. *Paleobiology*, **16**, 77–93.

Mai, D.H. (1989). Fossil remains of *Castanopsis* (D.Don) Spach (*Fagaceae*), and their importance to the European laurel-oak-forests. (In German). *Flora*, **182**, 269–86.

Meyer, J. (1987). *Plant galls and gall inducers*. Gebrüder Borntraeger, Berlin.

Pentecost, A. 91985). *Alnus* leaf impressions from a postglacial tufa in Yorkshire. *Annals of Botany,* **56,** 779–82.

Potonié, H. (1893). Die Flora des Rothliegenden von Thuringen. *Abhandlungen der Koniglich Preussischen Geologischen Landesanstalt,* **9,** 1–298.

Potter, J.R. and Dilcher, D.L. (1981). Biostratigraphic analysis of Middle Eocene floras of western Kentucky and Tennessee. In *Biostratigraphy of fossil plants: successional and palaeoecological analysis,* (ed. D.L. Dilcher and T.N. Taylor), pp. 211–25. Dowden, Hutchison, and Ross, Stroudsburg.

Roskam, J.C. (1992). Evolution of the gall-inducing guild. In *Biology of insect-induced galls,* (ed. J.D. Shorthouse and O. Rohfritsch), pp. 34–49. Oxford University Press, Oxford.

Ross, A.J. and Jarzembowski, E.A. (1993). Arthropoda (Hexapoda: insecta) In *The fossil record 2,* (ed. M.J. Benton), pp. 363–426. Chapman & Hall, London.

Scott, A.C., Stephenson, J., and Chaloner, W.G. (1992). Interaction and coevolution of plants and arthropods during the Palaeozoic and Mesozoic. *Philosophical Transactions of the Royal Society of London,* **335B,** 129–65.

Scudder, S.H. (1886). Systematic review of fossil insects. *Bulletin of the United States Geological Survey,* **5,** 9–129.

Shorthouse, J.D. (1986). Significance of nutritive cells in insect galls. *Proceedings of the Entomological Society of Washington,* **88,** 368–75.

Shorthouse, J.D. and Rohfritsch, O. (1992). *Biology of insect induced galls.* Oxford University Press, Oxford.

Southwood, T.R.E. (1973). The insect/plant relationship—an evolutionary perspective. In *Insect/plant relationships* (ed. H.F. Van Emden), *Symposium of the Royal Entomological Society of London,* **6,** 3–30.

Southwood, T.R.E. (1985). Interactions of plants and animals: pattern and process. *Oikos,* **44,** 5–11.

Stephenson, J. (1991). Evidence of plant/insect interactions in the Late Cretaceous and Early Tertiary. Unpublished PhD thesis, University of London.

Stephenson, J. and Scott, A.C. (1992). The geological history of arthropod damage to plants. *Terra Nova,* **4,** 542–52.

Straus, A. (1977). Gallen, Minen und andere Frasspuren im Pliokän von Willerhausen am Harz. *Verhandlungen des Botanischen Vereins der provinz Brandenburg,* **113,** 41–80.

Strong, D.R., Lawton, J.H., and Southwood, T.R.E. (1984). *Insects on plants: community patterns and mechanisms.* Blackwell Scientific Publications, Oxford.

Swanton, E.W. (1912). *British plant galls.* Methuen, London.

Wooton, R.J. (1988). The historical ecology of aquatic insects: an overview. *Palaeogeography, Palaeoclimatology, Palaeoecology,* **62,** 477–92.

Appendix

The following collections were examined for evidence of galls. All material is on undetermined angiosperm leaves. Details of localities, etc., may be found in Stephenson (1991) and references cited in the text.

The specimens prefixed F are from the National Museum, Prague, Czechoslovakia, IU were registered in the Biology Department, Indiana University, Indiana and have been transferred to the University of Gainesville, Florida, PP, and UP are from the Geology Department, Field Museum of Natural History, Chicago, Illinois, and V, are housed in the British Museum (Natural History), London.

Modern comparisons have been made using material and literature housed in the Entomology Department of the Natural History Museum, London.

Gall types on fossil leaves

1. Gall type 1 (Fig. 26.1a and 26.3h)

Material. Cenomanian: F36675, F768, UP256, v53724, IU 15708–7518, IU 15713–7516, IU 15713–7517. Turonian: PP13895a. Maastrichtian: PP8532. Middle Eocene: PP5351, PP5647/PP5653a, V48813, V49689, V58436.

Description, Eocene. Small cone galls, very similar to Cretaceous forms except they are usually associated with the major veins. No exit pores are visible due to a secondary infection by fungi.

Description, Cretaceous. Very small galls, dispersed generally throughout the leaf lamina and are not usually associated with any veins. Originally smooth galls. Single, central exit pore.

Recent analogue. *Eriophyes ulmicola* Nalepa (Acari: Eriophyidae) on *Ulmus campestris* L.

Very similar to recent galls produced in response to the gall mite *Eriophyes ulmicola* which produces a pouch gall on the undersurface of the leaf and a cone gall on the corresponding upper surface. This suggests that the impression fossils represent the upper surface of the leaf.

2. Gall type 2 (Figs 26.1b and 26.3g)

Material. Cenomanian: F1008. Turonian: PP13895b, PP14026a. Maastrichtian: PP5176a. Lower Eocene: V24142. Middle Eocene: PP12107, V46365, V46407, V47206, V47530, V48022, V49631, V50089, V50598, V50936b.

Description. Small to medium cone galls, situated near the midrib or leaf margin, originally smooth. Single, central exit pore.

Superficially very similar to gall type 1, only larger and associated either with the midrib (PP5176a) or the margin.

Recent analogue: *Eriophyes* species (Acari: Eriophyidae).

3. Gall type 3 (Fig. 26.1c)

Material. Cenomanian: IU15713–7516b. Maastrichtian: PP8664.

Description. Small galls, associated with the midrib, in apical half of the leaf. Each has a single central exit pore, the gall wall being originally either smooth or rough.

Recent analogue. *Cystiphora* species (Diptera: Cecidomyiidae). These circular galls closely resemble Recent *Cystiphora* galls except they are always associated with the midrib. *Cystiphora* produce pustule galls, most noticeable on the under surface of the leaf which may be the impression preserved in the fossil specimens.

4. Gall type 4 (Figs 26.1d and 26.3l)

Material. Maastrichtian: PP13553.

Description. Small to medium galls, distributed throughout leaf lamina, none are associated with the midrib or margin. A peripheral thickening of the tissues surrounds each gall and a central single exit pore exists in many.

Recent analogue. *Moritziella intermedia* Perg. (Hemiptera/Homoptera: Phylloxeridae) on *Carya*. These are virtually indistinguishable from Recent examples of galls formed by *Moritziella intermedia*. This insect produces pustule galls which are most noticeable on the upper surface of the leaf, dispersed over the entire surface and not associated with any veins or the leaf margin.

5. Gall type 5 (Fig. 26.1h)

Material. Turonian: PP14026. Maastrichtian: PP10498 and PP10587 pt. and cpt. Middle Eocene: V49728, V50185.

Description. Medium-sized irregular galls, found near the margin in the basal portion of the leaf, each gall has an irregular shaped wall which may have had flanges or extensions growing out over the lamina. Exit pores can not be seen.

Recent analogue. Species of Eriophyidae (Acari). A precise comparison to Recent galls is not possible but they are very similar to many produced by gall mites of the family Eriophyidae.

6. Gall type 6 (Figs 26.1f and 26.3b)

Material. Cenomanian: IU15703–4082, UP348.

Description. Medium-sized spot galls, distributed anywhere on the leaf lamina but not generally found in the more apical or basal areas, they tend not to be associated with any veins. Galls were originally smooth with a single central exit pore.

Recent analogue. *Myzus certus* (Walker) (Hemiptera/Homoptera: Aphididae) on *Stellaria media* L. and *Trioza ocoteae* Lzr. (Hemiptera/Homoptera: Psyllidae) on *Ocotea acutifolia* (Nees.) Very similar to Recent homopteran spot galls of the genera *Myzus* and *Trioza* especially when viewed from the upper surface.

7. Gall type 7 (Fig. 26.1i)

Material. Cenomanian: IU15703–7523. Turonian: PP13471.

Description. Medium to large pouch galls, distributed throughout the leaf lamina, between the veins but often overlapping them. The course of the veins is not affected by the smaller galls. Originally smooth walled, each has a single central exit pore.

Recent analogue. *Dasineura pustulans* (Rubsaamen) (Diptera:Cecidomyiidae) on *Filipendula ulmaria* L.

8. Gall type 8 (Fig. 26.1e)

Material. Cenomanian: F1861. Maastrichtian: PP5176b.

Description. Medium-sized ball galls, distributed either on or immediately adjacent to the midrib. Originally smooth walled with a single, peripheral exit pore.

Recent analogue. Various galls produced by the Cecidomyiidae (Diptera) or Cynipidae (Hymenoptera). Similar to many Recent cecidomyiid leaf galls including *Lasioptera populnea* Wachtl. and *Harmandia loewi* (Rubsaamen) both on *Populus tremula* L. but they also resemble the smaller female cynipid galls produced by *Cynips disticha* Htg. and *Cynips divisa* Htg. on *Quercus*.

9. Gall type 9 (Figs 26.1j,k and 26.3d)

Material. Cenomanian: PP14154. Maastrichtian: PP9196.

Description. Very large galls. On large leaves they are situated on either the midrib or a major vein, on smaller, lanceolate-like leaves their size obscures any signs of association with a particular structure of the leaf. Original gall wall either rough or smooth, single exit pore appears central.

Recent analogue. Various species of Cynipidae (Hymenoptera). Most similar to Recent *Diplolepis eglanteriae* (Hartig) on *Rosa* but also resemble many *Pontania* species, for example, *P. viminalis* L., *P. pedunculi* (Htg.), and *P. joergenseni* Enslin all on *Salix* species and female *Pediaspis aceris* Gmelin on *Acer pseudoplatanus* L. All belonging to the Cynipidae (Hymenoptera).

10. Gall type 10 (Figs 26.1l and 26.3e)

Material. Middle Eocene: V47982.

Description. Very small oval galls, distributed throughout the lamina except the apical end. Usually associated with the major veins and the largest are associated with the midrib. Originally smooth with lengthwise striations. No exit pores are visible.

Recent analogue. *Dasineura ulmariae* (Bremi) (Diptera: Cecidomyiidae) on *Filipendula ulmaria* (L.) Maxim. Very similar to Recent galls produced by *Dasineura ulmariae* which tend to lie on the major veins on the undersurface of the leaf. *Dasineura ulmariae* produces covering galls where the tissue around the larva is developed to form protective lips, this may be the reason why no exit pores are visible in the fossil material.

11. Gall type 11 (Fig. 26.1g)

Material. Middle Eocene: PP8932.

Description. Small cone galls. Identical to gall type 1 except the host plant is a

monocotyledon. They are distributed throughout the leaf lamina and not associated with any veins. The galls were originally smooth with a central, single exit pore.
Recent analogue. *Eriophyes ulmicola* Nalepa (acari: Eriophyidae) on *Ulmus campestris* L.

12. Gall type 12 (Figs 26.2a and 26.3m)

Material. Middle Eocene: V46409, V49425, V50320.
Description. Small pit galls, found mainly in the apical section of the leaf unassociated with the veins. May form clusters. Originally smooth walled.
Recent analogue. Species of Eriophyidae (Acari). No exact Recent equivalent exists.

13. Gall type 13 (Fig. 26.2c)

Material. Middle Eocene: PP15860.
Description. Small to medium oval galls. Unusual galls distributed throughout lamina with no particular association with veins. Originally smooth or perhaps with lengthwise striations. Position of exit pore/s unclear.
Recent analogue. None. These galls resemble no recorded Recent galls.

14. Gall type 14 (Fig. 26.2b)

Material. Middle Eocene: PP12481, PP12515.
Description. Medium-sized hollow galls, distributed throughout the leaf lamina and overlap each other. There appears to be no association with either the veins or margin although this may be obscured. Up to 200 individual galls may be present on a single leaf. They were originally smooth with a peripheral thickened ridge and a pronounced central exit pore.
Recent analogue. Female *Neuroterus numismalis* Olivier (Hymenoptera: Cynipidae) on *Quercus*. Similar to winter survival galls produced by Recent female *Neuroterus numismalis*, the toughened walls of the galls may have enhanced preservation potential.

15. Gall type 15 (Fig 26.2d)

Material. Middle Eocene: PP15197, V50154.
Description. Medium-sized galls, located midway along the leaf and always associated with the midrib. Originally smooth gall wall, may have had a single exit pore.
Recent analogue. *Dasineura urticae* (Perris) (Diptera: Cecidomyiidae) on *Urtica*. *Dasineura urticae* normally produces covering galls on the midrib or larger secondary veins, the exit pore tends to be rather central in keeping with the fossil specimens.

16. Gall type 16 (Figs 26.2e and 26.3i)

Material. Middle Eocene: V49947.
Description. Medium-sized galls, found in the basal half of the leaf, mainly along the midrib which has been distorted by the galls. Originally hairy and no exit pores can be distinguished.
Recent analogue. *Eriophes padi* (Nalepa) (Acari: Eriophyidae) on *Prunus spinosa* L. and *Hartigiola annulipes* (Hartig) (Diptera: Cecidomyiidae) on *Fagus silvatica* L. Very similar to Recent galls produced by both *Hartigiola annulipes* and *Eriophes padi* which distort the midrib as they grow.

17. Gall type 17 (Figs 26.2g and 26.3c)

Material. Middle Eocene: PP7704.
Description. Medium-sized galls, distributed throughout the leaf lamina, not associated with the midrib but may lie on the margin. Irregular gall wall which was originally smooth. No exit pores are visible.
Recent analogue. *Xerophylla caryaeglobuli* Walsh (Hemiptera/Homoptera: Phylloxeridae) on *Carya Xerophylla* produces ball galls on the undersurface of the leaf which create a characteristic 'bubble' effect on the corresponding upper surface. The galls in specimen PP7704 represent the upper surface gall marks. They are also very similar to marks on a leaf figured by Givulescu (1984, Pl.III, Fig. 5) which are incorrectly described as feeding traces of the ichnogenus *Phagophytichnus*.

18. Gall type 18 (Figs 26.2h and 26.3a)

Material. Middle Eocene: V25192, V48051, V49448b.
Description. Medium-sized spot galls, located throughout the leaf lamina but not usually associated with either the veins or the margin. Originally smooth walled with a peripheral thickening and centrally raised exit pore.
Recent analogue. Female *Neuroterus quercusbaccarum* L. (Hymenolptera: Cynipidae) on *Quercus*. Galls are produced on the undersurface of the leaf.

19. Gall type 19 (Fig. 26.2i)
Material. Middle Eocene: V40285 pt. & cpt.
Description. Medium to large irregular galls, found along the midrib and completely distorting it. The central part of the gall is typically cone-like with a single central exit pore. Around this spreads an irregular thickening along the midrib and onto the lamina on either side. Originally smooth in the centre, it may have had hairs on the outer flanges.
Recent analogue. *Iteomyia capreae* (Winnertz) (Diptera: Cecidomyiidae) on *Salix*, *Rabdophaga salicis* (Schrank) (Diptera: Cecidomyiidae) on *Salix* and *Massalongia rubra* (Keiffer) (Diptera: Cecidomyiidae) on *Betula* and various other recent cecidomyiid galls.

20. Gall type 20 (Figs 26.2f(a) and 26.3j)

Material. Late Palaeocene: V59878. Middle Eocene: V52584.
Description. Medium to large galls, found throughout the lamina but not associated with the veins or margin. Central portion contains a possible exit pore and is cone shaped. This is surrounded by a number of extensions which lie on top of the leaf lamina in a stellate arrangement.
Recent analogue. Dehisced *Pauropsylla montana* Uichanco (Hemiptera/Homoptera: Psyllidae) on *Ficus variegata* Blume. The inner cone-like section of the fossil specimens represents the thickening of the entry hole on the undersurface of the leaf. The flanges correspond to the walls of the gall which remain after the insect has escaped.

21. Gall type 21 (Fig. 26.2f(b)

Material. Middle Eocene: V52584.
Description. Medium ball galls, distributed throughout the leaf lamina and not associated with the veins or margin, may form clusters. Originally smooth walled. No exit pores visible.
Recent analogue. *Pauropsylla montana* (Hemiptera/Homoptera: Psyllidae) on *Ficus variegata*. These galls could possibly represent undehisced galls produced in response to *Pauropsylla*, but they differ from the Recent form in their lack of hairs.

22. Gall type 22 (Fig. 26.2l)

Material. Middle Eocene: V49689.
Description. Large irregular galls, usually associated with either the midrib or margin, they may be clustered or plurilocular. Originally smooth walled with an irregular margin, a number of exit pores may exist centrally.
Recent analogue. None. They do not resemble any recorded Recent galls.

23. Gall type 23 (Figs 26.2m and 26.3f)

Material. Middle Eocene: PP5647/PP5653b.
Description. Large ovoid galls, distributed along the length of the lamina approximately 10 mm from either margin and usually associated with the basal side of the secondary veins. Gall wall originally cerebroid and no exit pores are distinguishable.
Recent analogue. *Monarthropalpus buxi* (Geoffroy) (Diptera: Cecidomyiidae) on *Buxus sempervirens* L. Similar to Recent pustule galls produced on the undersurface of the leaf by *Monarthropalpus buxi*.

24. Gall type 24 (Figs 26.2k and 26.3k)

Material. Late Palaeocene: V48404, V49272, V49448a, V50013.
Description. Large to massive galls, distributed throughout leaf lamina, but not

usually near the margin and not associated with the veins. The central pouch-like portion is surrounded by a thickening of the lamina forming a 'halo' effect on the fossil leaf. No exit pores are visible.

Recent analogue. None. There are no recorded Recent galls which resemble these fossil specimens. The presence of gall type 18 (probably occurring on the leaf undersurface) on specimen V49448(b) suggests that type 24 is found on the upper surface of the leaf.

25. Gall type 25 (Fig. 26.2j)

Material. Middle Eocene: V48524.

Description. Very large/massive ball gall, which may be associated with a large secondary vein which it overlaps in the basal portion of the leaf. Originally rough walled, no exit pores are visible.

Recent analogue. Female *Cynips quercusfolii* L. (Hymenoptera: Cynipidae) on *Quercus*. Largest fossil gall recorded. Very similar to Recent galls produced by female *Cynips quercusfolii* which associate with secondary veins and have a rough gall wall.

Index

Systematics Association Publications

1. Bibliography of key works for the identification of the British fauna and flora, *3rd edition* (1967)†
   Edited by G.J. Kerrich, R.D. Meikle, and N. Tebble
2. Function and taxonomic importance (1959)†
   Edited by A.J. Cain
3. The species concept in palaeontology (1956)†
   Edited by P.C. Sylvester-Bradley
4. Taxonomy and geography (1962)†
   Edited by D. Nichols
5. Speciation in the sea (1963)†
   Edited by J.P. Harding and N. Tebble
6. Phenetic and phylogenetic classification (1964)†
   Edited by V.H. Heywood and J. McNeill
7. Aspects of Tethyan biogeography (1967)†
   Edited by C.G. Adams and D.V. Ager
8. The soil ecosystem (1969)†
   Edited by H. Sheals
9. Organisms and continents through time (1973)†
   Edited by N.F. Hughes
10. Cladistics: a practical course in systematics (1992)
    P.L. Forey, C.J. Humphries, I.J. Kitching, R.W. Scotland, D.J. Siebert, and D.M. Williams

†Published by the Association (out of print)

Systematics Association Special Volumes

1. The new systematics (1940)
   Edited by J.S. Huxley (Reprinted 1971)
2. Chemotaxonomy and serotaxonomy (1968)*
   Edited by J.G. Hawkes
3. Data processing in biology and geology (1971)*
   Edited by J.L. Cutbill
4. Scanning electron microscopy (1971)*
   Edited by V.H. Heywood
   Out of print
5. Taxonomy and ecology (1973)*
   Edited by V.H. Heywood

6. The changing flora and fauna of Britain (1974)*
   Edited by D.L. Hawksworth
   Out of print
7. Biological identification with computers (1975)*
   Edited by R.J. Pankhurst
8. Lichenology: progress and problems (1976)*
   Edited by D.H. Brown, D.L. Hawksworth, and R.H. Bailey
9. Key works to the fauna and flora of the British Isles and north-western Europe, *4th edition* (1978)*
   Edited by G.J. Kerrich, D.L. Hawksworth, and R.W. Sims
10. Modern approaches to the taxonomy of red and brown algae (1978)
    Edited by D.E.G. Irvine and J.H. Price
11. Biology and systematics of colonial organisms (1979)*
    Edited by G. Larwood and B.R. Rosen
12. The origin of major invertebrate groups (1979)*
    Edited by M.R. House
13. Advances in bryozoology (1979)*
    Edited by G.P. Larwood and M.B. Abbot
14. Bryophyte systematics (1979)*
    Edited by G.C.S. Clarke and J.G. Duckett
15. The terrestrial environment and the origin of land vertebrates (1980)
    Edited by A.L. Panchen
16. Chemosystematics: principles and practice (1980)*
    Edited by F.A. Bisby, J.G. Vaughan, and C.A. Wright
17. The shore environment: methods and ecosystems (2 Volumes) (1980)*
    Edited by J.H. Price, D.E.G. Irvine, and W.F. Farnham
18. The Ammonoidea (1981)*
    Edited by M.R. House and J.R. Senior
19. Biosystematics of social insects (1981)*
    Edited by P.E. Howse and J.-L. Clément
20. Genome evolution (1982)*
    Edited by G.A. Dover and R.B. Flavell
21. Problems of phylogenetic reconstruction (1982)*
    Edited by K.A. Joysey and A.E. Friday
22. Concepts in nematode systematics (1983)*
    Edited by A.R. Stone, H.M. Platt, and L.F. Khalil
23. Evolution, time and space: the emergence of the biosphere (1983)*
    Edited by R.W. Sims, J.H. Price, and P.E.S. Whalley
24. Protein polymorphism: adaptive and taxonomic significance (1983)*
    Edited by G.S. Oxford and D. Rollinson
25. Current concepts in plant taxonomy (1983)*
    Edited by V.H. Heywood and D.M. Moore
26. Databases in systematics (1984)*
    Edited by R. Allkin and F.A. Bisby

27. Systematics of the green algae (1984)*
    Edited by D.E.G. Irvine and D.M. John
28. The origins and relationships of lower invertebrates (1985)‡
    Edited by S. Conway Morris, J.D. George, R. Gibson, and H.M. Platt
29. Infraspecific classification of wild and cultivated plants (1986)‡
    Edited by B.T. Styles
30. Biomineralization in lower plants and animals (1986)‡
    Edited by B.S.C. Leadbeater and R. Riding
31. Systematic and taxonomic approaches in palaeobotany (1986)‡
    Edited by R.A. Spicer and B.A. Thomas
32. Coevolution and systematics (1986)‡
    Edited by A.R. Stone and D.L. Hawksworth
33. Key works to the fauna and flora of the British Isles and north-western Europe, *5th edition* (1988)‡
    Edited by R.W. Sims, P.Freeman, and D.L. Hawksworth
34. Extinction and survival in the fossil record (1988)‡
    Edited by G.P. Larwood
35. The phylogeny and classification of the tetrapods (2 Volumes) (1988)‡
    Edited by M.J. Benton
36. Prospects in systematics (1988)‡
    Edited by D.L. Hawksworth
37. Biosystematics of haematophagous insects (1988)‡
    Edited by M.W. Service
38. The chromophyte algae: problems and perspectives (1989)‡
    Edited by J.C. Green, B.S.C. Leadbeater, and W. Diver
39. Electrophoretic studies on agricultural pests (1989)‡
    Edited by Hugh D. Loxdale and J. den Hollander
40. Evolution, systematics, and fossil history of the Hamamelidae (2 Volumes) (1989)‡
    Edited by Peter R. Crane and Stephen Blackmore
41. Scanning electron microscopy in taxonomy and functional morphology (1990)‡
    Edited by D. Claugher
42. Major evolutionary radiations (1990)‡
    Edited by P.D. Taylor and G.P. Larwood
43. Tropical lichens: their systematics, conservation, and ecology (1991)‡
    Edited by D.J. Galloway
44. Pollen and spores: patterns of diversification (1991)‡
    Edited by S. Blackmore and S.H. Barnes
45. The biology of free-living heterotrophic flagellates (1991)‡
    Edited by D.J. Patterson and J. Larsen
46. Plant-animal interactions in the marine benthos (1992)‡
    Edited by D.M. John, S.J. Hawkins, and J.H. Price

47. The Ammonoidea: environment, ecology, and evolutionary change (1993)‡
    Edited by M.R. House
48. Designs for a global plant species information system‡
    Edited by F.A. Bisby, G.F. Russell, and R.J. Pankhurst
49. Plant galls: organisms, interactions, populations‡
    Edited by Michèle A.J. Williams

* Published by Academic Press for the Systematics Association
† Published by the Palaeontological Association in conjunction with the Systematics Association
‡ Published by the Oxford University Press for the Systematics Association